Engineering Geology for Society and Territory – Volume 4

Giorgio Lollino • Andrea Manconi
Jacques Locat • Yu Huang
Miquel Canals Artigas
Editors

Engineering Geology for Society and Territory – Volume 4

Marine and Coastal Processes

Springer

Editors
Giorgio Lollino
Andrea Manconi
Institute for Geo-hydrological Protection
National Research Council (CNR)
Turin
Italy

Jacques Locat
Department of Geology
Laval University
Quebec, QC
Canada

Yu Huang
Department of Geotechnical Engineering
Tongji University
Shanghai
China

Miquel Canals Artigas
Department of Stratigraphy, Paleontology
and Marine Geosciences
University of Barcelona
Barcelona
Spain

ISBN 978-3-319-08659-0 ISBN 978-3-319-08660-6 (eBook)
DOI 10.1007/978-3-319-08660-6
Springer Cham Heidelberg New York Dordrecht London

Library of Congress Control Number: 2014946956

Cover Illustration: Mouth of the Galeso submarine spring, Mar Piccolo, Taranto, Italy. The spring is located at 18 meters below the sea level, with a discharge 800 liters/sec. A special structure was projected and realized in order to drive the fresh water the surface without pumping systems, and without altering the hydrogeological regime. *Photo*: Giorgio Lollino.

Printed on acid-free paper

Springer is part of Springer Science+Business Media (www.springer.com)

Foreword

It is our pleasure to present this volume as a part of the book series of the Proceedings of the XII International IAEG Congress, Torino 2014.

For the 50th Anniversary, the Congress collected contributions relevant to all themes where the IAEG members have been involved, both in the research field and in professional activities.

Each volume is related to a specific topic, including:

1. Climate Change and Engineering Geology;
2. Landslide Processes;
3. River Basins, Reservoir Sedimentation and Water Resources;
4. Marine and Coastal Processes;
5. Urban Geology, Sustainable Planning and Landscape Exploitation;
6. Applied Geology for Major Engineering Projects;
7. Education, Professional Ethics and Public Recognition of Engineering Geology;
8. Preservation of Cultural Heritage.

The book series aims at constituting a milestone for our association, and a bridge for the development and challenges of Engineering Geology towards the future.

This ambition stimulated numerous conveners, who committed themselves to collect a large number of contributions from all parts of the world, and to select the best papers through two review stages. To highlight the work done by the conveners, the table of contents of the volumes maintains the structure of the sessions of the Congress.

The lectures delivered by prominent scientists, as well as the contributions of authors, have explored several questions ranging from scientific to economic aspects, from professional applications to ethical issues, which all have a possible impact on society and territory.

This volume testifies the evolution of engineering geology during the last 50 years and summarizes the recent results. We hope that you will be able to find stimulating contributions, which will support your research or professional activities.

Giorgio Lollino

Carlos Delgado

Preface

A better living in the future environment is the common wish of the whole of mankind. With the rapid development of coastal and marine resource exploitation and utilization worldwide, coastal and marine engineering geology has become a key component in human activities. Many engineering geological problems have been encountered in the process of planning, building and servicing in coastal and marine environments.

This session focuses on coastal management and shore protection, taking into account storm-related events and relative sea level change, planning of waste disposal, works for the remedy of coastal pollution, sea-floor pipeline engineering, slope stability analysis, and tsunami propagation and flooding. The session aims to provide a forum for engineering geologists who play a pivotal role in investigating, managing, using, exploiting and protecting the coastal and marine domains so that they can share their overall views, creative ideas and advanced knowledge. This is also an opportunity to generate profitable interactions with other experts, such as civil, forest and agrarian engineers, geomorphologists, oceanographers, geologists, geophysicists, biologists, land planners and environmentalists.

This volume contains more than 40 peer reviewed technical papers related to coastal and marine engineering geology on the topics of: (1) Coastal and offshore geohazards; (2) Coasts at threat: causes and consequences of coastal settlements and maritime transportation; (3) Monitoring and measurement of seabed dynamic processes; (4) Relative land subsidence in transitional coastal environment: causes, effects, quantification and monitoring; (5) Submarine mass movements: hazards and risk assessment. This collection of papers covers recent case histories, theoretical advances, laboratory and field-testing, and design methods beneficial to practitioners, researchers and other professionals. This volume will also showcase the rapidly evolving technologies used in investigating coastal and marine environments.

We would like to thank all authors and participants for sharing their ideas and results in the area of coastal and marine engineering geology. We are also appreciative of the conference organizers for their hard work in putting up so many interesting sessions. We wish to acknowledge the help from all the reviewers in advising and refining the contributions. The reader may find that in some of the papers they are left on their appetite, which is largely due to the guidelines limiting the number of pages to a maximum of 6. Readers are encouraged to directly contact the authors if they wish to have more information.

Finally, we hope that the proceedings will reflect the success of the XII IAEG Congress, which is the 50th Anniversary of IAEG.

Contents

Consiglio Nazionale delle Ricerche
Istituto di Ricerca per la Protezione Idrogeologica

The Istituto di Ricerca per la Protezione Idrogeologica (IRPI), of the Italian Consiglio Nazionale delle Ricerche (CNR), designs and executes research, technical and development activities in the vast and variegated field of natural hazards, vulnerability assessment and geo-risk mitigation. We study all geo-hydrological hazards, including floods, landslides, erosion processes, subsidence, droughts, and hazards in coastal and mountain areas. We investigate the availability and quality of water, the exploitation of geo-resources, and the disposal of wastes. We research the expected impact of climatic and environmental changes on geo-hazards and geo-resources, and we contribute to the design of sustainable adaptation strategies. Our outreach activities contribute to educate and inform on geo-hazards and their consequences in Italy.

We conduct our research and technical activities at various geographical and temporal scales, and in different physiographic and climatic regions, in Italy, in Europe, and in the World. Our scientific objective is the production of new knowledge about potentially dangerous natural phenomena, and their interactions with the natural and the human environment. We develop products, services, technologies and tools for the advanced, timely and accurate detection and monitoring of geo-hazards, for the assessment of geo-risks, and for the design and the implementation of sustainable strategies for risk reduction and adaptation. We are 100 dedicated scientists, technicians and administrative staff operating in five centres located in Perugia (headquarter), Bari, Cosenza, Padova and Torino. Our network of labs and expertizes is a recognized Centre of Competence on geo-hydrological hazards and risks for the Italian Civil Protection Department, an Office of the Prime Minister.

Part I

Coastal and Offshore Geodisasters

With the rapid development of coatal and marine resources exploitation and utilization in the world, many geodisaster problems have been encountered in coastal and offshore engineering. This part is aimed at providing a forum for all participants interested in contributing a presentation related to coastal and offshore geodisaster problems due to earthquakes, tsunamis, storm surges, floods, avalanches, landslides, erosion, sea level rise, human activities, and so on. Suggested themes for presentations including all aspects of geodisaster items (e.g., phenomenon, mechanism, test, analysis, computation, prevention, protection, and mitigation) in costal and offshore engineering geology.

On Soil Layer Damage Due to Gas Escape

1

Lu Xiaobing, X. D. Chen, X. H. Zhang, and S. Y. Wang

Abstract

Dissociation of natural gas hydrate (NGH) in seabed can product amounts of gas. If the soil layer over NGH layer is permeable, gas will escape. Gas escape can on one hand lead to the damage of seabed, on the other hand lead to the decrease of the density of sea water and so threatens the structures nearby. In this paper movement and expansion behavior of gas in water was first studied, then the damage of seabed due to gas escape was studied. Effects of gas pressure, thickness of soil layer were investigated. The expansion angle of gas in the water was obtained. The characteristics of the hole size induced by gas escape with gas pressure and soil thickness were also obtained. These results can be as references of deeper research and practice.

Keywords

Gas hydrate • Gas escape • Seabed damage

1.1 Introduction

NGH is a new potential source of energy. More and more researchers study the exploitation and exploration of NGH recently. Most of previous attentions are paid on the exploitation of NGH (Hisashi et al. 2002). However, NGH dissociation can lower the strength of NGH sediment and induce kinds of hazards such as landslide of stratum (Lu et al. 2008, 2010, 2011; Locat and Lee 2002).

Large amount of NGH dissociation can causes leakage of gas from seabed (Gilles et al. 1999). If the soil layer over the NHG layer is permeable, gas can percolate through this over layer and causes not only the damage of seabed, but also the formation of bubbles in sea water. Bubbles will flow upwards and decreases the density of sea water, which can lead to the damage of offshore structures such as platforms and ships (Zhang 1998; Sultan et al. 2004). Thus it is important to study the hazards caused by the leakage of gas dissociated from NGH.

1 m^3 NGH can product 164 m^3 methane gas after dissociation. If there is no leakage of gas, pore pressure with 50 MPa can form (Briaud and Chaouch 1997). The over layer will damage once the pore pressure is over its weight and tensile strength. For example, layered fracture or outburst can be observed (Zhang et al. 2011).

In this paper experiments were carried out to observe the diffusion of bubbles. The damage characteristics of the soil layer caused by the leakage of gas were studied also. Effects of pressure and thickness of the over layer were mainly considered.

1.2 Experiment of Gas Diffusion

The aim of this paper is to study the mechanism and phenomenon that gas percolates through the over layer and diffuses in the sea water during flowing upwards. Although gas produced from dissociation of gas hydrate is in a relatively large zone and passes the hole in the over layer with a variable pressure at field conditions, the model box adopted

L. Xiaobing (✉) · X. D. Chen · X. H. Zhang · S. Y. Wang
Institute of Mechanics, Chinese Academy of Sciences, Beijing 100190, China
e-mail: xblu@imech.ac.cn

G. Lollino et al. (eds.), *Engineering Geology for Society and Territory – Volume 4*,
DOI: 10.1007/978-3-319-08660-6_1,

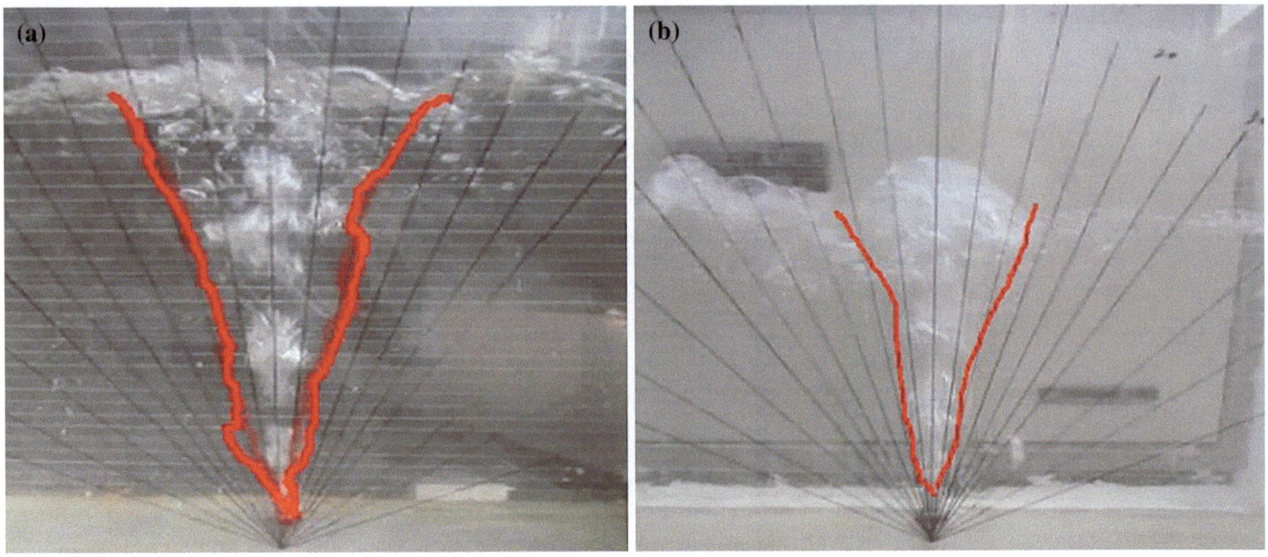

Fig. 1.1 Photo of the bubble expansion (hole diameter 3 cm, water depth 35 cm) **a** Gas pressure is 7.5 kPa, **b** Gas pressure is 25 kPa

here is with a size is length × width × height = 50 cm × 10 cm × 50 cm and gas flowed directly out from a hole at the bottom of the box to model the gas flow from the dissociation zone. It is difficult to model the whole process from dissociation of gas hydrate to seepage and damage of over layer. Though the adopted device is not designed by the similar law, it is enough to obtain the required results if only we control the main parameters (pressure, water depth, soil layer depth). Meantime, this choice can make the experiments be carried out fast and convenient. Controlling device of gas pressure and gas pump was used to output gas flow with different pressures.

A hole was set at the bottom of the box. Gas with designed pressure can pass through the hole into water. Two diameters of the holes, 3 and 6 cm, are chosen to control the output flow rate and pressure at the hole because the pressure of the gas source is fixed. Three depths of water layer, 25, 35 and 40 cm, were adopted to investigate the effects of water depth. Gas pressures at the hole were controlled as 2.5, 5, 7.5, 12.5, 15, 17.5 and 20 kPa. Camera was used to record the bubbles' movement during flowing upwards.

Diffusion of gas under different conditions is similar (Fig. 1.1). Bubble flow is with a diffusion angle 5°–10° near the bottom, and a little bigger than 10° near the surface. Depth of water and hole diameter have little effects on the diffusion angle in the range of water depth in our experiments.

1.3 Critical Bubble Content Which can Cause a Structure Settle

Assuming it is the critical state when a structure just settles in the water with bubbles, that means, the weight of the structure is just equal to the floating force, so

$$\rho_c = G/V_s \tag{1.1}$$

in which ρ_c is the water density in the critical state, V_s is the total volume of the structure, G is the weight of structure and load on it.

The weight of the bubble is neglected because it is too small relative to the weight of water, so for any water with density ρ:

$$\rho = \frac{\rho_w V_w}{V_w + V_g} \tag{1.2}$$

Let $V_g/(V_g + V_w)$, volume percentage of bubbles corresponding to density ω is:

$$\omega = \rho/\rho_w \tag{1.3}$$

So volume percentage of bubbles corresponding to critical density ω_c is:

$$\omega_c = \rho_c/\rho_w \tag{1.4}$$

in which V_w is the water volume, V_g the bubble volume, ρ_w water density.

The diffusion angle is assumed as 10° by the above experiments. The area with bubbles is assumed as a cone (Fig. 1.1). Water depth is h, then the radius of the zone with bubbles at the surface is x = h·tan10°. Assuming the total gas volume in this zone is a, solubility of methane gas hydrate in sea water is β.

Volume of the circular cone is

$$V = \frac{\pi r^2 h}{3} = \frac{\pi \tan^2 10 \cdot h^3}{3} \tag{1.5}$$

According to the critical density obtained in the above section, the total gas volume which can causes the settlement of structures can be obtained as follows considering it contains the bubbles in water and the dissolved gas in water.

$$a = [(1 - \omega_c)\beta + \omega_c]V \tag{1.6}$$

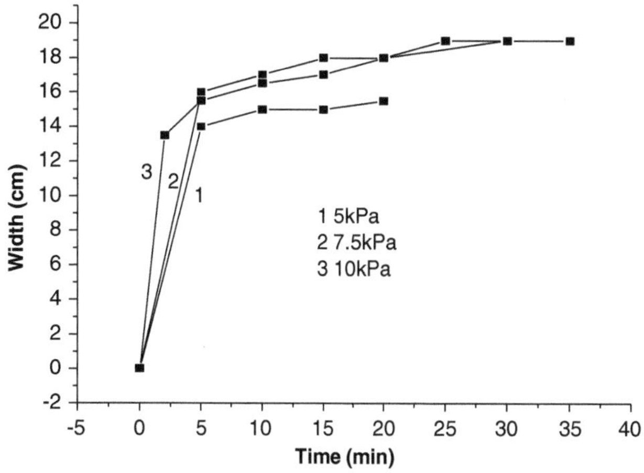

Fig. 1.2 Width of the damage zone

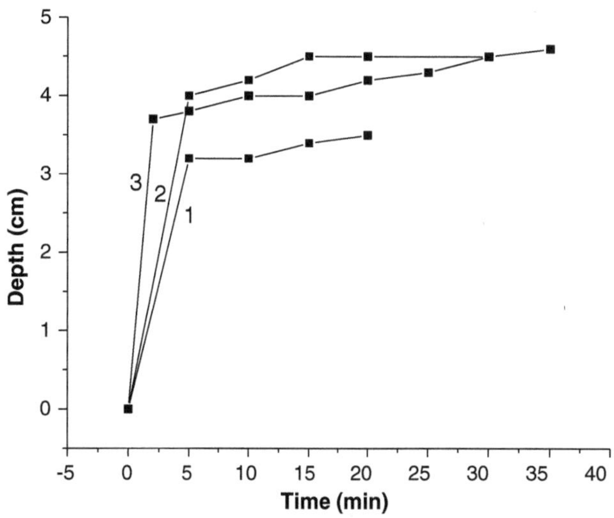

Fig. 1.3 Depth of damage zone

1.4 Experiment on the Soil Layer Damage Caused by Seepage Gas

1.4.1 Experimental Setup

The aim in this section is to study the damage of soil layer caused by the seepage of dissociated gas from NHG, so gas flowing with different pressures from a hole into the soil layer is used to model the gas seepage from NGH sediment into water but not NGH sediment. The silty sand with dry density of 1600 kg/m^3 was used.

Initial water content of the soil was $w = 6.35\ \%$, thickness of soil layer is adopted as 10, 15 and 25 cm respectively. Output pressures were 2.5, 5, 7.5, 10 and 12.5 kPa respectively. Soil layer was first prepared to the designed thickness, then water was percolated slowly from a hole at the bottom of the box till the water level was 20 cm over the surface of the soil layer. Experiments were carried out after the model stayed statically for some time. The course and damage forms of soil layer were observed. The size of the damage zone was recorded.

1.4.2 Experimental Results of Sediment Damage

(a) Results when the thickness of soil layer is 10 cm

It is observed that bubbles get out from the surface of the soil layer when the pressure is 2.5 kPa and gas seepage causes the damage of the soil layer. The damage zone is cone type whose width and depth are 9 and 2 cm respectively. The damage zones are similar under other pressures. Damage zone expands fast at first and then becomes stable gradually. The duration to arrive at the stable state increases with the rise of pressure. The width and depth of the damage zone under different pressures are shown in Figs. 1.2 and 1.3. The size and expansion speed of the damage zone increase with the rise of pressure. Obviously, gas flows faster when the pressure is larger, so the effects of erosion and damage are stronger.

(b) Results when the thickness of soil layer is 15 cm

It is observed that after the valve is opened, bubbles are shown at the surface of soil layer just above the gas hole. Water becomes unclear gradually. After some time, grains are taken out by bubbles. The damage zone increases with pressure and is larger than that under the thickness of soil layer 10 cm (Figs. 1.4 and 1.5). The reason is that the seepage line and the diffusion zone increase with the rise of the thickness of soil layer.

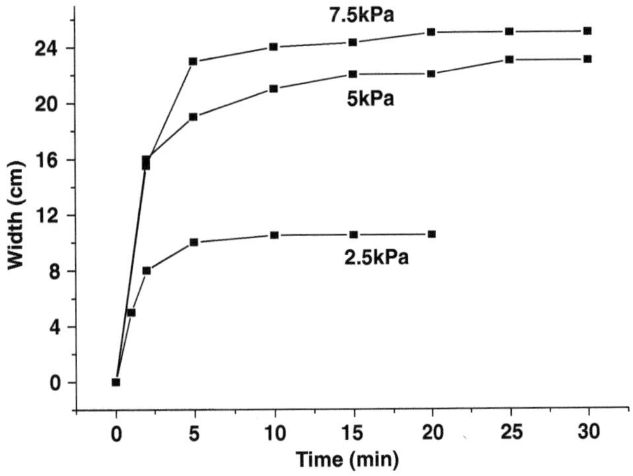

Fig. 1.4 Width of damage zone

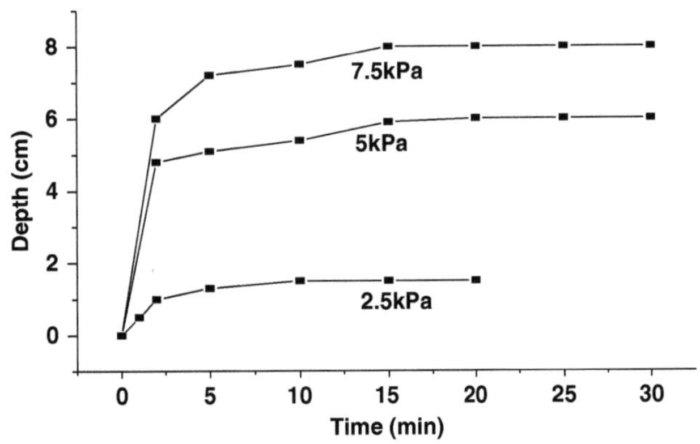

Fig. 1.5 Depth of damage zone

(c) Results when the thickness of soil layer is 25 cm

Under pressure of 2.5 kPa, there is no bubble is observed even 30 min after the valve is opened. Soil layer is not damaged. Under pressure of 5 kPa, a few grains are taken into water by bubbles. The soil layer is damaged gradually but the damage zone is small. The width and depth of the damage zone are 7 and 3 cm respectively. Shown in Table 1.1 are the width and depth under the pressure 7.5 kPa. When the gas pressure is given, the resistance is larger at larger depth of soil layer. Gas can not take grains away from soil layer when the thickness is over some value at given pressure. Thus no damage zone can be observed. For a given gas pressure, there exists a thickness of soil layer corresponding to maximum damage zone. Less than this thickness, damage zone increases with the increase of thickness. Otherwise, damage zone decreases.

Table 1.1 Depth and width of damage zone at pressure of 7.5 kPa and 25 cm thick

Pressure = 7.5 kPa		
Time (min)	Width (cm)	Depth (cm)
0	0	0
5	10	4
10	11	4.5
15	13	4.5
20	13	4.5
25	13	4.5

The changes of the width and depth of damage zone with gas pressure are shown in Fig. 1.6. It can be seen that the width and depth is linear to the gas pressure when the pressure is less than some value. After that the width and depth keep almost stable. The duration to arrive at maximum damage zone and the required pressure to cause the damage of soil layer increase with the thickness of soil layer. The depth of the damage zone first increases with soil thickness and then decreases because the seepage pressure can cause larger zone be damaged when the soil thickness is not large enough to induce the seepage pressure decrease obviously. However, with the increase of soil thickness, seepage pressure decreases largely so the erosion power of seepage decreases.

1.5 Discussion

During passing through the soil layer, gas has momentum and so can take some fine grains out from the soil layer. With the loose of fine grains, soil skeleton becomes unstable. Some part near the surface becomes catastrophic and the soil grains are taken away by the water flow and bubbles. During flowing upwards, velocity of water flow and bubble decrease gradually. Grains settle gradually and at last an eroded hole is formed.

With the increase of the thickness of soil layer, the resistance that the gas must overcome and the seepage route become large, meanwhile the duration for gas to pass through the soil layer becomes longer, thus the required pressure is large. For example, when the soil layer is 10 cm thick, gas can pass through the soil layer in about several

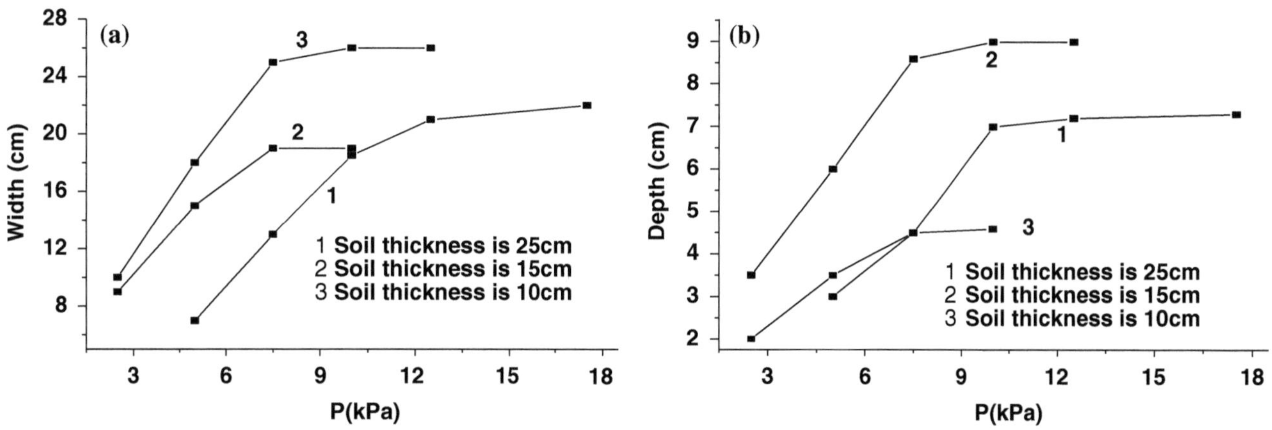

Fig. 1.6 Development of the depth of damage zone with gas pressure **a** Width versus pressure, **b** Depth versus pressure

seconds under 2.5 kPa; while the thickness of soil layer increases to 15 cm, gas can be observed after 1 min; when the soil layer is 25 cm thick, no gas is observed to pass through the soil layer, and it is observed to pass through only if the pressure is increased to 5 kPa.

1.6 Conclusions

Movement and expansion behavior of gas in the water and the damage of the soil layer due to gas escape were studied. Main conclusions obtained under the experimental conditions of this paper are as follows:

(1) Expansion angle of the escape bubbles in the water is about 5°–10°.
(2) Width and depth of the damage zone is linear to the gas pressure when the pressure is less than some value.
(3) Thickness of soil layer is the main factor to determine the size of damage zone. There is a critical thickness of soil layer, less than which the size of the damage zone increases with the thickness.

Acknowledgments This paper is supported by the National Science Foundation (No. 51239010, No. 11102209; No. 11272314).

References

Briaud JL, Chaouch A (1997) Hydrate melting in soil around hot conductor. ASCE J Geotech Geoenviron Eng 123(7):645–653
Gilles G, David G, Aleksandr M (1999) Characterization of in situ elastic properties of gas hydrate-bearing sediments on the Blake Ridge. J Geophys Res 104(B8):17781–17795
Hisashi OK, Sridhar N, Song F, Duane HS (2002) Synthesis of methane gas hydrate in porous sediments and its dissociation by depressurizing. Powder Technol 122:239–246
Lu XB, Wang L, Wang SY et al (2008) Study on the mechanical properties of THF hydrate deposit. In: Proceedings of the 18th international offshore and polar engineering conference, Vancouver, pp 57–60
Lu XB, Wang L et al (2010) Instability of seabed and pipes induced by NGH dissociation. In: Proceedings of the 20th international offshore and polar engineering conference, Beijing, pp 110–114
Lu XB, Zhang XH, Wang SY (2011) Strong failure of seabed induced by gas hydrate dissociation In: 7th international conference gas hydrate, Edinburgh
Locat J, Lee HJ (2002) Submarine landslide, advances and challenges. J Canada Geotechnol 39:93–212
Sultan N, Cochonat P, Foucher JP et al (2004) Effect of NGHs melting on seafloor slope instability. J Marine Geol 213:379–401
Zhang Youxue (1998) Experiment simulations of gas-driven eruptions: kinetics of bubble growth and effect of geometry. Bull Volcanol 59:281–290
Zhang XH, Lu XB, Zheng ZM (2011) Layered fracture and outburst due to dissociation of hydrate. J Petrol Sci Eng 76:212–216

2 Littoral Drift Barriers and the Problem of Proving Accelerated Recession

Max Barton and Sally Brown

Abstract

Any barrier to longshore drift has the potential to cause downdrift erosion. Understanding this problem means that any new construction of a groyne field or breakwater should be accompanied by monitoring to record changes in beach volumes and profiles as well any erosion of the coastline. In the past, such care may not have happened, especially in those cases involving boundaries between authorities responsible for the coast, or where the ground liable to be eroded was not considered to be of high value. This paper proposes a way by which previous groyne construction, or other coastal works, may be deemed responsible for increasing the rate of soft cliff erosion beyond a value which could be ascribed to being the result of a natural variation in recession rate.

Keywords

Longshore drift • Marine erosion • Recession • Groyne • Breakwater

2.1 Introduction

A set-back in the line of an undefended cliffed coast with respect to an updrift defended coast is a very common coastal feature (Fig. 2.1), ranging from tens to hundreds of metres in length. Where an undefended coast has previously had a large littoral drift volume of beach material and a new littoral drift barrier, such as a breakwater or groyne field, it acts as a barrier leading to a reduction in sediment downdrift. Naturally it is held responsible for the recession of the downdrift coast. Numerous examples of this have occurred on previously undeveloped coastlines (Anderson et al. 1983 and Komar 1983). An example of where doubt cannot exist is provided by the Orissa coast of India (Mohanty et al. 2012) where two groynes are protecting a harbour mouth: much accumulation is taking place updrift of one groyne and severe erosion taking place downdrift of the other. The erosion is clearly at the expense of the material being held up on the updrift side of the harbour.

By contrast, where the coastline has had a long history of development, with coastal defences constructed at various times and much interference with beach volumes, locally or within the sediment cell, it becomes problematic as to the extent to which observed downdrift erosion can be ascribed to the influence of any new coastal defence works. The purpose of this paper is to examine how we may be able to assess whether or not new, or recently extended coastal defences are responsible for an acceleration in the downdrift rate of recession.

2.2 Reasons for Proving Accelerated Recession

Our reasons for evaluating the extent to which new defence works are the principal cause of accelerated recession may be (i) for legal reasons where the liability for land loss is under investigation (Maddrell and Gowan 2001), (ii) planning and coastal management, especially in respect of the design of new works, (iii) determination of future land values and (iv) as a guide to the growth of the defended

M. Barton (✉) · S. Brown
Faculty of Engineering and the Environment, University of Southampton, Southampton, UK
e-mail: M.E.Barton@soton.ac.uk

G. Lollino et al. (eds.), *Engineering Geology for Society and Territory – Volume 4*,
DOI: 10.1007/978-3-319-08660-6_2, © Springer International Publishing Switzerland 2014

coast as an artificial headland and subsequent growth of an embayment downdrift. For all of these purposes, it will be necessary to take into account all of the factors which influence the rate of erosion and hence the recession of the coastline. These factors can be listed in terms of their time of operation as follows:

- **Historical**—hard and soft constructive activities up and downdrift, beach mining (particularly for sand and gravel) and the time lapse needed for the restoration of equilibrium following the updrift changes.
- **Present**—the periodicity of cliff movements, changes in geology as recession proceeds, the delayed response of clay cliffs to past toe erosion, groundwater fluctuations and the incidence of storm conditions.
- **Long Term**—wave climate and offshore bathymetric changes, sea-level rise and other responses to climate change.

2.3 Methodology

The previous rate of recession will be obtained from Ordnance Survey maps, aerial photographs, traditional ground surveying and modern differential GPS surveys. Recession is calculated with respect to the cliff edge which acts as the clearest and most unequivocal feature of the coastline. The error range will come from a combination of surveying and reading errors, the deformation of historical maps and unrectifiable photographic errors following survey practice and literature. The objective is to compare the observed recession since the date of construction ($R_{observed}$) with the recession that would have been obtained had the preceding recession rate continued in the absence of the sea defence works: the "maintained" value ($R_{maintained}$). Where the observed recession, allowing for the error range ($\pm E$), is clearly greater than the maintained recession, also allowing for the errors, then we can refer to the difference as the "excess" recession (R_{excess}) as follows:

$$R_{excess} = [R_{observed}(\pm E)] > [R_{maintained}(\pm E)] \qquad (2.1)$$

The value of the maintained recession is taken from a sufficiently long time period leading up to, and not after, the construction commenced. It is important that the maintained value covers a time period sufficiently long to obtain a full representation of all the "current" factors influencing the rate of erosion as listed above. The historical factors should be excluded unless there are reasons to consider they may still be active and likewise the long term factors can be also be excluded where over the time period selected, their influence is very small. The exact length of the time period will depend on the dates of available data and its consistency but short periods are liable to be overly reflective of infrequent events. From studies of various soft cliff localities in the UK, it has been observed that a reliable time period is approximately 30–50 years (Brown 2008). It is appreciated that despite a thorough check of causes of past retreat rates, these retreat rates may not continue in the future, but without detailed, costly modelling, it is the most simple, quick and best available method.

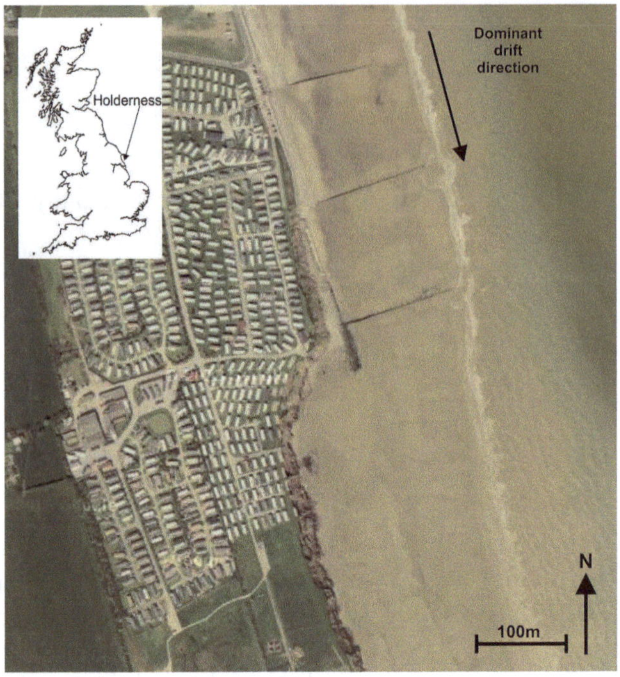

Fig. 2.1 Aerial view of the set-back downdrift of the defence works at Hornsea on the Holderness coast of Yorkshire, UK, where a 1,544 m groyne field was extended by 290 m (including a small shore parallel extension). The *inset* map shows the location of the Holderness coast (Photograph from Google Maps)

2.4 Applications at Holderness

2.4.1 Hornsea

Following the original piecemeal construction along 1,544 m of coast at Hornsea over the period 1906–1954, there is a clear excess retreat over what would have been the maintained rate of recession. However the extension of the groyne field by 290 m as shown in Fig. 2.1, completed in 1977, does not give an excess over the maintained value for a measurement period of 37 years. It appears probable that the retreat being experienced from the original construction was still influencing the downdrift coast and the small extension shown was not sufficient to make a major impact (Brown et al. 2012).

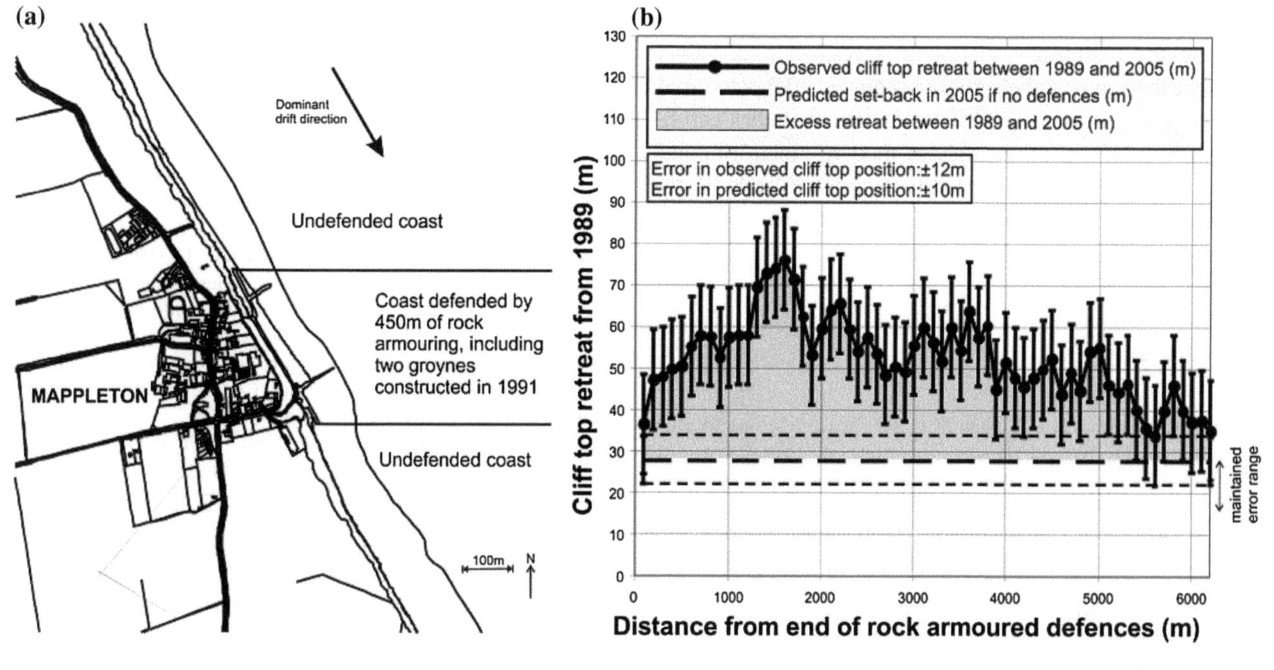

Fig. 2.2 **a** and **b** Mappleton defences and set-back down-drift of defence to illustrate excess retreat to 2005, if the maintained rate had continued after 1989

2.4.2 Mappleton

The village of Mappleton located 3.1 km downdrift of Hornsea provides a clear example where excess retreat, as defined by Eq. 2.1, has taken place. Before protection, the coastline was subject to parallel retreat. In 1991, 450 m of shore parallel rock armouring and two rock groynes were constructed (Fig. 2.2a), creating a set-back down-drift. Retreat rates downdrift increased from 1.7 ± 0.6 m/year (1952–1989) to 3.3 ± 0.8 m/year (1989–2005) after defence construction. Assuming the retreat rate from 1952–1989 was maintained, this resulted in an average excess retreat of 25 ± 12 m up to 4 km downdrift (Fig. 2.2b).

The increased rate of erosion at Mappleton became the subject of a land tribunal from 1998–1999 (Lands Tribunal 1999) where the landowners requested compensation for the excessive land lost. The outcome was that the defences were not responsible for the 7 years of excess erosion examined. However, as discussed in Sect. 2.2, coastal retreat has a natural temporal variation and a much longer period than 7 years after construction is required for proper assessment of the influence of the defences. Our figures show that over a period of 14 years, it is clear that excess retreat had occurred. Ideally, the assessment should extend over a longer period to provide a full representation of all the factors influencing erosion but in practical terms this could aggravate the hardship suffered by an impoverished landowner seeking redress for the land loss.

2.5 Discussion and Conclusion

Various authors have discussed the temporal and spatial variation in rates of cliff erosion on the Holderness coast (Maddrell et al. 1999; Pethick 1996 and Quinn et al. 2009), a major factor of which is the variation in storm surge activity and periodicity of retreat due to geotechnical reasons. It is for this reason that an appropriately long length of time must be allowed for the assessment of recession in order to obtain an average which reflects this degree of natural variation. The high rate of erosion which occurred in the short period of 7 years in the case of the Mappleton inquiry, allowed the Tribunal to point to the high variability of shoreline erosion as a factor which overrode any effects due to downdrift erosion (Lands Tribunal 1999). The contribution which downdrift erosion due to the groyne construction made to the severity of the cliff erosion was effectively ignored. Depletion of beach material by the construction of barriers on a coast where there is a high rate of longshore drift and where the total beach volume is not adequately replaced by such cliff erosion is a common cause of serious shoreline retreat worldwide. This inevitable process cannot be ignored in the design of defence schemes. Indeed, in recent years set-backs have been seen as a feature of the protection works to create emerging stable embayments (e.g. Hsu et al. 2008).

During the 20th Century the influence of sea-level rise would have been minimal compared with the other factors that influenced coastal retreat. With climate change, sea-level rise this century is expected to be more than the last, thus changing the temporal variation in recession rate. Hence historical shoreline analysis to assess downdrift erosion may become less appropriate. However historical studies can be augmented with the full panoply of modern monitoring and modelling techniques now readily available (such as the system described by Mohanty et al. 2012) to evaluate sediment movement and rates of recession.

References

Anderson GL, Hardaway CS, Gunn JR (1983) Beach response to spurs and groins. In: Weggel JR (ed) Proceedings of coastal structures '83. Waterway, Port, Coastal & Ocean Division of the American Society of Civil Engineers, pp 727–739

Brown S (2008) Soft cliff retreat adjacent to coastal defences, with particular reference to Holderness and Christchurch Bay. PhD Thesis, University of Southampton. http://eprints.soton.ac.uk/72899/

Brown S, Barton ME, Nicholls RJ (2012) The effect of coastal defences on cliff top retreat along the Holderness coastline. Proc Yorks Geol Soc 59(1):1–13

Hsu JRC, Benedet L, Klein AHF, Raabe ALA, Tsai C-P, Hsu T-W (2008) Appreciation of static bay beach concept for coastal management and protection. J Coast Res 24(1):198–215

Komar PD (1983) Coastal erosion in response to the construction of jetties and breakwaters. In: Komar PD (ed) CRC handbook of coastal processes and erosion. CRC Press Inc, Boca Raton, pp 191–204

Lands Tribunal (1999) Compensation in the matter of a notice of reference between John David Foster Earle, Susan Anne Fletcher Earle (Claimants) and East Riding of Yorkshire Council (Compensating Authority) regarding Grange Farm, Great Cowden, Yorkshire LCA/143/95

Maddrell RJ, Home R, Thurston N, Rennie D (1999) Impacts of changes to the bathymetry and wave energy on rates of coastal erosion. In: Proceedings of the 34th MAFF conference of river and coastal engineers, Keele University, pp 4.4.1–4.4.13

Maddrell RJ, Gowan D (2001) Legal implications of coastal structures. In: Allsop NWH (ed) Breakwaters, Coastal structures and coastlines. Thomas Telford, London, pp 461–470

Mohanty PK, Patra SK, Bramha S, Seth B, Pradhan U, Behera B, Mishra P, Panda US (2012) Impact of groins on beach morphology: a case study near Gopalpur Port, East Coast of India. J Coast Eng 28:132–142

Pethick J (1996) Coastal slope development: temporal and spatial periodicity in the Holderness cliff recession. In: Anderson MG, Brooks SM (eds) Advances in hillslope processes. Wiley, Chichester, pp 897–917

Quinn JD, Philip LK, Murphy W (2009) Understanding of the recession of the Holderness coast, east Yorkshire, UK: a new presentation of temporal and spatial patterns. Q J Eng Geol Hydrol 42:165–178

Field Test on Stiffened Deep Mixed Columns

3

Guan-bao Ye, Yong-sheng Cai, and Qing Liu

Abstract

A composite column, referred to as Stiffened Deep Mixed (SDM) column, can be used to improve the soft soils. It consists of a Deep Mixed (DM) column inserted by a small-sized precast concrete pile (called core pile as well) in center. The past study (citation) shows that the SDM column can provide high bearing capacity as compared with the pile with the same material of the core pile and same configuration of the SDM column under the same geological conditions. Therefore, the technique of SDM columns has a significant effect on saving construction material and budget. The previous projects indicate that the SDM column can reduce 20–40 % of budget at least. However, its mechanical behaviors have not been well understood yet. To investigate the mechanical behaviors of SDM column, a series of plate loading tests including nine Prestressed High Strength Concrete (PHC) piles and 30 SDM columns in total are conducted in field. Axial strains of two columns are measured during tests by the Fiber Bragg Grating (FBG) Sensor Technology. The axial force distributions of the core piles can be obtained based on the axial strains. The test results show the bearing capacity of SDM column is extremely high and the side resistant ratio between SDM columns and PHC piles is approximately in a range from 1.3 to 2.5.

Keywords

Stiffened deep mixed (SDM) column • Plate loading test • Core pile • Bearing capacity

3.1 Introduction

Deep Mixed (DM) column has been widely used to enhance the bearing capacity and increase stability and reduce settlement of the soft soils around the world (Porbaha 1998; Bergado et al. 1999). However, the low strength and stiffness of DM columns limit its application, especially for certain projects being sensitive to deformation and requiring high bearing capacity on soft soils (Bruce et al. 1998).

A new kind of composite column, called concrete-cored Deep Mixed column or Stiffened Deep Mixed (SDM) column, was proposed in last two decades and provides a new economic alternative to improve the soft soils. The previous projects indicate that the SDM column can reduce 20–40 % of budget at least. This composite column consists of DM column inserted by a precast concrete pile (core pile) in center. The core pile with higher strength and stiffness carries most of the load from the embankments or surcharges. The load carried by the core pile which is, in turn, transmitted to DM column around through their interfaces and then to soil (Dong et al. 2002; Wu et al. 2004). The interface shear strength between the DM column and the core pile, which increases linearly with the unconfined compressive strength of DM column, is sufficient to ensure the core pile and soil-cement can work together (Ding et al. 2010).

G. Ye · Y. Cai (✉) · Q. Liu
Key Laboratory of Geotechnical and Underground Engineering of Ministry of Education, Tongji University, 1239 Siping Road, Shanghai 200092, China
e-mail: cys036@163.com

G. Ye · Y. Cai · Q. Liu
Department of Geotechnical Engineering, Tongji University, 1239 Siping Road, Shanghai 200092, China

G. Lollino et al. (eds.), *Engineering Geology for Society and Territory – Volume 4*,
DOI: 10.1007/978-3-319-08660-6_3, © Springer International Publishing Switzerland 2014

Table 3.1 Soil properties

No.	Soil layer	Thickness (m)	Void ratio	Cohesion (kPa)	Friction angle (°)	Compression modulus (MPa)
1	Clay	4.30	1.038	15.7	12.3	3.62
2	Sandy silt	12.90	0.782	4.5	28.6	11.5
3	Fine sand	7.70	0.674	2.5	33.2	17.4
4	Clay	7.60	0.908	29.7	14.8	6.2
5	Fine sand	8.00	0.730	3.1	31.4	14.2

This paper presents a series of plate loading tests of the SDM columns in field which have a long DM column and a short concrete core pile in center. The bearing capacity of the SDM columns are evaluated and the force axial force distributions of the core piles are calculated based on the test results of the axial strains.

3.2 Project Description

The field tests are conducted in a project of 16 residential apartments in Jiangsu Province, China. Pile foundation is adopted for their foundations and the desired bearing capacity is from 2,600 to 4,000 kN with the floors of the apartments from 17 to 26.

The geological condition in the area of the project is investigated and Table 3.1 shows the properties of the subsoil layers.

Layer 3 is selected as the bearing stratum for the pile foundation. In original scheme, Prestressed High Strength Concrete (PHC) piles are selected, but cannot meet the required bearing capacity if the piles seated on Layer 3. Thus, the pile length has to be increased to reach a deeper bearing stratum so that they can satisfy the desired bearing capacity. However, the increasing pile length would increase the budget. After that, the SDM column foundation with the core pile of PHC pile is recommended to replace the original scheme. Before construction, a series of field tests are conducted to determine the pile parameters such as pile length and diameter of both core pile and DM column.

3.3 Field Test

3.3.1 Procedures of Field Test

For feasibility study, plate loading tests are conducted to compare the bearing capacities of PHC piles and SDM columns. Nine PHC piles with different length are constructed individually in different areas as well as 28 SDM columns. The partial test schemes are listed in Table 3.2 in detail. The slow maintained load test method is adopted in the test. For SDM columns, the load plate covers the plan area of both core pile and DM column.

After feasibility study, two additional SDM columns consisting of the DM columns with a diameter of 800 mm and length of 14 m and the PHC piles with a diameter of 500 mm and length of 11 m are installed for the further study on mechanical behavior by employing Fiber Bragg Grating (FBG) Sensor Technology. Hill and Meltz (1997) indicated that "A fiber Bragg grating (FBG) is a periodic perturbation of the refractive index along the fiber length which is formed by exposure of the core to an intense optical interference pattern."

FBG sensors are embedded in the core piles of SDM columns. To prevent the sensors from damage of construction, two grooves with roughly 1 cm-deep and 0.5 cm-wide along the pile shaft are cut symmetrically using grooving machine. The FBG sensors are pasted in the grooves at certain positions, i.e., 0.7, 2.7, 4.7, 6.7, 8.7 and 10.7 m from the pile top, respectively. Then seal the grooves with epoxy resin so that FBG sensors can be well protected.

3.3.2 Test Results and Discussion

3.3.2.1 Feasibility Study

The bearing capacities of both PHC piles and SDM columns are summarized in Table 3.2. For SDM columns, only several representative columns are presented. The values of the ultimate bearing capacity of SDM columns are in a range from 4050kN to 5500kN. All of them are greater than those of 28 m-long PHC piles and close to those of the PHC piles with length of 33–38 m. It indicates that the existence of DM column surrounding the core pile can significantly increase the total amount of side resistance and the load can transfer from the core pile to the DM column and then to the surrounding soil effectively. Moreover, the Load-settlement curves (Q-s curves) in Fig. 3.1 indicate that the settlement of SDM columns is close to those of PHC piles.

In addition, the average side resistances of PHC piles and SDM columns are calculated while ignoring the tip resistance as the side resistance contributed the most part of the bearing capacity. The side resistances of SDM column ranging from 124.0 to 164.1 kPa are significantly greater than those of PHC piles with a range from 65.5 to 98.4 kPa, and the side resistance ratio between SDM columns and PHC piles is approximately from 1.3 to 2.5.

Table 3.2 Test scheme and results

No.	PHC pile			DM column		Ultimate bearing capacity (kN)	Side resistance (kPa)
	Length (m)	Diameter (mm)	Thickness (mm)	Length (m)	Diameter (mm)		
1	28	500	125			2880, 3520	65.5, 80.0
2	33	500	125			4,800	92.6
3	34	500	125			4,440, 4,680	83.1, 87.6
4	36	500	125			4,440	78.5
5	37	500	125			4,840, 5,720	83.3, 98.4
6	38	500	125			5,720	95.8
7	12	400	95	16.5	800	4,950	164.1
8	13	400	95	17.0	800	4,960	151.8
9	13	400	95	16.5	800	4,050	124.0
10	13	500	125	18.0	800	4,950	151.5
11	14	400	95	18.0	800	5,150	146.4
12	14	400	95	17.5	800	5,040	143.2
13	14	400	95	18.0	800	5,400	153.5
14	14	500	125	18.0	800	5,500	156.3
15	15	500	125	18.0	800	5,500	145.9

Note 1. No. *1–6* represent PHC piles and No. *7–15* represent SDM columns
2. For some SDM columns, no significant failure occurs and the max load applied is considered as the ultimate bearing capacity
3. For SDM column, the core pile length is considered as the pile length when calculating the side resistance, because the axis force is small below the core pile (Ding et al. 2010)

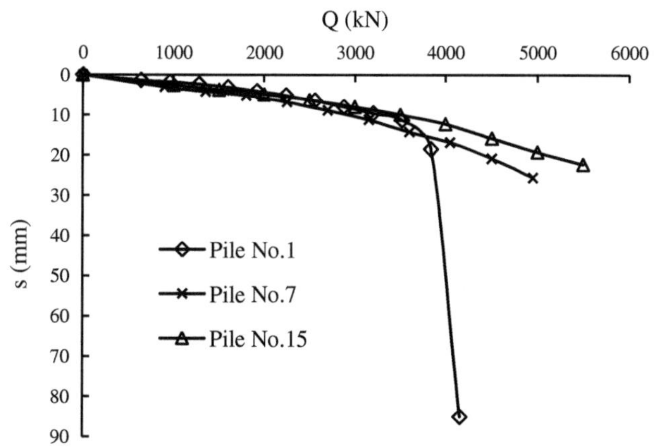

Fig. 3.1 Q-s curves of plate loading test

Generally speaking, the test results indicate that the bearing capacity of SDM column is extremely high and the settlement is small.

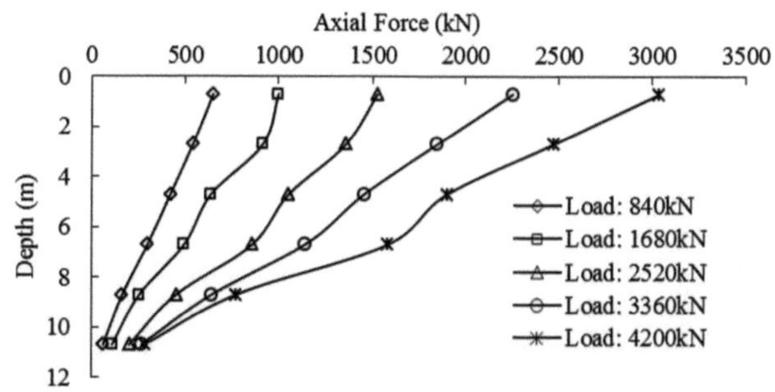

Fig. 3.2 Axial force distribution of core pile

Table 3.3 Axial force distribution on top of SDM column

Load (kN)		840	1,680	2,520	3,360	4,200
Core pile	Force (kN)	699	1,114	1,687	2,420	3,269
	Percentage(%)	83	66	67	72	78
DMcolumn	Force (kN)	141	566	833	940	931
	Percentage(%)	17	34	33	28	22

3.3.3 Axial Force Distribution

The max load applied on two additional SDM columns is 4200 kN and no significant failure occurs, so the bearing capacity of single pile meets requirement. The axial force distribution of the core pile at different load steps is shown in Fig. 3.2. It shows that the axial force decreases rapidly and almost linearly along the pile length from the top to the pile base. Based on least squares method, the axial force of core pile on top can be calculated. Then, the axial force of DM column on top can be obtained, shown in Table 3.3. It indicates that the core pile carries the most part of load from 66 to 83 % and the percentage increases with the applied load applied excluding 840 kN load.

3.4 Conclusions

Based on the analyses on the plate loading tests of the SDM columns in field, the following conclusions can be drawn:

- Bearing capacity of SDM column is extremely high because the DM column around the core pile can significantly increase the total amount of side resistance.
- The side resistance ratio between SDM columns and PHC piles is approximately from 1.3 to 2.5.
- The core pile of SDM columns carries most part of the external load from 66 to 83 % and the axial force of core pile decreases almost linearly along pile length.

Acknowledgments The authors appreciate the financial support from the National Natural Science Foundation of China (NSFC) (No. 51078271) for this work.

References

Porbaha A (1998) State of the art in deep mixing technology. Part I—basic concepts and overview. Ground Improv 2:81–92

Hill KO, Meltz G (1997) Fiber Bragg grating technology fundamentals and overview. J Lightwave Technol 15(8):1263–1276

Bergado DT, Ruenkrairergsa T, Taesiri Y, Balasubramaniam AS (1999) Deep soil mixing to reduce embankment settlement. Ground Improv 3(1):1–18

Bruce DA, Bruce MEC, DiMillio AF (1998) Deep mixing method: a global perspective. Civil Eng 68(12):35–41

Ding Y-J, Li J-J, Liu E (2010) Load transfer mechanism of reinforced mixing pile. J Tianjin Univ 43(6):530–536 (in Chinese)

Dong P, Chen Z-Z, Qin R (2002) Use of Concrete-cored DCM Pile in Soft Ground. Chin J Geotech Eng 24(2):204–207 (in Chinese)

Wu M, Dou Y-M, Wang E-Y (2004) A study on load transfer mechanism of stiffened DCM pile. Chin J Geotech Eng 26(3):432–434 (in Chinese)

Measurement Analysis of Bohai Bank Deep Foundation Pit Project in Tianjin Soft Soil Area

4

Yi-qun Tang, Jing-jing Yan, Zhi-jun Sun, and Jie Zhou

Abstract

In this paper, monitoring results of Bohai Bank deep foundation pit in Tianjin has been summarized and analyzed. The characters of ground settlement caused by excavation and lateral movement of the bracing have been analyzed. The temporal and spatial variation rules of deformation pit have been studied. Making use of related work of the soil loss theory, the ground settlement has been calculated inversely. The study shows that: the pit foundation deformation rule of this project is similar to the rules in Shanghai soft soil area. Spatial effect is obvious in pit deformation process. The deformation of the pit is sensitive to depth and stage of excavation. In the position where has the maximum lateral movement, the ratio of the horizontal distance to the depth of excavation is 0.85. The settlement values calculated by soil loss theory are larger than measured values. The quickly changing of the settlement caused by excavation does not happen at the same time as the deformation of bracing. The main factor that affects the settlement may be the addition load caused by surroundings, while deformation of bracing is affected by active earth pressure on bracing.

Keywords

Soft soil area • Pit excavation • Spatial effect • Soil loss theory

4.1 Introduction

With the development of high-rise buildings and utilization of underground spaces in soft soil areas, the deformation problems in deep foundation pit excavation has received extensive attention. Real-time monitoring and analyzing on success or failure pits will contribute to the development of foundation pits (Li 2007). The forecast methods of foundation pit deformation can be divided into two: theoretical analysis and field data analysis. The former includes empirical theory, which is continuously developed such as Peck's theory, and numerical analysis, which refers mainly to the finite element analysis (Peck 1969; O'Rourke 1981; Chen et al. 2007). As for the later, many scholars have studied dozens of deep foundation pits in Hangzhou and Shanghai soft soil areas and got lots useful results (Liu et al. 1999; 'Li et al. 2007). However, in Tianjin, there only limited study

Y. Tang (✉)
Key Laboratory of Geotechnical Engineering and Underground Engineering of Ministry of Education, Tongji University, 1239 Siping Road, 200092 Shanghai, People's Republic of China
e-mail: tangyiqun2@mail.tongji.edu.cn

Y. Tang · J. Yan · Z. Sun · J. Zhou
Department of Geotechnical Engineering, Tongji University, 1239 Siping Road, 200092 Shanghai, People's Republic of China
e-mail: yjj351@126.com

Z. Sun
e-mail: sunzhijun2006@126.com

J. Zhou
e-mail: zjie@umich.edu

Y. Tang
Joint Research Center of Urban Environment and Sustainable Development, Tongji University, 1239 Siping Road, 200092 Shanghai, People's Republic of China

G. Lollino et al. (eds.), *Engineering Geology for Society and Territory – Volume 4*,
DOI: 10.1007/978-3-319-08660-6_4,

has been done in this field. This paper fully analyzes the monitoring results of Bohai Bank deep foundation pit, which is in Tianjin soft soil area and gives recommendations for excavation of deep foundation pits in Tianjin.

4.2 Project Overview

Bohai Bank deep foundation pit project, which is located at the intersection of Liuwei Road and east Haihe Road, has the Haihe River on its south and metro line 9 on its north. The minimum distance between metro tunnel and the pit is only 12 m, and as for the Tianjin Custom Building that on its northwest, it is 15 m. A high-rise building that on its east is also under construction. The position and shape of this project is shown in Fig. 4.1. The total structure area is 187,000 m^2, while the area of underground structure is 137,000 m^2. The main building is 250 m high with 50 floors and the podium building is 24 m high with 4 floors. The project has 3 layers of basement with 50,000 m^2 and the area of pit is about 16,600 m^2. The main excavation depth of podium building is 16.7 m while the main building is 18.8 m. Diaphragm walls are used as bracing and the depth is 33 m and 34.5 m respectively.

4.2.1 Engineering Geological Condition

The soil properties of this project are shown in Table 4.1. The third layer is clay, whose high water content and high void ratio lead to the high compressibility. The bottom depth of this layer is about 4.2–5.2 m, and the uneven layer thickness may affect the settlement of the pit. The sixth layer is silt clay, which also with high water content plays a significant role on pit deformation because it is the thickest layer in the excavation range. The depth of groundwater is 4 m and generally changes with the season in a range of 0.5–1.0 m.

4.2.2 Foundation Pit Construction Plan and Deformation Monitoring Scheme

The project excavation area is very large and the surrounding conditions are complex. In order to control the settlement and insure the safety of surrounding buildings, step by step excavation is needed. Pit should be excavated in the set time and install bracing in time, especially for the bottom, which should be installed reinforced concrete floor right after the excavation to decrease the exposure time.

Here are principles of monitoring distribution works of surrounding roads: First, arrange settlement-monitoring points along the bracing wall. Considering the pit shape and surrounding environment, there are 8 monitoring sections

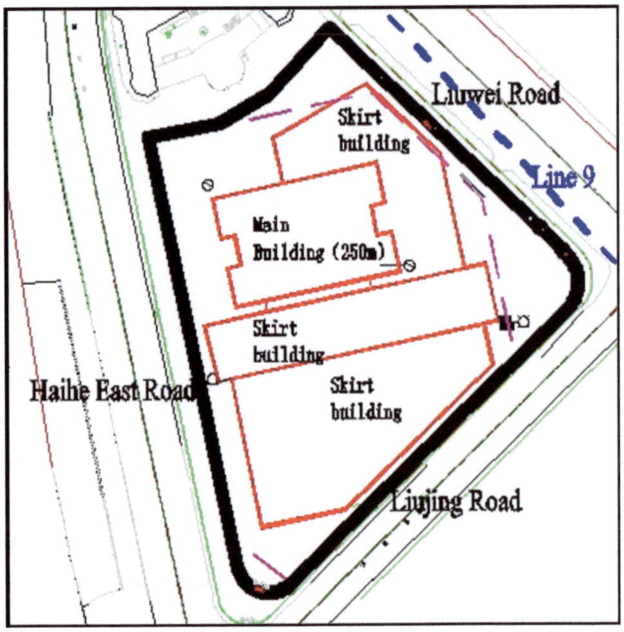

Fig. 4.1 Foundation pit plan

have been set, each of which has 5 monitoring point with 5 m interval, shown in Fig. 4.2.

Secondly, arrange inclinometer points on supporting structure. Because of the complexity of surrounding environment, 11 inclinometer points have been installed on the diaphragm wall, which is the supporting structure of the pit and has small deformation. The location of inclinometer points is shown in Fig. 4.3.

4.3 Analysis of Site Deformation Monitoring Data

4.3.1 Settlement Feature Around Pit

The settlement caused by excavation often has adverse effects on foundations of roads and buildings that close to the pit. The adverse effects mainly display as uneven settlement of buildings or collapse, even worse. This paper mainly focuses on the settlement analysis of surrounding roads.

The settlement curves of monitoring sections by the end of pit excavation are shown in Fig. 4.4, which indicates that the real settlement curves are more complex than those on theory. This is because that there are many factors such as non-homogeneous properties of soil and the vehicle dynamic load may influence the pit settlement, making the actual settlement curves be irregular compared to the ones based on theory. However two main settlement forms of pit deformation theory—triangle settlement form and groove settlement form (Heish and Ou 1998; Clough and O'Rourke

Table 4.1 Physical and mechanical indexes of soil

Soil layer	Soil type	Water content (%)	Void ratio	Soil thickness (m)	Soil weight (kN/m^3)	Internal friction angle φ (°)	Cohesion c (kPa)	Coefficient of permeability K (cm/s)	
								Kv	Kh
3	Clay	35.13	1.01	1.50	19.0	15.30	15.30	8.57×10^{-8}	2.62×10^{-7}
4	Silty clay	27.28	0.77	2.5	19.7	15.10	19.58	2.97×10^{-7}	1.48×10^{-6}
6	Silty clay	30.11	0.84	8.5	19.3	22.22	14.59	2.18×10^{-6}	2.36×10^{-6}
7	Silty clay	26.35	0.76	1.5	20.1	19.50	12.75	1.00×10^{-7}	1.00×10^{-7}
8	Silty clay	23.65	0.67	8.5	20.2	19.61	16.89	1.76×10^{-7}	2.23×10^{-7}
9	Silty sand	17.77	0.53	5.0	20.8	34.21	4.20	1.83×10^{-4}	3.15×10^{-4}
10	Clay	28.23	0.81	4.0	19.7	13.71	27.72	1.00×10^{-7}	3.59×10^{-7}
11	Silty sand	19.28	0.59	5.0	20.3	31.94	3.68	4.07×10^{-5}	3.63×10^{-4}
12a	Silty clay, clay	23.13	0.65	3.0	20.7	23.43	23.91	2.14×10^{-7}	4.30×10^{-6}
12b	Silty sand, silty clay	21.66	0.62	7.0	19.7	35.47	6.09	5.22×10^{-5}	2.26×10^{-4}

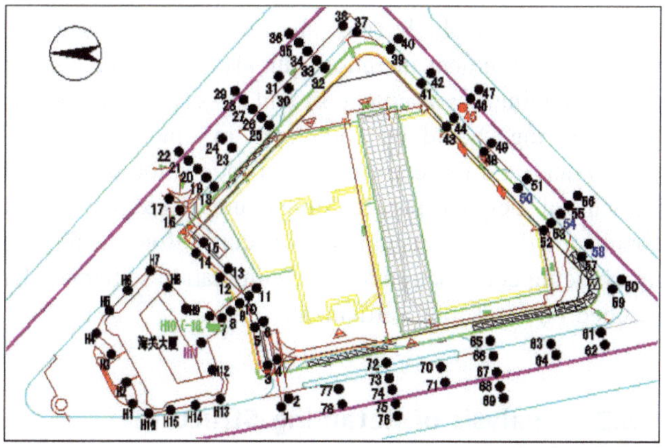

Fig. 4.2 Pit settlement monitoring point layout

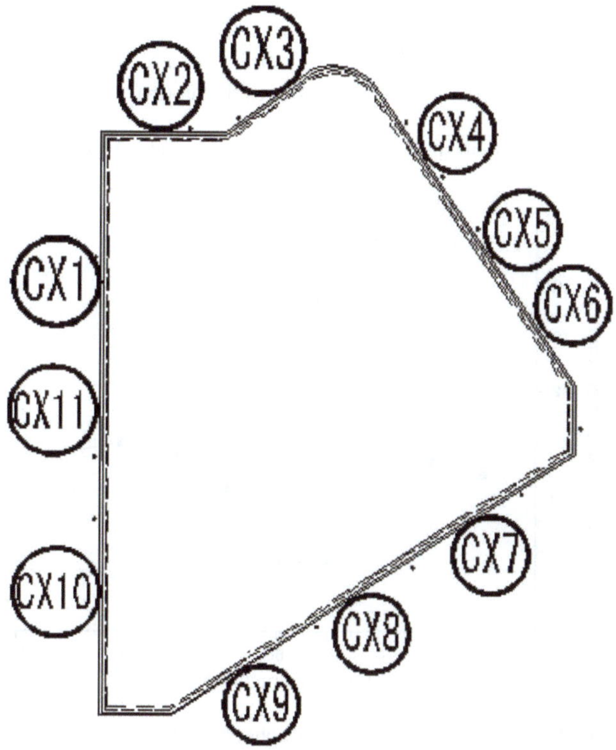

Fig. 4.3 Diaphragm wall inclinometer point plan

1990)—still can be seen and the former one takes more percentage compared with the latter. Taking two settlement profiles of Liujing Road for example, the largest settlement position is about 15 m to the pit and the two curves apparently are neither triangle form nor groove form. This is possibly caused by vehicle dynamic load and additional load from upper structure construction, which could make larger settlement. It can be speculated that the additional load caused by surrounding environment is an important factor that influences the pit deformation.

From Fig. 4.4, the largest settlement is generally at the position, where is 10 m away from the pit. This is consistent to the conclusion drawn by Liu and Wang (2009) that the largest settlement position is about 8–12 m to the wall for ultra-deep foundation pit.

The lateral settlement monitoring curves along diaphragm wall are shown in Fig. 4.5. It can be clearly seen that pit settlement has corner effect, which means that the settlement close to the corner of the wall is smaller than the settlement near the middle of the wall. Liu and Wang (2009) proposed that the deeper the pit is, the more obvious the corner effect is. Therefore, three-dimension is better than two-dimension in pit numerical simulation because the former one can reflect the corner effect. On the other side, the corner effect should be considered in choosing the bracing form in order to better control the settlement of excavation.

Settlement duration curves of some representative points are shown in Fig. 4.6, from which it can be found that the settlement mainly occurred in the first step excavation and the rest seem much more stable. According to the statistic of

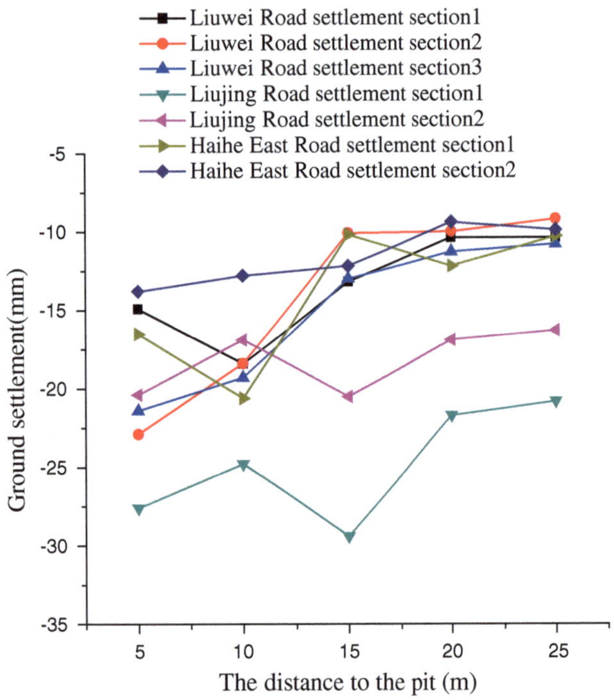

Fig. 4.4 Settlement curves of monitoring sections after pit excavation

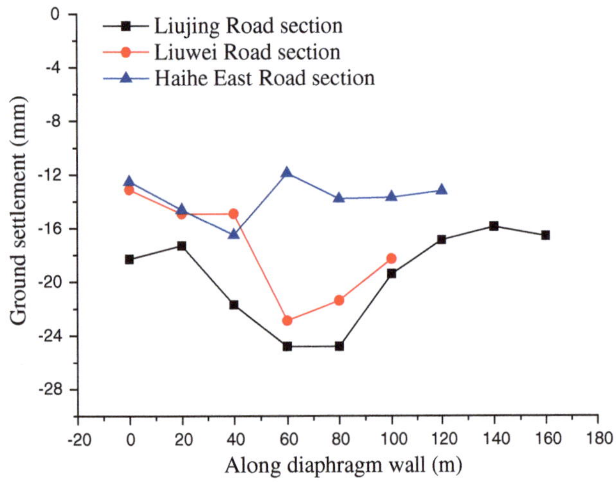

Fig. 4.5 Settlement curves along diaphragm wall

all monitoring points, the settlement by the end of first step excavation accounts for 60 to 90 % of total settlement, and the average is 85 %. It is also shows that the settlement increases quickly in initial stage, and after the completion of the first layer bracing, the settlement rate decreases significantly. The excavation before the first layer bracing completes is similar to cantilever excavation, which can lead to triangular settlement form (Liu and Wang 2009). This gives a reasonable explanation for why the triangular settlement form is more than groove settlement form in Fig. 4.4. In general, the pit support scheme is reasonable in controlling the settlement of pit for the largest settlement is less than

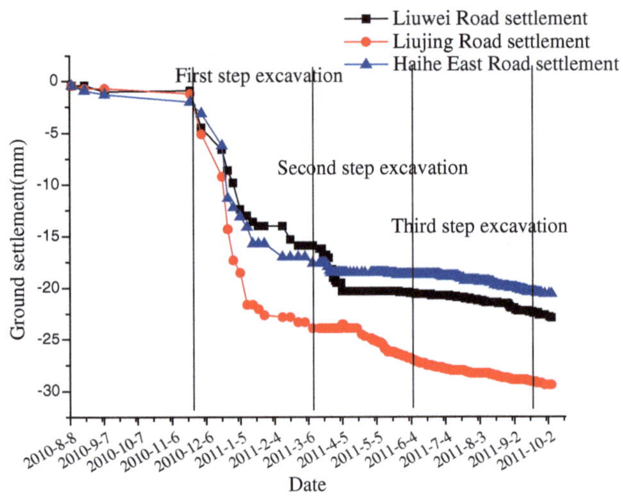

Fig. 4.6 Settlement duration curves of representative points

4 cm. But it should be pay attention to initial stage, in which the settlement increases quickly so that the changes of surrounding environment should be cared and the first layer bracing should be completed in time. Yuan et al. (2004) analyzed the deformation of foundation pit and settlement of neighboring buildings and made the similar conclusion that settlement mainly occurs before the finish of first bracing construction and the cumulative settlement has reached 86 % of total settlement.

4.3.2 Analysis of Retaining Structure Deformation

This project arranged 11 inclinometer points on retaining structure, shown in Fig. 4.3. The maximum lateral displacement and its position are shown in Table 4.2.

According to Table 4.2, the position of maximum displacement (MD) is around the depth of 15 m, accounting for 0.85 of the excavation depth. Considering the position of inclinometer points, it can be found that the deformation of retaining structures is closely related to the geographical location. The lateral displacement on Haihe East Road by CX1, CX10 and CX11 exceed 40 mm and because of the spatial effect, the displacement value of CX1 is smaller than that of CX10 and CX11. CX2 and CX3 are also influenced by the spatial effect. Shown by the monitoring results of CX4, CX5 and CX6 on Liuwei Road, the maximum lateral displacement is about 40 mm. The deformation on Liujing Road is about 35 mm shown by CX7, CX8 and CX9, which is smaller than the other roads. The reason for this may be a project under upper structure construction, whose pit is earlier excavated so that the soil has a trend to the opposite making the decrease of active earth pressure. The side on Haihe East Road is the longest side of the pit, which can

Table 4.2 Displacement data statistics of inclinometer points

Displacement	CX1	CX2	CX3	CX4	CX5	CX6	CX7	CX8	CX9	CX10	CX11
MD/mm	40.4	28.7	34.6	41.8	39.9	40.8	34.2	35.6	34.4	44	46.6
PMD/m	16.2	15.2	15.7	14.7	14.7	13.7	19.2	14.2	13.7	14.2	14.2
DE/m	19.5	17.4	17.4	17.4	17.4	17.4	17.4	17.4	17.4	17.4	17.4
Ratio	0.83	0.87	0.90	0.84	0.84	0.79	1.10	0.82	0.79	0.82	0.82

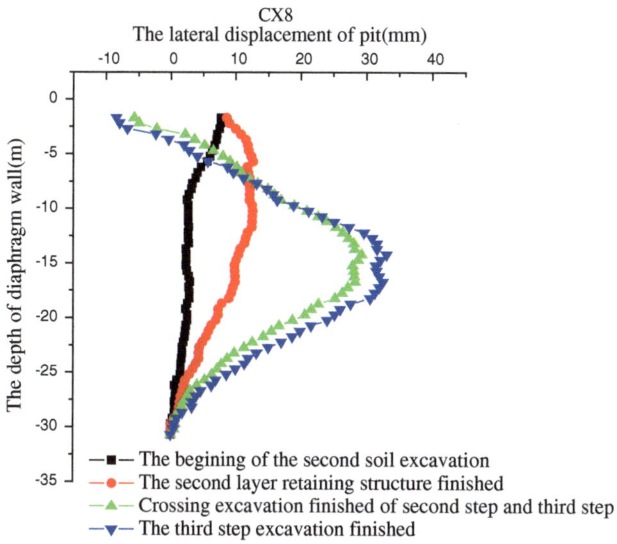

Fig. 4.7 Measured value of CX8

influence the deformation of retaining structures. In general, the maximum deformation of retaining structures is closely related to the surrounding environment. The environment can influence the active earth pressure on retaining structures, which can affect the displacement of the structures.

The measured value of CX8 is shown in Fig. 4.7. The typical changing process of lateral displacement in different excavation stages can be seen in this figure. First, the top of the structure tilts slowly, followed by the falling of the maximum deformation position and the diminishing of upper deformation. The retaining structure turns into a parabolic shape and the position of maximum deformation moves down with the displacement increasing. From Fig. 4.7, it can be seen that the deformation rate is not stable and most deformation happens in a certain time. The maximum rate occurs during the crossing excavation period of the second step and the third step. During this period, the position of maximum displacement has the same depth with the excavation. It then moves down with the excavation until reaching the depth of 15 m, along with a slightly increasing in deformation value. All of this indicates that the pit deformation is sensitive to the changing of excavation stages.

4.4 Effect of Retaining Structure Deformation to Ground Settlement

In order to study the effect of retaining structure deformation to ground settlement, relevant method of soil loss theory was used. The measured data of structure deformation is used to reckon the ground settlement,and then the predicting results was compared to the measured data so that the theoretical relationship could be analyzed. On the other hand, the measured data of structure deformation and ground settlement was analyzed in space and time in order to find the main factors affecting the deformation and settlement.

4.4.1 Soil Loss Theory

Soil loss theory uses the principle that structure lateral displacement is related to ground settlement and adopts the finite element method or elastic foundation beam method to calculate ground settlement (Liu and Wang 2009).

The finite element method or elastic foundation beam method is more suitable for simulating two-dimensional pit than three-dimensional pit because of the spatial effect. In this paper, the measured data of structure deformation is used to reckon the ground settlement. In this way, the shortage of finite element method and elastic foundation beam method can be avoided and the rationality of soil loss method could be tested by the certain project at the same time. The process of calculation is shown in Fig. 4.8.

4.4.2 Calculation Procedures

The calculation in accordance with the following steps:

(1) Use the measured lateral displacement data to calculate the surrounded area S_w of retaining structure by approximate integral method.
(2) The above area is multiplied by the coefficient of m, which is a correction coefficient of the area. Considering the measured data, it can take 1.0 according to Clough and O'Rourke (1990). In real pit project that using diaphragm wall, the lateral displacement is 1.4

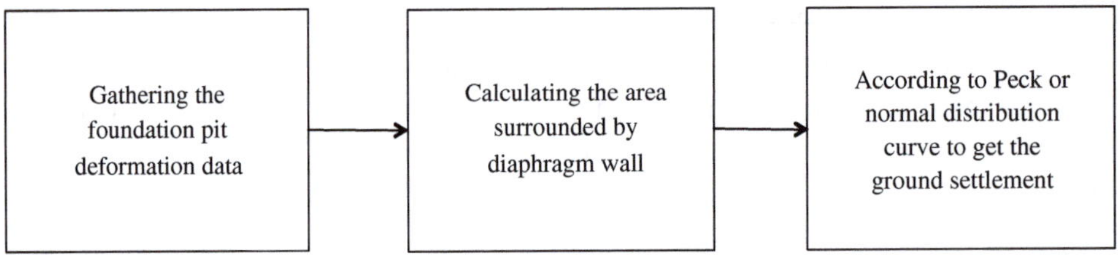

Fig. 4.8 Calculation progress

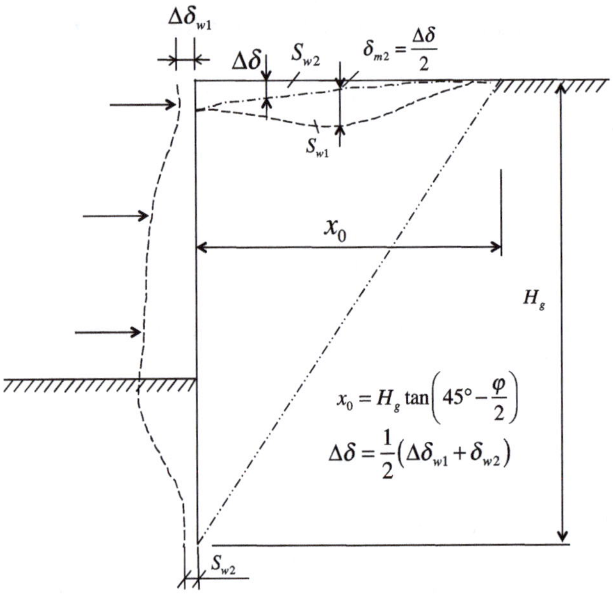

Fig. 4.9 Ground settlement calculation schematic. $\Delta\delta_{w1}$ is the top displacement of the wall; $\Delta\delta_{w2}$ is the bottom displacement of the wall; settlement area S_w can be divided into S_{w1} and S_{w2}. S_{w1} can be calculated by normal distribution curve rule and S_{w2} calculated by the triangle similar theory

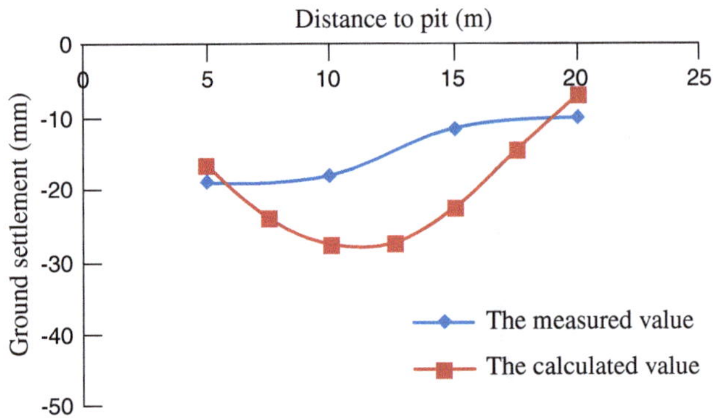

Fig. 4.10 Measured and calculated curves on Liuwei road

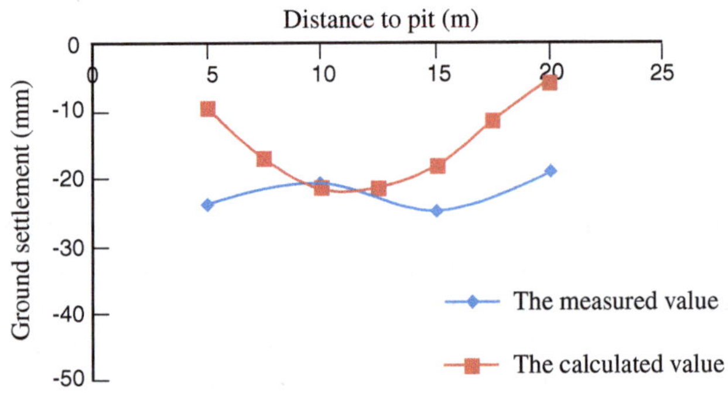

Fig. 4.11 Measured and calculated curves on Liujing road

times of the ground settlement. So S_w would be divided by 1.4 which means that m will take 0.7.

(3) The method of calculating ground settlement is shown in Fig. 4.9.

First, calculate x_0, and the internal friction angle φ could use the weighted average value of soil in the depth of retaining wall. After calculating the settlement of S_{w1} and S_{w2} separately and then summing them up, the final settlement can be got.

4.4.3 Results and Analysis of Calculation

The ground settlement curves of Liujin Road and Liuwei Road were calculated. The calculated area used average value of the road. The measured curve and calculated curves of Liujin Road and Liuwei Road are shown in Figs. 4.10 and 4.11.

From Figs. 4.10 and 4.11, it can be seen that the calculated curves and measured curves have some crosses, but there are also some differences. In Fig. 4.10, the calculated values are mostly larger than the measured values, while in Fig. 4.11, the maximum calculated value is similar to the maximum measured value, but the shape of curve is quite different. This may because of the limit of the soil loss theory and the factors that may influence the ground settlement in real project.

The basic principle of the soil loss theory is that soil movement area of lateral displacement and ground settlement is related. But the influence distance in calculation is smaller than the real. For example, the calculated influence distance of this pit is 22.34 m which accounts for 1.3 times of the depth of excavation. From many previous studies it is known that the influence distance of pit in soft soil area is almost 2 times of the depth of excavation. This means that the ground settlement calculated by soil loss theory would be larger than the real settlement. It can be seen in Fig. 4.11 that the measured values are larger than the calculated values because of the under upper structure construction, which causes more additional load and leads to larger settlement than usual.

4.4.4 Analysis of Measured Data

According to the analysis of measured data, there is a time interval between the development of retaining structures deformation and the development of ground settlement. They are sensitive to different stages of excavation. The main ground settlement is occurred in the period of the first step excavation while the main retaining structures deformation is occurred in the crossing excavation period of the second step and the third step. The retaining structures deformation seems not to be relevant to ground settlement. On the Haihe East road, the structure deformation is large while its ground settlement is small. On the Liujing Road, the ground settlement is large while its structure deformation is small.

Considering the local environment, there is a project under upper structure construction. The pit of that project is earlier than this pit; the soil in the middle of the road has a trend to the opposite pit so that the active earth pressure is relatively small which could decrease the deformation of retaining structures. And at the same time, additional load caused by vehicles could make larger settlement than usual. So this can explain why the ground settlement is large while retaining structure deformation is small.

Through the above analysis, it can be deduced that the main factor for ground settlement is the additional load caused by surrounding environment, and the main factors affecting retaining structures deformation are the retaining structures type and the changes of active earth pressure on the structures.

Overall, the maximum lateral displacement of the pit is not more than 5 cm and ground settlement don't exceed 4 cm, which proves that the choice of retaining structures has obvious effect on controlling lateral displacement and ground settlement.

4.5 Conclusion

The measured record of Bohai Bank deep foundation pit in Tianjin has been summarized and analyzed. Making use of related work of the soil loss theory, the ground settlement has been calculated inversely and the relationship between retaining structures deformation and ground settlement caused by pit excavation are analyzed. According to the analysis, the following conclusions and recommendations are given.

(1) The ground settlement and retaining structures deformation rule of this project are similar to the rules in Shanghai soft soil area. The largest settlement is in the position of about 10 m to pit while the position of maximum displacement is located at the depth of 15 m.
(2) Most settlement occurs before the first layer bracing completing and the excavation at this period is similar to cantilever excavation which will lead to triangle settlement form take more percentage compared with groove settlement form. The pit has spatial effect, so for numerical simulation, the three-dimensional model will be more accurate.
(3) The lateral displacement rule is coincident to the typical retaining structure deformation law. It has a rapid development period, which means that pit deformation is sensitive to changes of excavation stages. The position of maximum displacement is around the depth of 15 m, accounting for 0.85 of the excavation depth.
(4) The ground settlement calculated by soil loss theory is larger than the real settlement. There is a time interval between the development of retaining structures deformation and the development of ground settlement. And the retaining structures deformation seems not to be relevant to ground settlement in this project. The main ground settlement is occurred in the period of the first step excavation, which accounted for 85 % of the whole settlement. The main retaining structures deformation occurs in the crossing excavation period of the second step and the third step.
(5) Through the analysis, it can be deduced that the main factor affecting ground settlement is the additional load caused by surrounding environment, and the main factors affecting retaining structures deformation are the retaining structures type and changes of active earth pressure on the structures.

Acknowledgments The work presented in this paper was supported by the National Key Technologies R&D Program of China (Grant No. 2012BAJ11B04) and Shanghai Leading Academic Discipline Project, Project Number: B308.

References

Chen W, Wu CD, Huang JF et al (2007) Theory, application and tests of 3D analysis method for deep foundation pits. Chin J Geotech Eng 29(5):729–735 (In Chinese)
Clough GW, O'Rourke TD (1990) Construction induced movements of insitu walls. In: Proceedings of the ASCE conference on design and performance of earth retaining structures. Geotechnical Special, New York

Hsieh PG, Ou CY (1998) Shape of ground surface settlement profiles caused by excavation. J Canadia Geotech 35(6): 1004–1017

Li GM (2007) Measured record analysis method applies in predicting of monitoring deep foundation excavation. Dissertation, Tianjin University (In Chinese)

Li L, Yang M, Xiong JH (2007) Analysis of the deformation characteristics of deep excavations in soft clay. China Civ Eng J 40(4):66–72 (In Chinese)

Liu GB, Wang WD (2009) Excavation engineering handbook. China Architecture & Building, Beijing (In Chinese)

Liu XW, Shi ZY, Yi DQ (1999) Deformation characteristics analysis of braced excavation on soft clay. Chin J Geotech Eng 21(4):456–459 (In Chinese)

O'Rourke TD (1981) Ground movement caused by braced excavation. J Geotech Geo-environ Eng 107(9):1159–1178

PECK RB (1969) Deep excavation & tunneling in soft ground. State-of-the-art-report. In: Proceedings of the 7th international conference on soil mechanics and foundation engineering

Yuan JY, Wu Q, Ai ZY (2004) Analysis on deformation of foundation pit enclosure and settlement of neighboring building in case of top-down technology. Undergr Space 24(1):44–47 (In Chinese)

The Coastal Geological Environment and Blue–Green Economic Chain in China

5

Yaoru Lu, Qi Liu, and Zhanfei Gu

Abstract

The characteristics of coastal zones are controlled by the geological structure and properties of rock and soil mass, and accordingly they can be classified into several categories. In this paper, six important problems confronted in coastal geological environment were discussed: (1) earthquake and tsunami; (2) seawater invasion; (3) coast erosion and sedimentation; (4) coastal subsidence; (5) climate—geological disaster chain control; and (6) sea level change. In the West Coast Economic Zone with the core being Fujian Province, China, some coastal cities, such as Fuzhou, Xiamen, Quanzhou, are affected by the Minjiang, Jiulongjiang and Jinjiang rivers. This paper discusses the geological environment of these cities, and the significant connection with the river ecology. A conceptual framework was proposed for developing a sustainable ecological river and a water-front city with ecological civilization.

Keywords

Coastal zone • Geological environment • Ecological problems • Blue economy and green economy

The sea surrounded coastal zones are always important economic development areas in the world. However, the economic development of the coastal areas will inevitably restricted by coastal geological environment. In this paper, this issue will be taken in discussion.

5.1 The Basic Geological Characteristic of the Coastal Zone

The rise of geological structure, sedimentation and fault, fold, has important influence on the formation of the coastal zone, coastal zone accordingly can be divided into: ① rising coastal zone; ② subsidence coastal zone; ③ fracture coastal zone; ④ fold coastal zone. Some were affected by the fracture along the river mouth area or by the fracture parallel to the coast. In the coastal zone, the corner of two groups of faults, gulf is apt to develop due to the mixing action of Marine and river erosion. Of course, there are also bay formed by folding.

Lu Yaoru, male, born in 1931, academician of chinese academy of engineering, professor, mainly research on the hydrogeology and engineering geology. Liu Qi, female, born in 1980, lecturer, mainly research on the water and rock interaction.

Y. Lu · Q. Liu (✉) · Z. Gu
Department of Geotechnical Engineering, Tongji University, Shanghai 200092, China
e-mail: liuqi472@163.com

Y. Lu
e-mail: yrlu@tongji.edu.cn

Y. Lu
Institute of Hydrogeology and Environmental Geology, CAGS, Shijiazhuang 050061, China

Y. Lu · Q. Liu
Joint Research Center of Urban Environment and Sustainable Development, Ministry of Education, Shanghai 200092, China

G. Lollino et al. (eds.), *Engineering Geology for Society and Territory – Volume 4*,
DOI: 10.1007/978-3-319-08660-6_5, © Springer International Publishing Switzerland 2014

The rock distribution of the coastal zone controls the important characteristic of the coast. The coastal zone usually are: ① rock coastal zone; ② muddy coastal zone; ③ sandy coastal zone. Rocky coast zone can be divided into: ① igneous rocks (granite, etc.) coastal zone; ② the volcanic rocks (basalt, etc.) coastal zone; ③ carbonate rocks coastal zone; ④ sand shale coastal zone; ⑤ migmatite coastal zone. The coastal zone of different lithology have obviously different geological environment. The delta foreland area of large rivers, such as the Yangtze River, Yellow River and the Pearl River etc., mostly are muddy and sandy coastal zone.

5.2 The Important Problems of Geological Environment

The geological environment problems are different for different geological structure conditions and lithology of the coastal zone. Here some common problems are emphasized (Chen 2005; Lu et al. 2010).

5.2.1 Earthquake and Tsunami

Ocean surrounded coastal zone is important economic development area, however, it is also geologically active area. Corresponding, some earthquakes can also cause tsunamis, for example, on July 28, 1978, the 7.8-magnitude earthquake devastated the Tangshan in China, killed 240,000 people. At that time earthquake monitoring and early warning was imperfect, and the earthquake was also a sudden. On December 26, 2004, Indonesia's Sumatra earthquake in the sea caused the tsunami, affected the coastal zones of a number of surrounding countries, killed nearly 300,000 people, and destroyed many towns, residential areas of these coastal zones. This earthquake and tsunami had some precursory phenomena, such as: ① abnormal steam increase in the ocean; ② a earthquake observatory of groundwater in Pingyuan county, Meizhou city, Guangdong province, the water column jetted out was about 50 meters above the surface; ③ the research institute of Beijing university of technology on December 19–22, 2004, before the earthquake, had received infrasonic wave; ④ in the east China sea of Jiangsu province, continental scientific drilling continental scientific drilling (CCSD) in the 5 km deep underground recorded the anomalous phenomenon of the inert gas He and Ar (Fig. 5.1). The earthquake precursory response of above several aspects are worthy of attention in the future. In mainland China, more than 10 provinces, autonomous regions and municipalities, at the same time of the earthquake, the hydrogeological monitor holes appeared abnormal phenomenon. This is coseismic effects (Fig. 5.2, Monitoring and Forecasting Department of China Earthquake Administration 2005). The precursory earthquake effects and coseismic effects of so large areas, explains the close relationship in atmospheric envelops (lithosphere, hydrosphere, atmosphere and biosphere) movement, also supply reference for the future earthquake prediction.

5.2.2 Seawater Invasion

In the coastal zone, due to the effect of tide, seawater is closely related to the change of dynamic conditions of groundwater. The gravity F_{sm-e} of the sun and moon for the earth can produce the tide on the surface of the earth solid, this is mainly due to the density change, which caused by the gravity of the earth surface layer solid material, corresponding the volume of pore fissure in the solid tide change, leading to the underground water level changes then liquid tide appear. Compare with the tide of sea water, the changes of groundwater level caused by the moon, also have a periodicity and lunar periodicity (Fig. 5.3). So, the moon for the earth tidal force is proportional to the mass of the moon, inverse proportional to the cub of the distance from the center of the earth to the center of moon. The mass of the sun is bigger than the moon, but the distance from the sun to the earth is much further, so the tidal forces small instead. Some calculation results are shown the biggest moon tide force is 2.17 times as the sun.

Combined with human exploitation of groundwater in the coastal zone, the seawater intrusion force, impact and destruction groundwater resources, also affect the people's life and economic development in the coastline. Take Dalian area for example, intensified seawater invasion, formed wide brackish water areas (Fig. 5.4).

Due to the artificial extraction of karst underground water, the seawater intrusion is intensified. Dalian area underground karst fresh water and brackish water and seawater chemical composition are shown in Table 5.1. The brackish water is made up of 76.78 % of karst water and 23.22 % seawater (Lu 1999).

5.2.3 Coastal Erosion and Sedimentation

The hydrodynamic state changes of the Coastal zone involves the status of the river discharge into the sea, including its kinetic energy and mud content, also involves the influence of the geological structure lifting movement and the change of sea level (tidal).

The scour erosion and sedimentation in coastal zone would bring some changes to coastal zone, thus affect the efficacy of waterway and port facilities. The coastal retreat

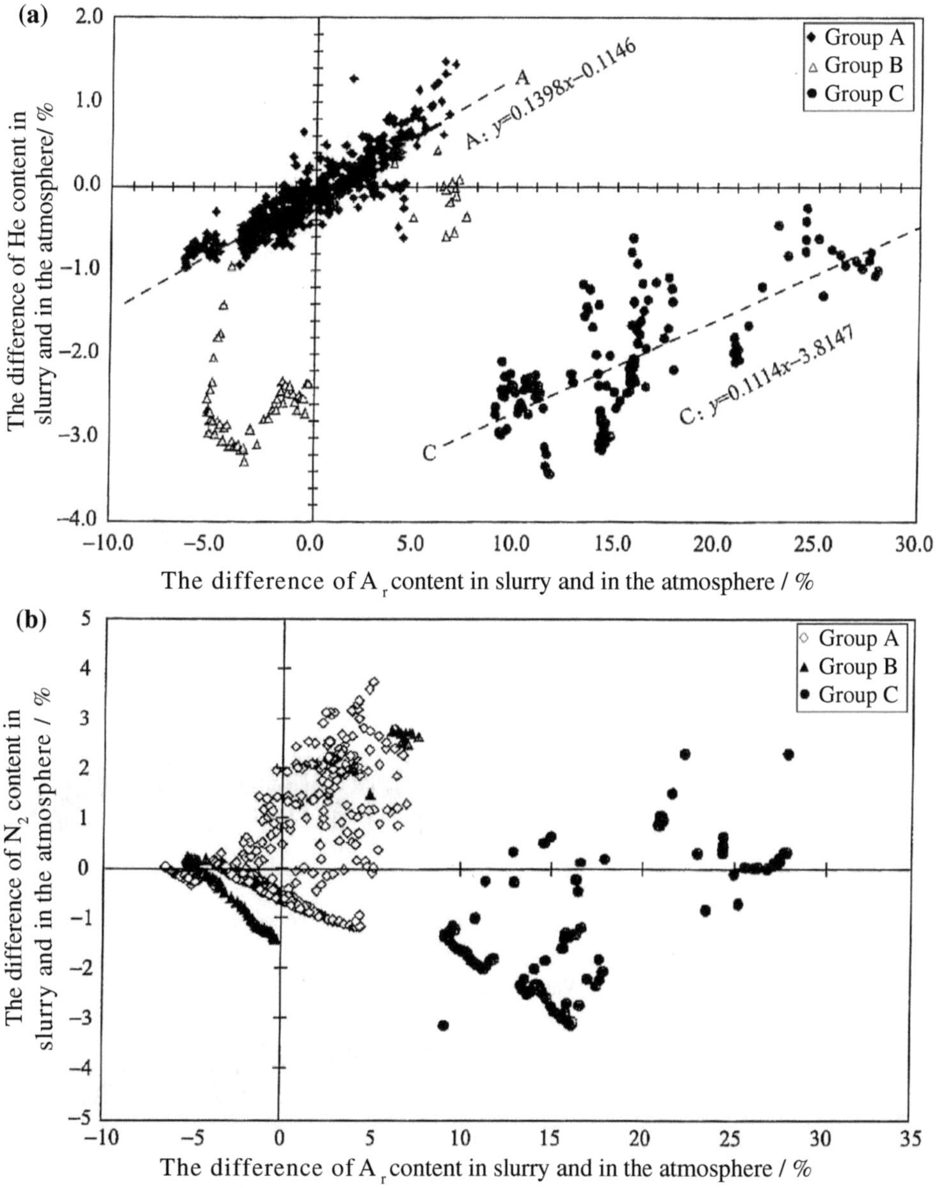

Fig. 5.1 Continental scientific deep drilling mud gas H_e and A_r and N_2 deviating from the air in the components (Zeng et al. 2005). **a** He-A, **b** N_2 -Ar

rate can be up to 0.2–7.7 m/a, and sedimentation makes the water back, the coastal retreat rate be up to 38–200 m/a at some areas. Some artificial sea reclamation does not belong to the category of the natural role. Excessive artificial sea reclamation bring adverse effect to the coastal environment.

In southeast China coastal area, in order to increase arable land area or the construction of pier, etc. artificial sea reclamation used more. The artificial landfill bring out some environmental problems: ① the land subsidence: blow sand spit, construction time is short, cannot make artificial filling soil better consolidation, which often leads to continuous land subsidence, impact the building later; ② increase the deposition: artificial landfill change water dynamic, easy to affect the river estuary, and sedimentation; ③ increase the flood threat: estuaries and artificial landfill affect the dynamic of land river, flood water into the sea not easily, increase the harm to the land; ④ environmental pollution, the river flow dynamic affected, some hysteresis current, make the terrigenous pollutants accumulation, adding water pollution to river mouth.

5.2.4 Coastal Land Subsidence

The coastal land subsidence is often made composite effect of natural and man. Human impact mainly is the extraction of groundwater and load of high-rise buildings, as well as the underground space development. The combination effect makes large areas land subsidence, then lead to great crisis to the coast city, where is easy to be affected by major

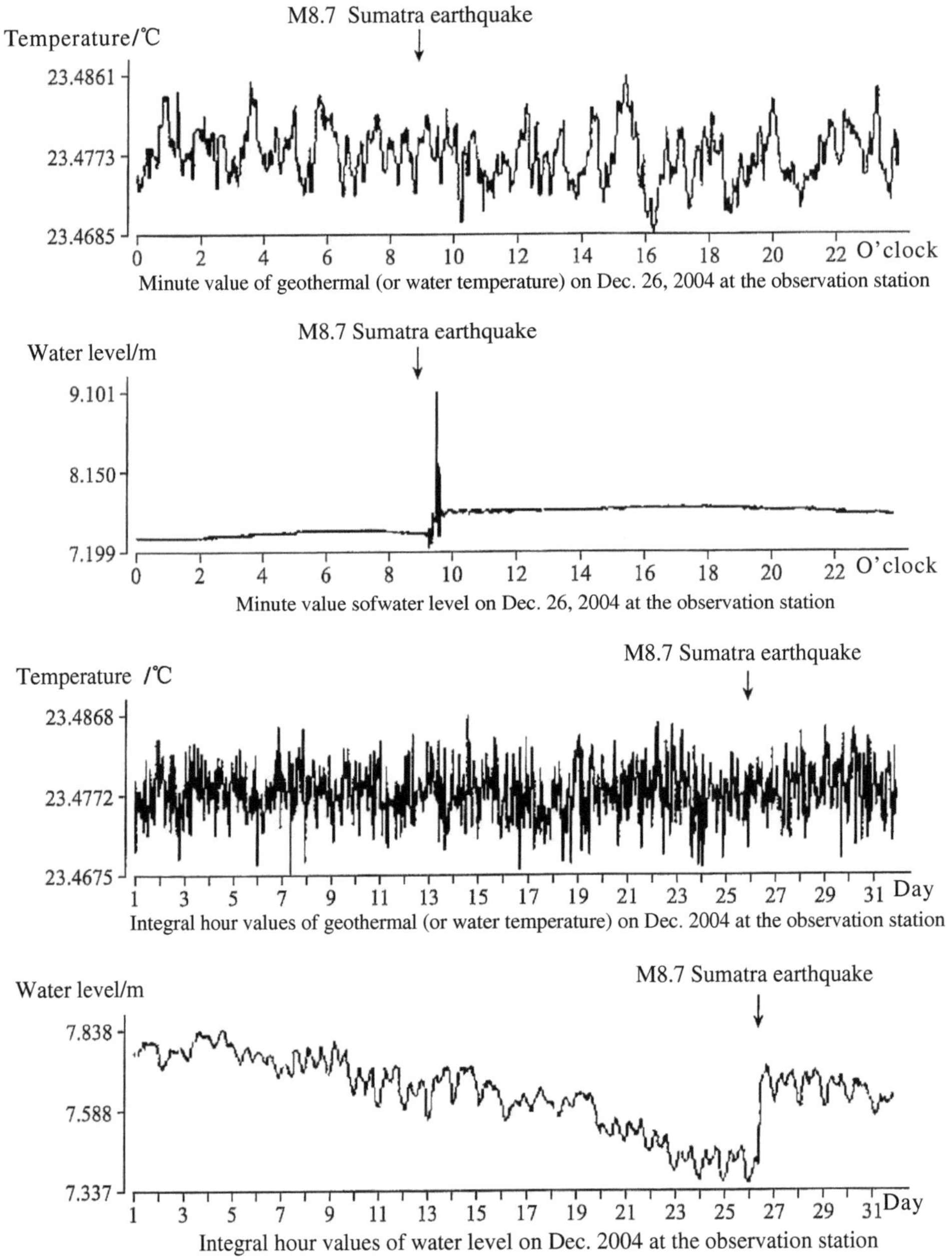

Fig. 5.2 Change curves of water level and temperature on Dec. 2004 at Liaocheng, Shandong province observation station

storm, typhoon, etc., making the city vulnerable in front of nature. The land subsidence area reached 60,000 km^2 with more than 200 mm depth in north coastal China. The land subsidence of the Yangtze River delta, like the Shanghai-Suzhou-Wuxi-Changzhou and other places has already linked together, with a 3 m subsidence center depth.

It should be stressed that the land subsidence caused by human activities such as pumping is not reversible. That is means, the ground water exploration controlling and artificial recharge are impossible to make the existing subsidence restored, could only reduce the sedimentation rate. In Venice, Italy, it is also hard to save its subsidence and the fate of the city, even with hundreds of thousands of root pile foundation. Many megacities on the coastal zone are all facing this threat.

5.2.5 Climate- Geological Disasters Chain Control

Coastal zone is affected by Marine climate disasters, and climate disasters will induce geological disasters. The dangers of climate disasters-geological disasters chain exist in many countries and cities in the world. The geological hazards brought landslide and debris flow, aggravated the ground subsidence and ground collapse. In particular, the

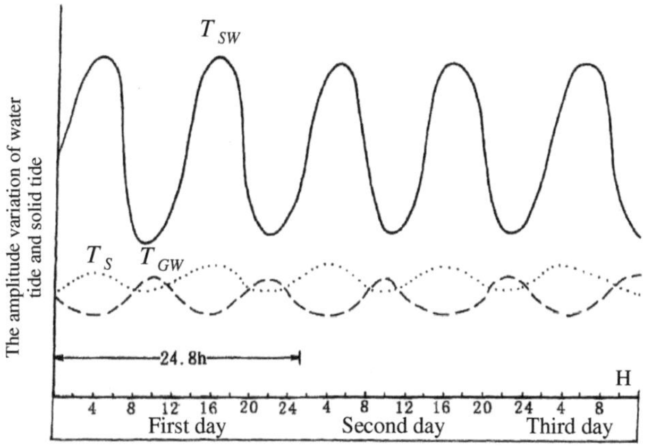

Fig. 5.3 Analysis of daily amplitude variation of earth tide caused by the sun and moon (Lu 1999). T_{SW} sea water tide; T_S solid tide; T_{GW} groundwater tide

geological disasters and seawater invasion and rainstorm water logging by overlapping phenomenon, which caused the loss is amazing. In the monsoon region of Fujian province, for example, the typhoon mainly concentrated in June to September each year, landing or influence of typhoon of Fujian province by more than 62 % of the total number of typhoon in China. Typhoons, storms and other disasters not only caused a lot of flooding in coastal zone, also caused many geological disasters and drought disasters. In Fujian and Zhejiang province, year after year people have to avoid the typhoon and the prevention and control of the impact of climate-geological disaster chain and establish corresponding early warning system, advance leave the danger zone of the people nearly ten thousands or million people. In Fujian province and its adjacent area for 530 years the probability of floods and drought conditions, the statistics as shown in Table 5.2.

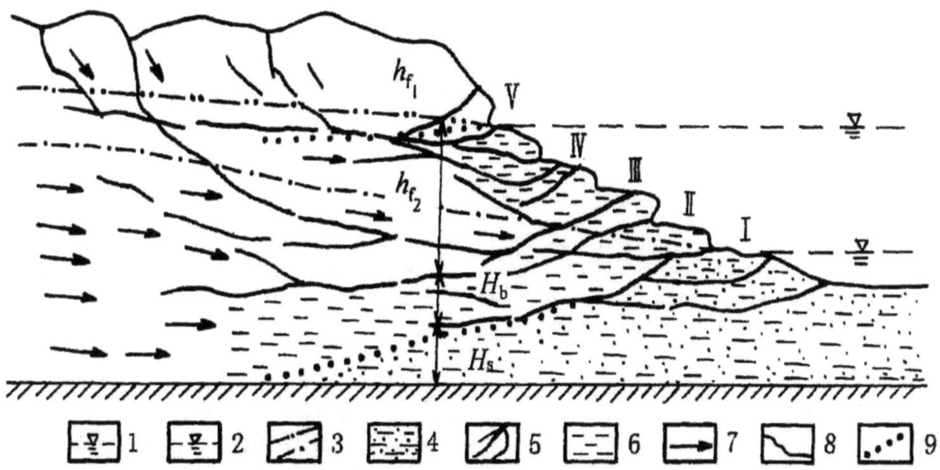

Fig. 5.4 Hydrodynamic profile of cave system caused by the sea water. *1* high sea level; *2* low sea level; *3* high and low groundwater; *4* salty groundwater; *5* karst channel; *6* brackish water; *7* karst fresh water and flow; *8* under the sea terrace; *9* the boundary of different water (variable); *I–V* marine terraces; H_b, H_s the water head value of brackish water and salt water; h_{f1}, h_{f2} respectively corresponding water head value of karst fresh water above and under sea level

Table 5.1 The correlation of different water in Dalian, Liaoning province

Water properties		Bohai sea water	Brackish water	Karst fresh water	Huanghai sea water
Cation (mg/L)	K^+	377.5	35.5	0.5	350.0
	Na^+	10000.0	1975.0	220.0	9500.0
	Ca^{2+}	392.62	407.53	5.47	365.29
	Mg^{2+}	1261.65	337.65	2.11	1187.79
	Total	12031.77	2755.68	228.08	11403.08
Anion (mg/L)	Cl^-	18487.14	4369.09	101.55	17425.08
	SO_4^{2-}	2541.90	589.62	33.83	2382.29
	HCO_3^-	151.43	207.93	368.96	124.30
	Total	21180.47	5166.64	504.34	19931.67
Total salinity (mg/L)		34,498	8,146	620	33,336

Table 5.2 The flood and drought disaster frequency from the year of 1470–2000 in west and east coast zone

Disaster areas	Disaster frequency (%)				
	Heavy flood areas	Light flood areas	All the year round	Light drought areas	Heavy drought areas
Wenzhou	11	23	40	16	10
Shangrao	17	23	36	17	7
Ganzhou	11	23	46	15	4
Fuzhou	14	27	30	21	8
Yongan	15	32	27	17	9
Zhangzhou	14	20	28	25	13
Taibei	99	38	39	9	6
Tainan	13	35	34	13	4

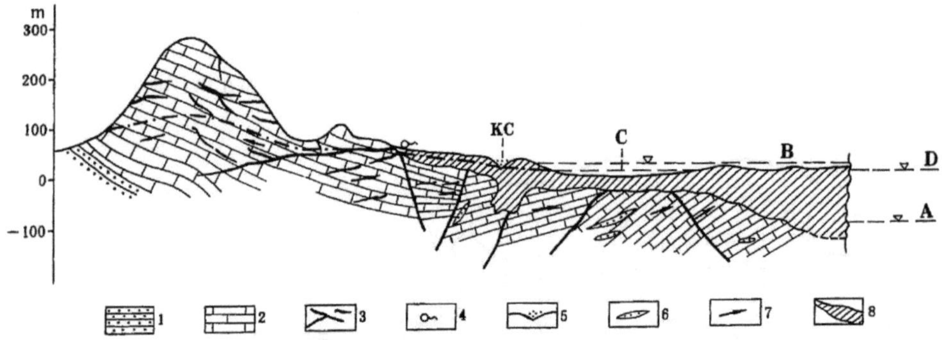

Fig. 5.5 Change schematic of karst hydrogeology and sea water level of the Xihu Lake of Hangzhou, Zhejiang province (Lu 2012). *1* non-carbonate rocks; *2* carbonate rocks; *3* karst cave and channel; *4* karst springs; *5* karst collapse; *6* underground karst; *7* underground karst water flow direction; *8* Quaternary strata; *A* the sea level of 15,000 years ago; *B* the sea level of 5,000–7,000 years ago; *C* water level of Xihu lake; *D* the modern sea level; *KC* karst collapse

5.2.6 Sea Level Change

In nature, the earth's crust evolution occurs during the process of sea level changes: seas change into mountains and mountains change into oceans. The Himalayas are the Tethys Sea 60 million years ago. The sea water area in Xihu Lake of Hangzhou, Zhejiang province can be up to front of the Lingyin temple about 7,000–8,000 years ago (Fig. 5.5). According to many research results (by the marine sediments has been fossils, such as *nakonokawense*, *Elphidiella kiangsuesis*, *Trochammina inflafa*), the last late pleistocene ice age, sea level is lower than the current sea level about one hundred meters.

The greenhouse effect caused by human activities makes the influence of human activity and natural evolution overlap. The situation gets more complicated. In nearly ten years, the sea level of Fujian province has raised 55 mm.

Above six geological environment problems of coastal zone, are not only isolated, but closely associated. So, it should be comprehensive researched.

5.3 Discussion Development of West Coast Economic Zone (WCEZ) and Geological Environment Problems

West coast economic zone take Fujian province as the core, and closely affected by coastal cities: Ningde, Fuzhou, Putian, Quanzhou, Xiamen, Zhangzhou as the six cities, and the Wenzhou of southern Zhejiang province and Chaozhou of northeast Guangdong, etc.

In terms of Fujian province, in addition to the above six cities, other inland cities, especially the main rivers in Fujian province, such as Minjiang river, Jinjiang river, Jiulongjiiang river, etc. connected with the six coastal cities, also related to Nanping, Sanming, Longyan city etc. That is to say, six coastal cities affect the development of many second level inland cities.

So, for the west coast economic zone, six coastal zones should study well the above six aspects of coastal geological environment problems, to better grasp its development. At the same time, should pay attention to the following three important aspects.

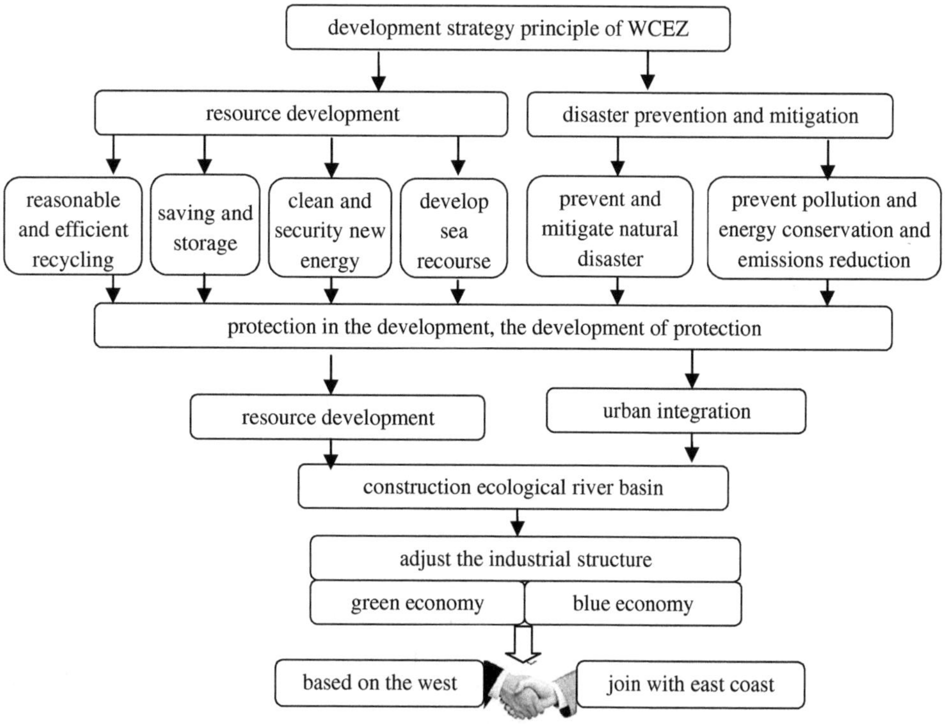

Fig. 5.6 The idea of patterning of development strategy principle of WCEZ

5.3.1 Pay Attention to Ecological Civilization, Construct Ecological River Basin

Give priority to the Minjiang river, Jinjiangriver, Jiulongjiang river, and the corresponding coastal three urban agglomerations: Fuzhou-Ningde, Quanzhou-Putian and Xiamen-Zhangzhou, we must better study on coastal geological environment and the river basin geological environment, better construct ecological, carry forward the ecological civilization, to build a harmonious environment.

5.3.2 The Strategic Principles of Development

The coastal zone cities in Minjiang and Jiulongjiang river basin, from the upstream to the estuary, should develop green economy, harmonious, natural and eco-friendly, with less pollution. Coastal zone city is the end of the green economy, also the starting city to develop blue economy to the ocean. We should develop Marine fishery, Marine mineral resources, and Marine tourism to better protect the blue ocean.

The core content of green and blue economy, mainly related to development resources reasonable and efficiently, and effective in disaster prevention and mitigation, the development of specific strategic principles summarized in Fig. 5.6, Focus on one word is: protection in the development, the development of protection.

About west coast economic development strategy, in summary: spanning development west coast economic zone in Fujian province as the core; Scientific development of high-end leading industry and the foundation industry of the people's livelihood and advanced farming; Building blue and green economy chain of land, rivers, sea, islands; balance urban and rural areas, optimization of urban agglomeration, harmonious channel environment, disaster prevention and benefit promotion; strengthen cross-strait cooperation, construction experimental area, relying on high-quality port group and communication rapid traffic network, promotion and develop a peace maritime silk road to all over the world.

5.3.3 The Way of Scientific Development

West coast development not only involved in the development of Fujian province, also not only involved southern Zhejiang, eastern Jiangxi and northeast of Guangdong province development problems. Actually, west coast development is involved in the whole of China mainland, how can have peaceful and harmonious country situation and rapid develop west coast economic zone in Fujian province as the core, and be able to walking together with east coast, make cross-strait peace and harmonious development, to help the peaceful reunification of the future. To reflect the development strategy, we should have corresponding ways and methods of scientific development (Fig. 5.7).

Constructing west coast economic zone with developed, advanced, rich and good ecological environment, it is to make the harmonious development of Fujian province and Taiwan better, make the bridgehead of Fujian province became a five harmonious situation—harmonious, peaceful,

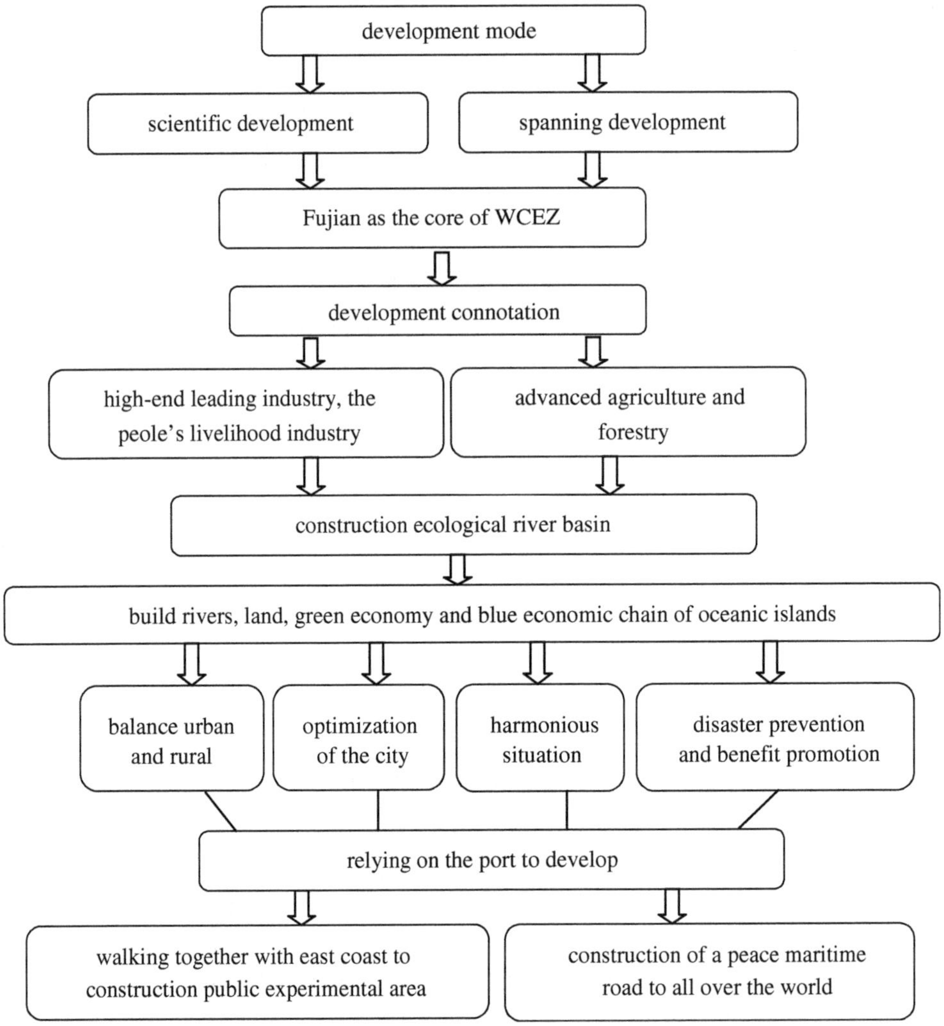

Fig. 5.7 The idea of patterning of scientific development way of WCEZ

smooth, friendly and happy situation, and plus the five special relationship, Fujian province will make significant contributions for development and unity of the both Fujian province and Taiwan province.

Acknowledgments Foundation support: the significant consulting project of chinese academy of engineering, Grant 41302220 from the national natural science foundation of china; Grant 20120072120032 from the specialized research found for the doctoral program of higher education of China.

References

Chen M (2005) The city Geo-environmental characteristics on chinese coastal zone and the major Geo-environment problem. China geological survey. In: Proceedings of geologic environment and city development [R]. China publishing house, Beijing, pp 8–14

Lu Y (1999) Research on the evolutions of Karst hydrogeological environments and their engineering impacts[M]. Science Press, Beijing, p 6

Lu Y, He K, Li X, et al (2010) Urban agglomeration region geological and ecological environment and sustainable development research on Shandong peninsula [M]. Geological Publishing House, Beijing

Lu Y (2012) Karst in china-a world of distinctive [M]. Higher Education Press, Beijing, p 9

Monitoring and Forecasting Department of China Earthquake Administration (2005) Sumatra, Indonesia 8.7- magnitude earthquake and its influence on the Chinese mainland in 2004[M]. The Earthquake Publishing House, Beijing

Zeng L, Xu Z, Luo L, et al (2004) Mw 9.3-magnitude sumatra-andeman earthquake and gas geochemistry anomalies in CCSD borehole [J]. J Rock 2005(2):513–520

Coastal Hazards Within Indigenous Settlements of Chukchi Peninsula

6

Alexey Maslakov and Gleb Kraev

Abstract

The paper contains results of surface cartographic and geocryological analysis of geotechnical stability within built-up coastal areas of Chukchi peninsula. Comparing topographic maps and plans of the years 1960s–1990s with extra-high resolution satellite imagery of the last several years revealed coastal retreat and ravines' growth. The deterioration of the geotechnical conditions may be caused by climate change and its consequences (e.g., expanding of open water period duration) or intensification of human activity. Based on materials of engineering surveys, including data on the properties of frozen soils, a stability forecast was made and recommendations of coastal protection were given.

Keywords

Coastal erosion • Shore retreat • Chukchi peninsula • Satellite imagery

6.1 Introduction

Arctic coastal retreat and thermal erosion acceleration are the result of climate change, its direct effect on permafrost temperature, and open water period expand, both being destabilizing factors for the coasts (State of Arctic Coast 2010). The average shoreline retreat rate in the Arctic is 0.57 m/yr (Lantuit et al. 2012). Depending on coastal morphology, sediments, and permafrost features, as well as the location in respect to prevailing winds and sea ice duration, the registered coastal retreat rate has reached 3–15 m/yr. However, only 5 % of the shoreline is covered by regular instrumental or remote studies (Fundamentals 1998; Lantuit et al. 2012). Moreover, even less data on coastal dynamics exist for Chukchi and Bering Sea coasts, which have always been the place of population concentration.

Coastal erosion is threatening both communities and industrial infrastructure. The plight of several communities in Alaska and Russian Arctic has been widely documented emphasizing the need to move to safer sites (Mackay 1986; State of Arctic Coast 2010; Kraev et al. 2011). Analysis of meteorological characteristics for the last decades (Artic and Antarctic Research Institute, http://www.aari.nw.ru) reveals that sea ice has started forming later in autumn, exposing Chukchi settlements to more frequent and more damaging storms. Besides the threat to buildings, many landfills, sewage lagoons and water sources are situated in locations where they can be damaged by erosion which could cause negative environmental impact.

Coastal erosion in the Arctic is not a new phenomenon and many Arctic communities have been dealing with it for years. However, there are no comprehensive global assessments of the vulnerability of Arctic communities and infrastructure to accelerated coastal erosion. The situation in some US and Canadian communities is so critical that they are considering immediate relocation (e.g., Shishmaref:

A. Maslakov (✉)
Geographical Department, Moscow State University, GSP-1, Leninskie Gory 1, 119991 Moscow, Russia
e-mail: alekseymaslakov@yandex.ru

G. Kraev
Center for Ecology and Productivity of Forests, Profsoyuznaya Street, 84/32, 117997 Moscow, Russia

http://www.shishmarefrelocation.com). In other cases (e.g., Tuktoyaktuk: Mackay 1986), phased retreat to a new location is an option which is now being considered.

6.2 Methodology

On the Russian side of the Bering Strait, coastal dynamics studies were initiated in 2010 within coastal indigenous settlements, built on permafrost. Several communities on the shores of Chuckhi and Bering seas within Chuckhi peninsula were chosen (Fig. 6.1). The coastal erosion rates are studied with the help of archival topographic maps and plans (Map of the planning 1967; Situation plan 1992; Topographic planning 1992), high resolution satellite imagery and modern geodetic studies. The use of various methods allows the reconstruction of coastal erosion rates for several decades.

Fig. 6.1 The settlements, considered in the study

6.3 Coastal Hazards Within the Considered Communities

6.3.1 Lorino Settlement

The settlement of Lorino is situated on the Middle-Pleistocene marine terrace composed by fine- to coarse-grained sands with low and medium ice content. The permafrost temperature within the settlement area is −3 to −4° C, the active layer thickness is 0.5–1.2 m. The community, being one of the largest indigenous settlements of Chukchi Peninsula (Lorino settlement, http://lorino1.narod.ru), has been experiencing the rates of erosion of up to 1 m/yr (Fig. 6.2) during the last 45 years (Kraev et al. 2011). The coastal bluff collapses during storms, when sea waves remove thawed grounds and erode frozen deposits at base of slope. Between storms exposed frozen deposits and ice thaw and slide to the base of bluff. Because of such coastal retreat mechanism, some engineering facilities have already been destructed within the settlement, and some of them—the fur farm, storehouses and the radio centre—are under the threat of demolition and relocation. During the study of coastal erosion rates, some fluctuations were revealed. In the period of 1967–1979 the shoreline retreat rate was 0.24 m/yr. In the next 13 years (1979–1992) coastal bluff within Lorino settlement was destructing with speeds exceeding 0.5 m/yr. From 1992 to 2010, a reduction of thermal abrasion is registered: the retreat rate has decreased to 0.2 m/yr. In last 3 years (2010–2013) coastal erosion has intensified again, exceeding retreat rates of 2 m/yr. Probably the main reasons of such retreat rate fluctuations within the studied coast

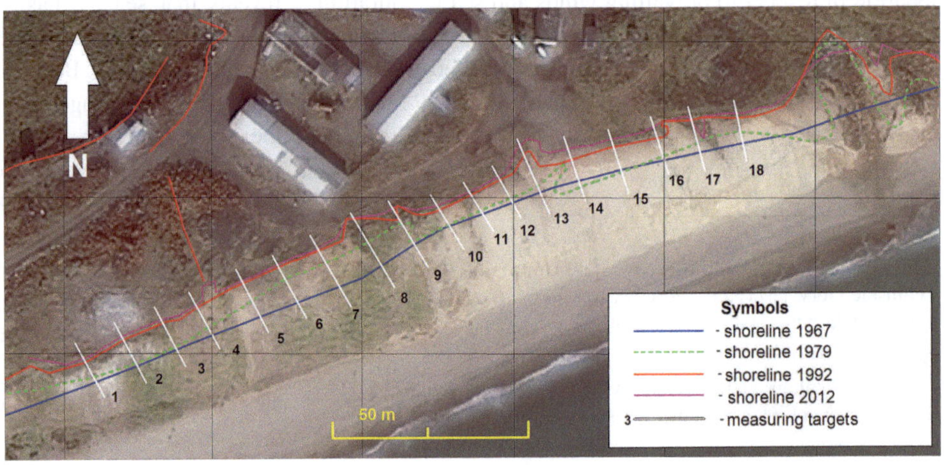

Fig. 6.2 Shore retreat within Lorino settlement for last 45 years

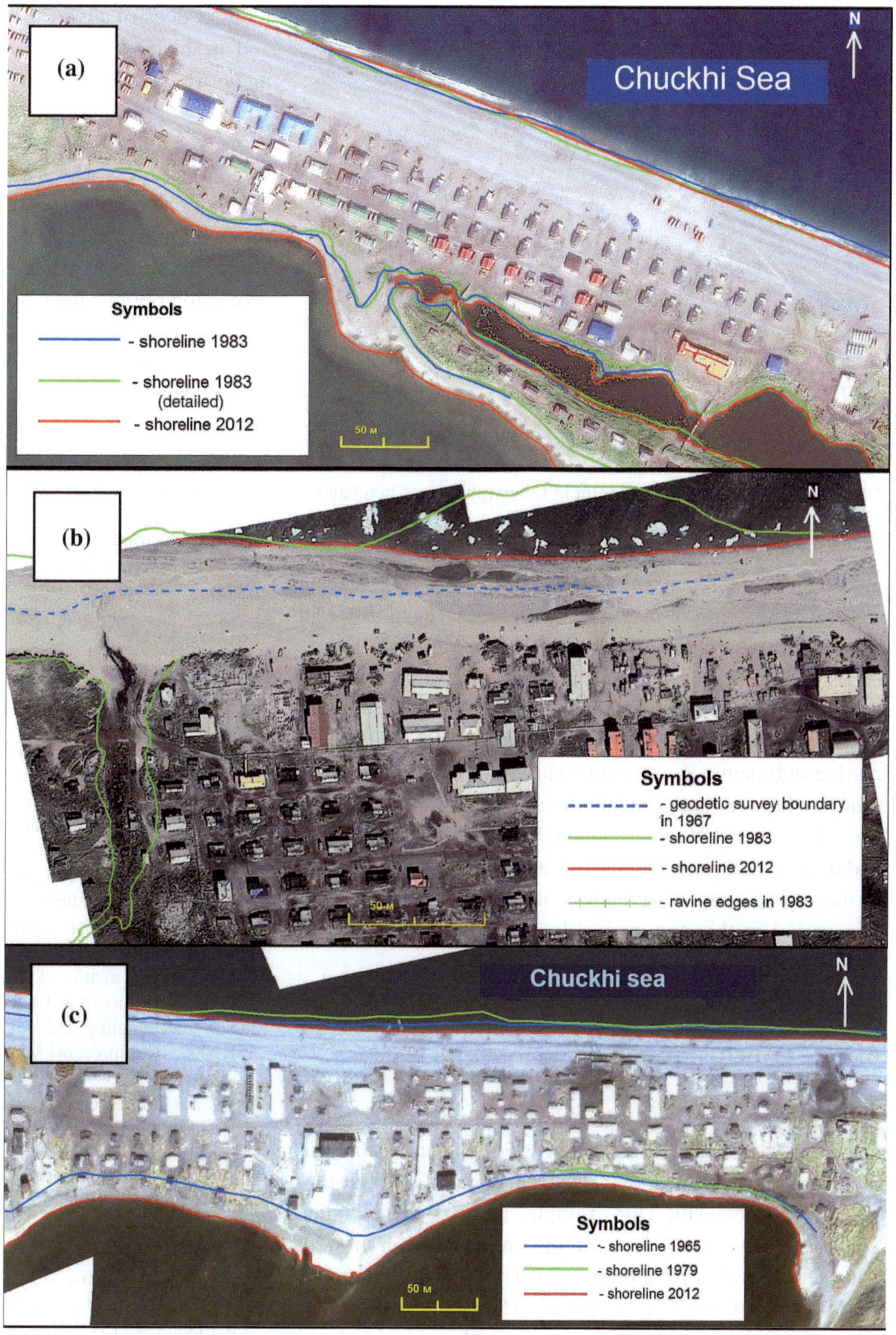

Fig. 6.3 Coastal dynamics within studied settlements: **a** Inchoun; **b** Neshkan; **c** Uelen

section are features of atmospheric circulation in different years (http://www.aari.nw.ru) and consecutive exposure of ground layers with various erosion resistance in the coastal bluff (Shur et al. 1984; Zhestkova 1982; Topographic mapping 1992). Besides, intensification of shore ravines growth was observed. New ravines formed in places, where thawing of ice bodies had been located. New erosion steps were found in older existing ravines.

6.3.2 Inchoun, Neshkan and Uelen Settlements

The communities Inchoun, Neshkan and Uelen are situated on pebble-sandy Holocene spits, which separate lagoons from Chuckhi Sea. Atmospheric circulation features are the same for these localities as well. The grounds within the built up areas are frozen: the permafrost temperature is −3 to −5 °C, the total ice content varies from 10 to 45 %; the active

Fig. 6.4 Consequences of fall storm in Uelen (photo by Cherepanova 2012)

layer thickness reaches 1.2 m (Materials of engineering 1967). Comparison of maps and satellite images of different years within the communities have shown varied dynamics of the seaward coastline and mostly accumulation of the lagoon coastline (Fig. 6.3). Lagoon shore accumulation was probably caused by the filling of the lagoon with aeolian and marine sediments. Sea coastline variations didn't reveal any trend in shore retreat or accumulation. These variations were caused, on the whole, by sediment balance on marine slopes and tidal fluctuations. Permafrost factors play secondary part in these processes. Nevertheless, the expanding of the ice-free period in this area causes longer wave impact on sea shores and higher storm frequency.

6.4 Conclusions

Permafrost thawing and increase of wave influence in the last years have caused destruction of engineering communications and facilities in majority of coastal communities of Chuckhi peninsula (for example, Fig. 6.4). For studied communities, a prior forecast of smooth acceleration of coastal erosion has been developed. The basic techniques to stabilize the situation are proposed. 'Hard' protection in the form of sea walls and revetments is costly and as long as funds are limited, the design and/or construction may not be adequate. Softer forms of protection such as beach nourishment are cheaper and have already been tested in some Arctic communities. In Barrow, Alaska, this form of protection was implemented (State of Arctic Coast 2010), but was terminated following the destruction of the dredge during a storm. The most economically effective method for this remote area is beach nourishment. Unfortunately, this method doesn't give a full guarantee for shore protection, and it should be used periodically (every year). In settlements (e. g., Uelen), where engineering equipment (excavators) for the enrichment of the beach is not available, shore revetments are already using, despite the high cost of these measures. Nevertheless, engineering and financial calculations of nourishment material volume and the cost of the works are necessary in the settlements, exposed to coastal hazards. In general, the success or failure of various protection options have not been well documented, if at all.

Comprehensive studies are required to be conducted in the other coastal settlements of Chuckhi peninsula. These studies have to include environmental, engineering, geocryological and geotechnical aspects. Suitable shoreline protection measures allow to avoid costly restore the consequences of coastal erosion.

Acknowledgements The work was financially supported by the program CALM, U.S. National Science Foundation OPP-0352957 grant.

References

Arctic and Antarctic Research Institute site. http://www.aari.nw.ru

Kraev GN, Maslakov AA, Grebenets VI, Kalyanto NL (2011) Engineering-geocryological problems on territories of the settlements of indigenous nations of Eastern Chukotka. Eng Geol J 3:52–58, PNIIIS, Moscow (in Russian)

Lantuit H, Overduin P, Wetterich S (2012) Arctic coastal erosion: a review. In: Proceedings of the Tenth International Conference on Permafrost, vol 4. The Northern, Salekhard, p 304

Mackay JR (1986) Fifty years (1935 to 1985) of coastal retreat west of Tuktoyaktuk, district of Mackenzie. In: Current Research, Part A, Geological Survey of Canada, Paper 86-1A, p 727–735

Map of the planning and development with the Lorino (1967) M.: 1:2000. Rosgiprozem. Magadan land management Expedition (in Russian)

Materials of engineering survey (1967) Object: General plan of Neshkan settlement/central manor farm "Vperyod"/Magadan expedition land management, Magadan, pp 30 (in Russian)

Official site, devoted to relocation of Shishmaref settlement and some other ones. http://www.shishmarefrelocation.com

Regional and Historical Geocryology World. In: Ershov ED (ed) (1998) Fundamentals of Geocryology. Part 3. Moscow State University, Moscow, p 575 (in Russian)

Shur L, Vasiliev AA, Weissman L, Zaikanov VG, Maksimov VV, Petrukhin N (1984) New results of observations of the destruction of the coast in the permafrost zone/coastal processes in permafrost. Nauka, Novosibirsk, p 12–19 (in Russian)

Situation plan (1992) Lorino. M. 1:1000. ChukotTISIz (in Russian)

State of Arctic Coast (2010) Scientific Review and Outlook, Helmholtz-Zentrum Geesthacht, Geesthacth, 2011, pp 170

Site of Lorino settlement. http://lorino1.narod.ru (in Russian)

Topographic mapping from Lorino (1992) M. 1:500. ChukotTISIz, (1992) (in Russian)

Zhestkova TN (1982) Formation of cryogenic soil structure. Nauka, Moscow, pp 217 (in Russian)

A Case Study on the Determination of the Strength of Seabed Sediments from SW Taiwan

7

Huai-Houh Hsu, Jia-Jyun Dong, Che-Ming Yang, Win-Bin Cheng, Shu-Kun Hsu, Jia-Hao Huang, and Yu-Jo Su

Abstract

In order to explore the natural energy resources in the offshore of south west (SW) of Taiwan, there has been systematical surveyed by geophysical and geomechanical methods. However, most of these methods put emphasis on estimating the potential quality of energy resources. There lacks of studying on the geological conditions of seafloor stability for the area across the continental slope and the shallower areas. A project is executed to investigate the seabed stability situations. The cored submarine mud samplers are performed laboratory tests to obtain engineering properties. On the other hand, the 2006 Pingtung earthquake had triggered numbers of submarine landslides in the offshore of SW Taiwan. By means of submarine bottom profiles acquired before and after this event, the back analysis approach can be used to evaluate the representative in situ strength and compares with the strength parameters determined by laboratory triaxial tests. The analysis result indicates the effective friction angle (φ') of the sliding surface is 14° with a cohesion (c′) of 6 kPa. According to the infinite slope stability theory, the landslide with a thickness of 30 m yields an undrained shear strength (c_u) of 37 kPa under the critical condition. These evaluated strength parameters are close to the results obtained from the isotropically consolidated undrained triaxial (CIU) tests ($\varphi' = 15.3°$, $c' = 19.4$ kPa).

Keywords

Earthquake • Submarine landslide • Strength parameters • Back analysis of slope stability • Triaxial tests

H.-H. Hsu (✉) · J.-H. Huang · Y.-J. Su
Department of Civil Engineering, Chienkuo Technology University, Changhua, Taiwan Republic of China
e-mail: hhhsu@ctu.edu.twgeolab; hsu@outlook.com

J.-J. Dong · C.-M. Yang
Graduate Institute of Applied Geology, National Central University, Jhongli, Taoyuan, Taiwan Republic of China
e-mail: jjdong@geo.ncu.edu.tw

W.-B. Cheng
Department of Environment and Property Management, Jinwen University of Science and Technology, New Taipei City, Taiwan Republic of China
e-mail: wbin@just.edu.tw

S.-K. Hsu
Department of Earth Sciences, National Central University, Jhongli, Taoyuan, Taiwan, Republic of China
e-mail: hsu@ncu.edu.tw

7.1 Introduction

Characterizing the shear strength of offshore sediments is essential to evaluate the stability problems related to the use of natural energy resources. Randolph (2004) reviewed practice and potential advances in characterizing the soft sediments typically encountered in deepwater offshore developments. Obviously, it is a challenge to investigate the stability of offshore with water depth approaches to hundreds meters. A project is carried out to put emphasis on developing the technology of offshore site characterization, especially focus on solving the stability problems of seabed. This project highlights the submarine landslides and liquefaction of the offshore sediments, which is highly related to the strength of sediments. The soil samplers obtained from

G. Lollino et al. (eds.), *Engineering Geology for Society and Territory – Volume 4*,
DOI: 10.1007/978-3-319-08660-6_7, © Springer International Publishing Switzerland 2014

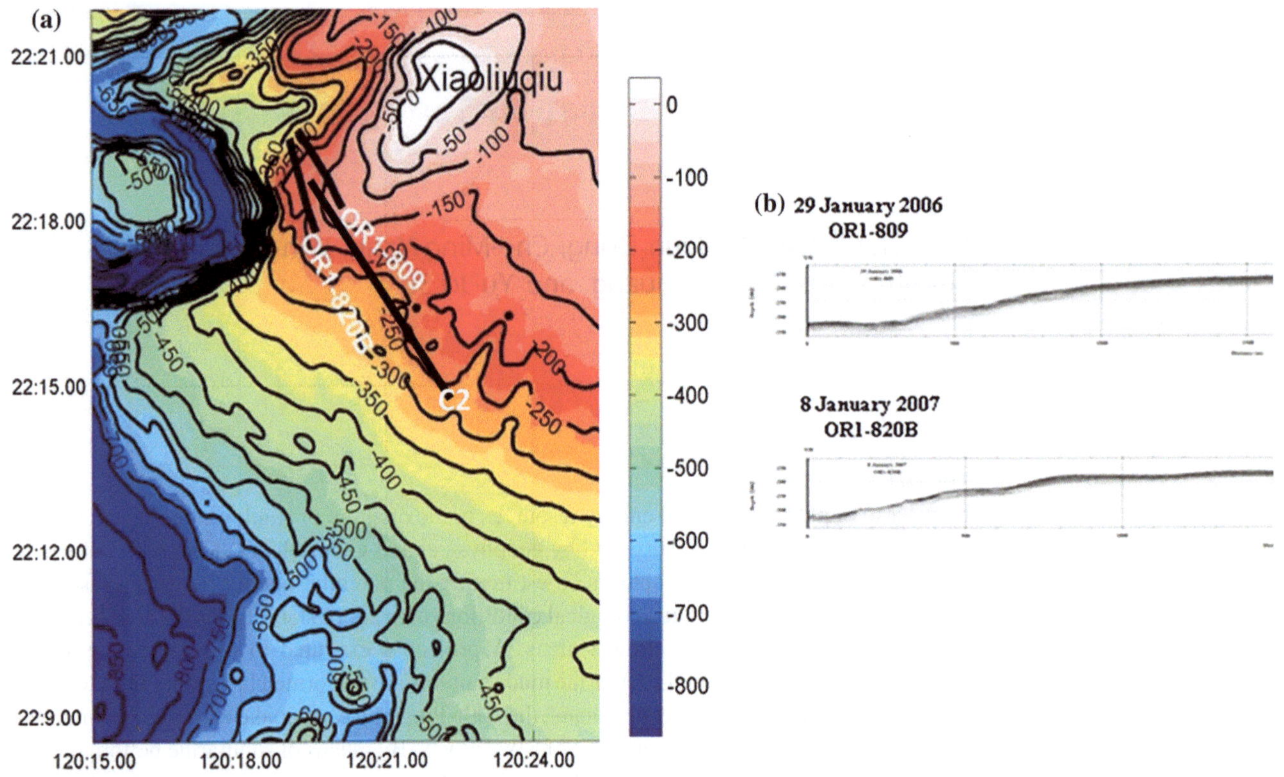

Fig. 7.1 **a** Detail bathymetry near the SW Xiaoliuqiu of offshore SW Taiwan; **b** chirp sonar images of the seabed before and after the 2006 Pingtung earthquake

the seabed will perform laboratory tests and the strength parameters of offshore sediments can be decided. Based on the testing results, the stability of slopes can therefore be assessed without difficulty.

Occurred in the offshore of SW Taiwan on 26 December 2006 with a magnitude of 7.0, the Pingtung earthquake had triggered numbers of submarine landslides (Hsu et al. 2008). This event provides an excellent opportunity to incorporate the back analysis approach to evaluate the representative in situ strength. In this study, two chirp sonar images (Fig. 7.1) of the seabed near the SW Xiaoliuqiu before (OR1-809, 29 January 2006) and after (OR1-820B, 8 January 2007) the Pingtung earthquake are adopted to identify the location of sliding surface. The back analysis results could be used to compare with the strength parameters determined by the triaxial tests.

7.2 Soil Properties

The submarine mud sampler (Core No.: C2) was cored at the water depth of 168 m by gravity sampling method near the SW Xiaoliuqiu (Fig. 7.1). A series of laboratory experiments were performed to establish reference properties. These experiments include basic physical property tests and the isotropically consolidated undrained triaxial (CIU) tests.

The physical properties of submarine mud are summarized in Table 7.1. Table 7.1 shows the Unified Soil Classification (USCS), natural water content (ω_n), mass unit weight (γ_m), void ratio (e), specific gravity (G_s) and Atterberg's limits (plastic limit and liquid limit). Because of the ω_n (= 42.9 %) is higher than the Atterberg's liquid limit (LL = 36.7 %), the submarine mud is a sensitivity soil and hard to be remolded. The void ratio (= 1.352) indicates the mud sampler is highly porous.

A series of CIU triaxial tests with excess pore pressure measurements were performed on the submarine mud. The triaxial specimens were sheared by axial compression (AC). A total of 3 CIU-AC triaxial tests were performed with combinations of 3 levels of confining pressure. Variables applied in this series of triaxial tests are summarized in Table 7.2. The main objective of these tests is to evaluate the strength. Test results show the effective friction angle (φ') is 15.3° and cohesion (c') is 19.4 kPa (Table 7.2).

7.3 Slope Stability Analysis

In this study, two chirp sonar images of the Pingtung earthquake are adopted to identify the location of sliding surface (Fig. 7.1). Back analysis utilizing this submarine landslide is worthwhile to obtain the representative shear

Table 7.1 Physical properties of submarine mud

Core no.	USCS	ω_n %	γ_m, kN/m^3	Void ratio, e	G_s	PL %	LL %
C2	ML	42.9	16.97	1.352	2.843	28.4	36.7

Table 7.2 Summary of the CIU-AC tests performed

Core no.	Confining pressure kPa	Strength parameters	
		c′, kPa	φ', deg.
C2	101,245, 479	19.4	15.3

strength of the sliding surface. It is indicated that the studied landslide resulted in a circular sliding surface (Fig. 7.2). The dimensions of the landslide body are 330 m in length and a maximum thickness of 30 m. Utilizing the widely used software for slope stability (SLIDE 5.0, by Bishop's simplified method), the strength parameters under the critical condition (i.e. safety factor = 1) can be back calculated. The input parameters (Table 7.3) included unit weight (γ = 17 kN/m^3) and seismic coefficient for pseudo-static analysis (k_h = 0.067, and k_v = 0.033) related to the Pingtung earthquake (Central Weather Bureau, Rec. No. 95107).

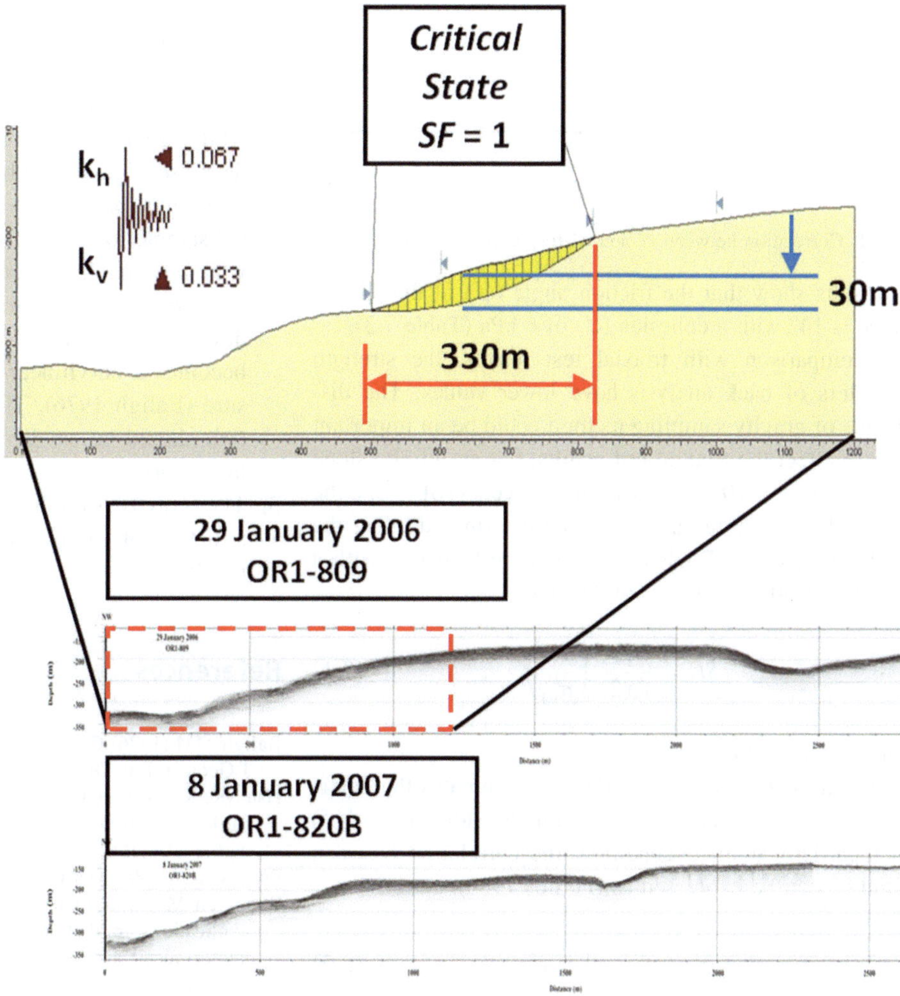

Fig. 7.2 Landslide profiles and back analysis of slope stability

Table 7.3 Summary of the back analysis performed

	Inputs						Results	
Method	Failure mode	γ kN/m^3	Seismic coefficients		Landslide body		Strength parameters	
			k_h	k_v	Length m	Thickness m	c′ kPa	φ' deg.
Bishop's simplified	Circular	17	0.067	0.033	330	30	6	14

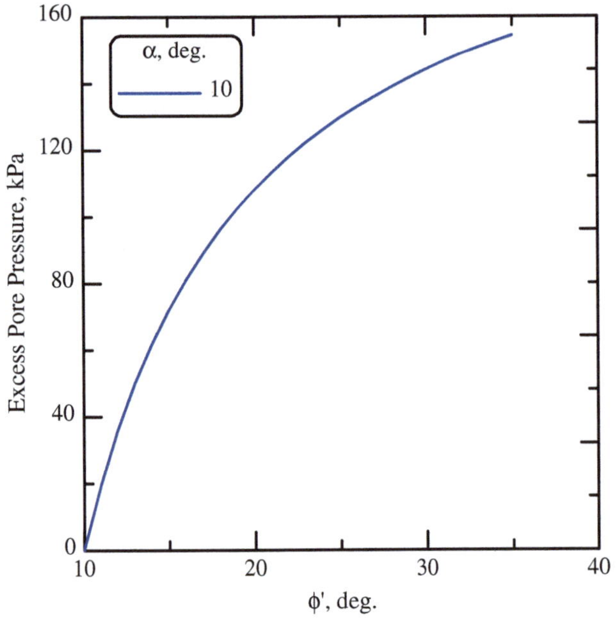

Fig. 7.3 Correlation between φ' and u^+ for an infinite slope

The results show that the friction angle (φ') of the sliding surface is 14° with a cohesion (c′) of 6 kPa (Table 7.3).

In comparison with triaxial test results, the strength parameters of back analysis have lower values. The disturbance of gravity sampling method could be an important factor to affect the triaxial test results. For an infinite slope (slope angle $\alpha = 10°$) at which failure occurs (SF = 1), the undrained shear strength (c_u) can be determined using the following equation (Poulos 1988) and the landslide with a thickness of 30 m yields c_u of 37 kPa:

$$SF = \frac{2}{\sin 2\alpha}\left(\frac{c_u}{\sigma'_{v0}}\right) \tag{3.1}$$

where σ'_{v0} is vertical effective stress.

The excess pore pressure (u^+) also dominates the determination of φ'. Concerning an infinite slope ($\alpha = 10°$) under partially drained condition, the correlation between φ' and u^+ can be expressed as (Poulos 1988):

$$\alpha = \varphi'\left(1 - \frac{u^+}{\sigma'_{v0}}\right) \; if \; c' = 0 \tag{3.2}$$

As shown in Fig. 7.3, the generating value of excess pore pressure on the shearing surface will affect the φ' to be derived.

7.4 Conclusions

The landslide trigged by 2006 Pingtung earthquake provides an opportunity to assess the slope stability of seabed. The prediction of effective friction angle (φ') by back analysis method is slightly lower than that of deciding from the CIU-AC triaxial test results. The conclusions can be made as follows:

1. The excess pore pressure on the sliding surface weakened the resistance shear strength. The Bishop's simplified method assumes there occurs no excess pore pressure, while the raise of excess pore pressure will cause to decrease φ' predicted by back analysis of slope stability.
2. The water pressure is approximately 1.65 MPa for the C2 sampler obtained from the water depth up to 168 m, which is much higher than the confining pressure of performing CIU-AC triaxial tests in the laboratory. The development of Mohr-Coulomb failure envelope becomes a curvilinear type under high confining pressure (Baligh 1976). The in situ effective friction angle (φ') should be smaller than that of determining in the laboratory.
3. The gravity sampling method obviously disturbs the structure of sampler and consequently affects the test results.

References

Baligh MM (1976) Cavity expansion in sands with curved envelopes. J Geotech Eng Div ASCE 102(11):1131–1146

Hsu SK, Kuo J, Lo CL, Tsai CH, Doo WB, Ku CY, Sibuet JC (2008) Turbidity currents, submarine landslides and the 2006 Pingtung earthquake off SW Taiwan. Terr Atmos Oceanic Sci 19(6):767–772

Poulos HG (1988) Marine Geotechnics. Unwin Hyman, London

Randolph MF (2004) Characterisation of soft sediments for offshore applications. In: da Fonseca V, Mayne (eds) Proceedings ISC-2 on geotechnical and geophysical site characterization. Millpress, Rotterdam

Assessment of Maritime Erosion Index for Ionian-Lucanian Coast

8

M. Greco and G. Martino

Abstract

It is assumed worldwide that the assessment of coastal erosion and induced risk can be carried out through concise indexes which take into account both morphological and anthropic characteristics of littoral system as well as energy intensity of wave climate. The classic literature proposes different indicators in order to describe erosion/progradation activities along the coast and climate wave conditions, generally expressed by wave energy. This means that the physical strength for coastal erosion is commonly assumed to coincide to the annual storm intensity and treated like an independent variable. In the present paper, the *maritime erosion index* has been proposed to be employed in risk assessment induced by continuous nearshore wave modelling and storm events corresponding to different boundary conditions. Such an index takes into account both erosion power and climate wave power. Sensitivity analysis has been carried out with reference to the case study of Ionian-Lucanian coast (Southern Italy) through the comparison with commonly used *storm intensity and exposure index*.

Keywords

Risk analysis • Erosion index • Wave climate • Coastal erosion

8.1 Introduction

The use of indexes and indicators is a synthetic and powerful tool in coastal planning and management as well as to define possible inundation scenarios in terms of both ordinary and extreme events and mitigation actions. Coastal inundation and erosion risks are strongly related: as erosion increases, inundation hazard increases as well due to the reduction of self-defence capability. The analysis of coastal erosion is generally made considering a set of synthetic indexes which take into account morphological, geological and sedimentological properties of the coastal system, short, medium and long term wave climate and socio-economic development and infrastructure layout. These indexes are also differentiated in driving force, pressure, state, impact and response (DPSIR) (OECD 2002). In such a conceptual model, wave climate intensity and erosion coastal rate play the role of independent variables listed in the set of cause and effect respectively, while the mutual interaction is obviously consistent. General assumption refers to, indeed, possible direct dependence between erosion rate and wave energy budget allowing us to consider decreasing erosion as the energy wave decreases and vice versa. The *storm exposure and intensity index* adopted by Ranieri (2010) works in such a way giving an assessment of wave climate activities here assumed as stress. Further, the energy wave

M. Greco · G. Martino (✉)
School of Engineering, University of Basilicata, Via Dell'Ateneo Lucano, 85100 Potenza, Italy
e-mail: giovanni.martino@unibas.it

M. Greco
e-mail: michele.greco@unibas.it

G. Lollino et al. (eds.), *Engineering Geology for Society and Territory – Volume 4*,
DOI: 10.1007/978-3-319-08660-6_8,

assessment is evaluated orthogonally to the shoreline neglecting the longitudinal components even though they are responsible for the sediment transport processes; hence, the index seems not to be able to give an exhaustive view of coastal erosion dynamics.

In the paper, a new physically based index, named *maritime erosion index*, has been proposed in the vulnerability rank analysis for coastal risk assessment. Index evaluation is performed on two homogeneous datasets referred to a relevant number of observations available for a 60 year time interval. Such data sets consist of hindcasting data and shoreline change data, observed on the Ionian-Lucanian coast. For the available data, both indexes have been computed and compared.

8.2 Maritime Erosion Index

The shoreline change is basically related to physical processes of erosion, transport and deposition of sediments, meteorology and wave climate. It also regards the energy of the wave field acting during storm events and/or a medium period.

The "maritime erosion index" considers the energy stress affecting the shore area, which characterizes the territorial system state and is finalized to vulnerability assessment. It is defined by the way of maximum annual erosion, T_E, and surface-wave parameters (H, T wave height and period respectively) corresponding to a specific physical condition, i.e. morphological, maximum, mean and/or with respect to a return period.

$$\vartheta = \frac{T_E}{\sqrt{[3]\frac{gH^2}{T}}} \tag{8.1}$$

The resulting coefficient, ϑ, is formally dimensionless, even though expressed in (s/year). T_E is the velocity of shoreline displacements (erosion or progradation) assumed positive when representing erosion. The denominator represents the *unit wave power* corresponding to the *mean available wave energy* per year from which the transport rate variations in the study area depends on. The more ϑ increases the less system integrity decreases leaving system vulnerability increasing as well.

Maritime erosion index, indeed, allows us to draw natural variation of sediment erosion, transport and deposition as well as shore protection system effectiveness, by referring to a mean annual wave climate or a morphological one.

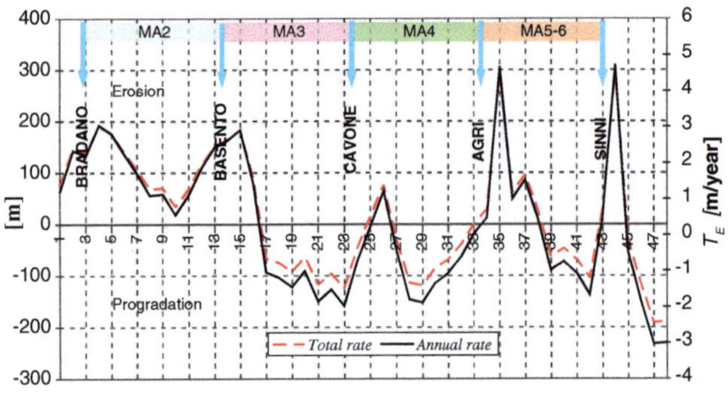

Fig. 8.1 Total rate (*left axis*) and annual erosion/progradation velocity, T_E, observed along the study littoral for the time interval 1949–2012

8.3 Study Area

The study area is located in Southern Italy and quite precisely inside the Taranto Gulf located in the middle part of the Mediterranean Basin. Such an area, corresponding to the Lucanian littoral, is included into a single physiographic unit and exhibits a coastline orientation at 30° N direction with quasi-parallel bathymetry. The coast is delimited by the Bradano River (NE) and the Sinni River (SO), and can be divided in 4 macro-areas (MAs), corresponding to physiographic sub-units. The observed mean annual erosion (or progradation) velocities, for the period 1949–2012, have been estimated through a multisource and multi-temporal approach (Guariglia et al. 2006) based on change detection analysis of satellite imagery and cartographic, aerial and LIDAR data. Figure 8.1 reports the observed erosion/progradation rate evaluated on 48 geo-referenced transects along the coast (sorted from NE to SW) both in terms of mean annual velocity, T_E and cumulative values at 2012.

In detail, Fig. 8.1 gives information about the territorial vulnerability degree in terms of erosion for the whole study area apart from MA 3. Such a phenomenon principally depends on local river dynamics and sediment transport changes due to extensive land defence and river settlement interventions made over 30 years (1960–990), whose effects are still going on.

8.4 Estimation of the Maritime Erosion Index

Due to the absence of way buoy data, wave climate analysis has been performed by means of SMB hindcasting updated method (Greco et al. 2004). Meteorological and wave

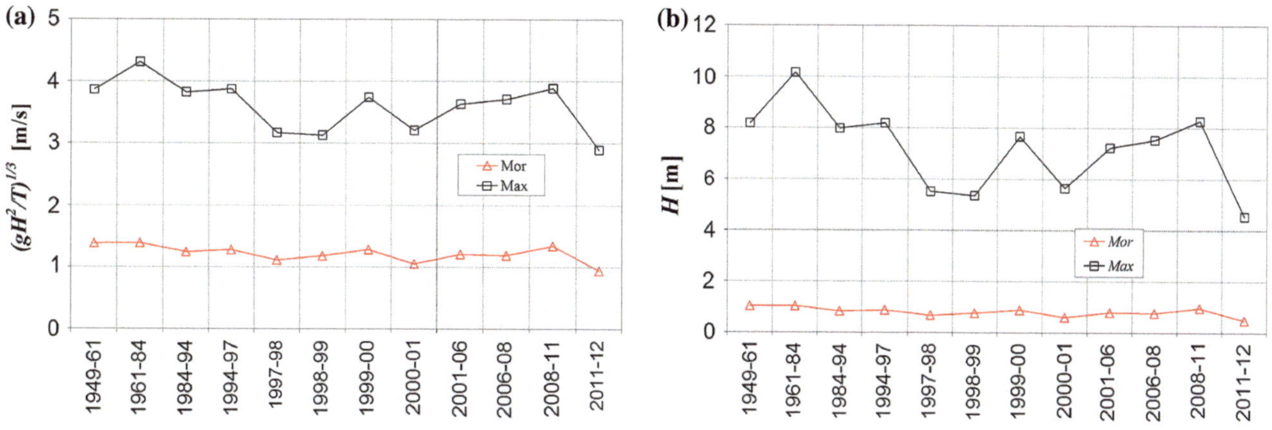

Fig. 8.2 Unit wave power (**a**) and wave height (**b**) referred to morphological and maximum wave conditions

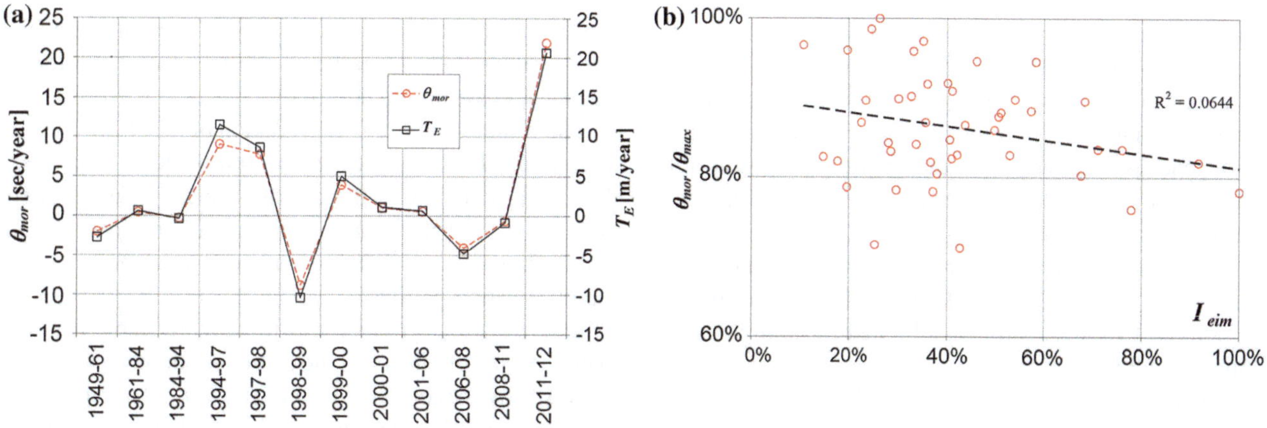

Fig. 8.3 Temporal variation of T_E and ϑ_{mor} (**a**) ($\vartheta_{mor}/\vartheta_{max}$) versus ($I_{eim}$) (**b**)

climate analyses have been performed with wind direction and speed data set in the time interval 1968–2012, collected at Marina di Ginosa meteorological station (id. N. 325 of the CNMCA network—National Center for Aeronautic Meteorology and Climatology of Italian Air Force) located in the Northern Taranto Gulf and representative of the study area wind climate.

Figure 8.2 synthetically shows the results of wave climate analysis for the whole dataset period expressed in terms of deep water morphological (*Mor*) and maximum (*Max*) unit wave power values (a) computed by morphological (H_{mor}) and maximum (H_{max}) wave heights (b).

Time distribution of the maritime erosion index (ϑ) all over the time interval as well as the observed rate T_E are reported in Fig. 8.3a. It shows a uniform time distribution which jointly read with Fig. 8.2a proves a slightly increasing erosion rate. On the other hand, from the combined reading of the Figs. 8.2b and 8.3a, a general decay of coastal erosion self-defence arises presenting an almost constant erosional trend versus a decreasing wave energy rate, that is the littoral has an increasing vulnerability.

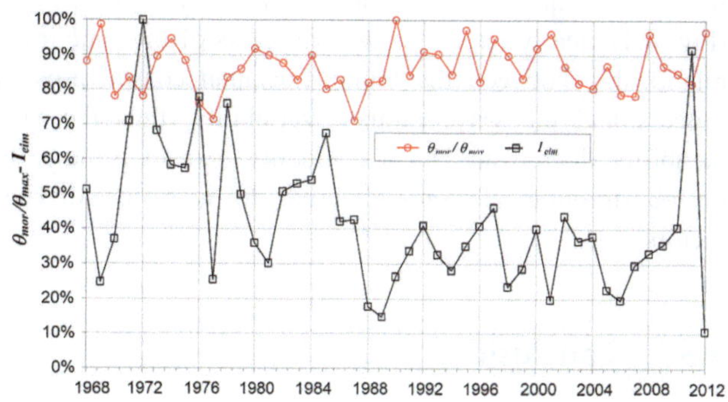

Fig. 8.4 Annual variation of ($\vartheta_{mor}/\vartheta_{max}$) and ($I_{eim}$) indexes

Moreover, comparison analysis between the proposed ϑ index and the *storm exposure and intensity index* (I_{eim}) (Ranieri 2010), defined by the ratio between the mean annual energy, E, and maximum one, E_{max}, is reported in Fig. 8.3b.

The distribution [($\vartheta_{mor}/\vartheta_{max}$) − I_{eim}] sketches a general absence of physical dependence among the indexes. In fact, time comparison between the two index distributions

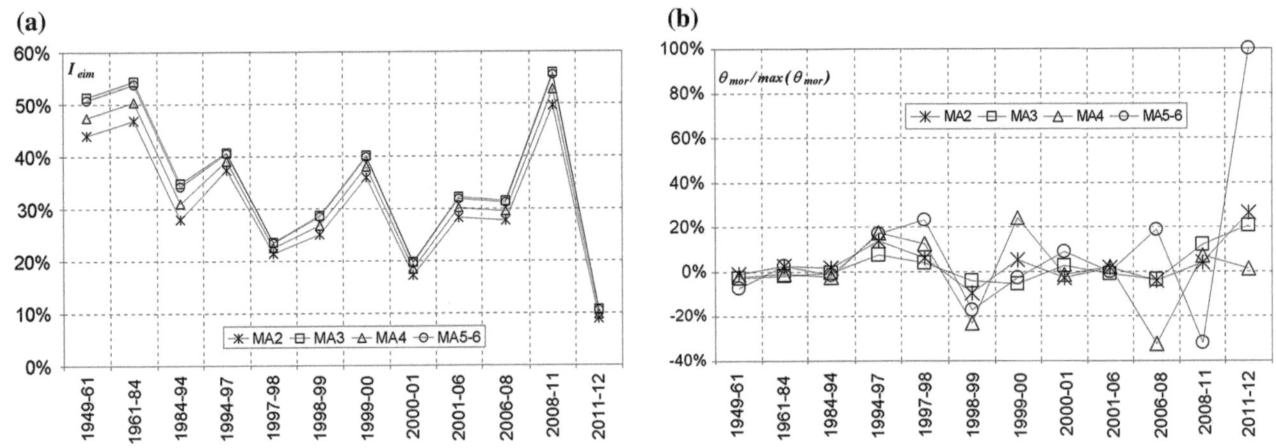

Fig. 8.5 Temporal variation of $(\vartheta_{mor}/\vartheta_{max})$ and (I_{eim}) per macro-area

(Fig. 8.4) presents a time variation of both indexes, in the observation period. The plot outlines a wide variation of I_{eim}, showing a smooth decreasing trend which supports a decreasing intensity of climate wave, versus a "quasi" homogeneous behaviour of the maritime erosion ratio which allows us to assume a steady unit wave power among the observation periods.

Furthermore, I_{eim} refers to a wide variability with respect to $(\vartheta_{mor}/\vartheta_{max})$ and I_{eim} spikes generally do not correspond to erosion rate excesses, outlining a limited capability to select critical conditions or events.

Finally, Fig. 8.5 underlines how ϑ_{mor}, normalized by the maximum one, seems to be more stable than I_{eim} even in terms of effects on shoreline, with reference to the average value of T_E observed per macro-area. In other words, the quasi-steady conditions of the $(\vartheta_{mor}/\vartheta_{max})$ (Fig. 8.4) clearly expresses the weakness of the functional relationship between the Ionian-Lucanian coastal erosion process and the wave climate variability, thus the vulnerability degree and local criticality depend principally on constraining elements external to maritime driving factors.

8.5 Conclusions

The vulnerability rank estimation in coastal areas can be expressed by indexes depending on wave climate parameters. The Authors recognize the *storm exposure and intensity index* (I_{eim}) as representative for the wave climate effects on the coastal dynamics and propose an alternative index named *maritime erosion index* (ϑ).

Such indexes and the comparison between them have been evaluated on a significant coastal area of the Ionian-Lucanian coast in Southern Italy.

The results outline a wide variation of the (I_{eim}), showing a smooth decreasing trend in time which supports a decreasing intensity of the climate wave, versus a "quasi" homogeneous behaviour of the maritime erosion ratio which allows us to assume a steady unit wave power among the observation periods.

The $(\vartheta_{mor}/\vartheta_{max})$ ratio seems to be more stable than I_{eim}, even in terms of effects on the shoreline. In other words, the quasi-steady conditions of the $(\vartheta_{mor}/\vartheta_{max})$ ratio clearly express the weakness of the functional relationship between coastal erosion processes and wave climate variability, considering the vulnerability level not directly affected by maritime driving factors.

References

Greco M, Mauro A, Martino G (2004) Semiprobabilistic integrated model for real sea forecasting, coastal environment. In: Proceedings of 5th international conference on environmental problems in coastal regions, Wessex Institute of Technology, UK, and University of Alicante, Spain, 26–28 Apr 2004, pp 321–330

Guariglia A, Buonamassa A, Losurdo A, Saladino R, Trivigno ML, Zaccagnino A, Colangelo A (2006) A multisource approach for coastline mapping and identification of shoreline changes. Ann Geophys 49(1):295–304

OECD (2002) Overview of sustainable development indicators used by national and international agencies. Statistics Working Papers

Ranieri G (2010) Il calcolo del rischio di erosione applicato al tratto di costa tra Margherita di Savoia e Barletta. In: XXXII Convegno Nazionale di Idraulica e Costruzioni Idrauliche, Palermo, 14–17 Sept 2010

An Inter-Comparison of Coastal Vulnerability Assessment Methods

9

G. Benassai, G. Di Paola, P. Aucelli, M. Passarella, and L. Mucerino

Abstract

In this paper a Coastal Vulnerability Assessment method (CVA) was compared with the well-established CVI method in order to obtain validation data and discuss the different results. The structure of the proposed method relies upon a key parameter, known as impact index I_i, which accounts for wave climate, bathymetry and sediment data and depends on the wave run-up height, the seasonal and long-term erosion, and the efficiency of coastal protection structures. The comparison between this CVA and the existing CVI method was done on a test site already examined. The calculation of the different contributions of each method was transformed in appropriate vulnerability scores in order to examine and discuss the different results.

Keywords

Coastal vulnerability • Vulnerability index • Wave run-up • Storm surge

9.1 Introduction

Coastal vulnerability, defined as the susceptibility of a coastal area to be affected either by inundation or erosion, affects the majority of coasts worldwide and is accountable for destruction of property and infrastructure. Therefore its evaluation is of major importance (Arun Kumar and Kunte 2012).

Among the different methods for assessing the coastal vulnerability, the authors recently suggested the Coastal Vulnerability Assessment method (CVA) (Benassai et al. 2013; Di Paola et al. 2011), who takes into account both the inundation of the inshore land and the beach retreat due to storm surge. The assessment relies upon a new parameter, the impact index, which depends on run-up height, seasonal and multi-year erosion index, coastal structures efficiency.

This method was compared with the most commonly used method for assessing the coastal vulnerability, that is the Coastal Vulnerability Index (CVI) (Gornitz et al. 1994), which was developed with minor modifications by Gaki-Papanastassiou et al. (2010). This approach attributes different classes of vulnerability related to climatic change (i.e. mean elevation, geology, coastal landform, shoreline, wave height and tidal range) to different coastal sections, defining the relative susceptibility of a coastline.

The CVA approach was compared with the previous coastal vulnerability index in the test site of the Sele coastal plain, which is highly sensitive to coastal inundation,

G. Benassai (✉)
Department of Engineering, University Parthenope, Centro Direzionale Is. C4, 80143 Naples, Italy
e-mail: benassai@uniparthenope.it

G. Di Paola
Department of Biosciences and Territory, University of Molise, C.da Fonte Lappone, 86090 Pesche (IS), Italy

P. Aucelli
Department of Science and Technology, University Parthenope, Centro Direzionale Is. C4, 80143 Naples, Italy

M. Passarella
University of Plymouth, Drake Circus, PL4 8AA, Plymouth, UK

L. Mucerino
Department of Earth Sciences, Genoa University, Corso Europa 26, 16132 Genoa, Italy

G. Lollino et al. (eds.), *Engineering Geology for Society and Territory – Volume 4*,
DOI: 10.1007/978-3-319-08660-6_9, © Springer International Publishing Switzerland 2014

particularly in the NW sector (Pappone et al. 2012). The calculation of the different contributions of each method was transformed in appropriate vulnerability scores which were summed to obtain CVA and CVI scores, in order to examine and discuss the different results.

9.2 Coastal Vulnerability Assessment Model

The model proposed in this paper for the assessment of coastal vulnerability is based on the methodology developed and suggested by Di Paola et al. (2011), where a new key parameter known as impact index CVA is properly used for coastal vulnerability evaluation.

The CVA can be calculated according to the following equation:

$$CVA = I_{Ru} + I_R + I_D + E + T \quad (9.1)$$

where I_{Ru} is an index associated to wave run-up distance, I_R is the short-term erosion index for the shoreline, I_D is the backshore coastal protection structures stability index, E is the beach erosion rate index and T is the tidal range.

Here the CVA is carried out by evaluating Eq. (9.1) without considering I_D and T index contributions (Benassai et al. 2013). For each index, the variable values have been applied to ranks 1, 2, 3 and 4 from "very low" to "high" (Table 9.1), by the simple addition of the single indexes, according to EUROSION project (Directorate General Environment European Commission 2004).

9.3 Test Site and Experimental Results

The test site for the inter-comparison experiment is the Sele plain, one of the widest alluvial coastal plains of Southern Italy, which extends from NW to SE in the Gulf of Salerno, in the Southern Tyrrhenian Sea (Fig. 9.1a).

In order to define a detailed mapping of the emerged and submerged beach, a number of 10 topographic and bathymetric profiles were performed, reported in Fig. 9.1b.

CVA was calculated on the basis of the evaluation of I_{Ru}, I_R, E over the ten profiles realized on the Sele coastal plain.

The wave run-up height index I_{Ru} assumes values that depend on the percentage associated to the maximum horizontal wave run-up distance on the beach (X_{max}) normalized with respect to the emerged beach width (L). X_{max} is retrieved through the wave run-up height, which depends on both beach and wave properties, which was evaluated

Table 9.1 Ranking of the single variables and of the whole index in CVA

Variables	Stability-1	Low-2	Moderate-3	High-4
I_R (%)	<15	15–30	30–50	>50
I_{Ru} (%)	<40	40–60	60–80	>80
E (m/year)	<0.5	0.5–1.0	1.0–2.0	>2.0
CVA	3	4–6	7–9	10–12

according to the empirical approach proposed in Stockdon et al. 2006.

The short-term erosion index I_R provides a measurement of potential beach retreat and is used for the dynamical calculation of the shoreline retreat based on the convolution method of Kriebel and Dean (1993). I_R values depend on the percentage associated to the maximum beach retreat (R_{max}) normalized with the beach width L.

Finally, the beach erosion rate index E was performed through the comparison of topographic maps, aerial photographs and multi-spectral satellite images taking into account beach variations prior to 1998 (1954–1975, 1975–1984 and 1984–1998) obtained by ISPRA and more recent data (2003–2009) reported in Table 9.2.

From the results of Table 9.2 it can be noted that profiles P1 and P2 show in the recent period (2003–2009) a moderate erosional trend of 0.7 m/year, profiles P3 and P4 show an erosional trend of 1.2 m/year, profiles P5, P6 and P7 show an erosional trend of 2.0 m/year, and finally the transects P8, P9 and P10, exhibit a clear stability with a mean accretion of 1 m in 6 years.

The average impact index CVA, obtained through the sum of aforementioned indexes, shows that beach profiles P8–P9–P10, P1–P2–P4–P5 and P3–P6–P7 exhibit low ($4 \leq CVA \leq 6$), moderate ($7 \leq CVA \leq 9$), and high ($10 \leq CVA \leq 12$) coastal vulnerability ranking, respectively. Experimental results clearly show that P7 is the most critical case among the ten considered transects, since it exhibits the highest CVA value and thus is the most exposed beach profile.

In synthesis, Fig. 9.2 reveals that profiles from P1 to P5 exhibit a moderate vulnerability, profiles P6 and P7 present a high vulnerability, while this is low for the last three profiles (P8, P9, P10).

Furthermore, Fig. 9.2 shows the different components of CVA index evidenced for each profile. With regard to the component I_{Ru} of the impact index, it shows that profiles from P1 to P7 exhibit a quite homogeneous value of I_{Ru}, while the profiles from P8 to P10 present a lower value. With regard to the I_R component, Fig. 9.2 shows that

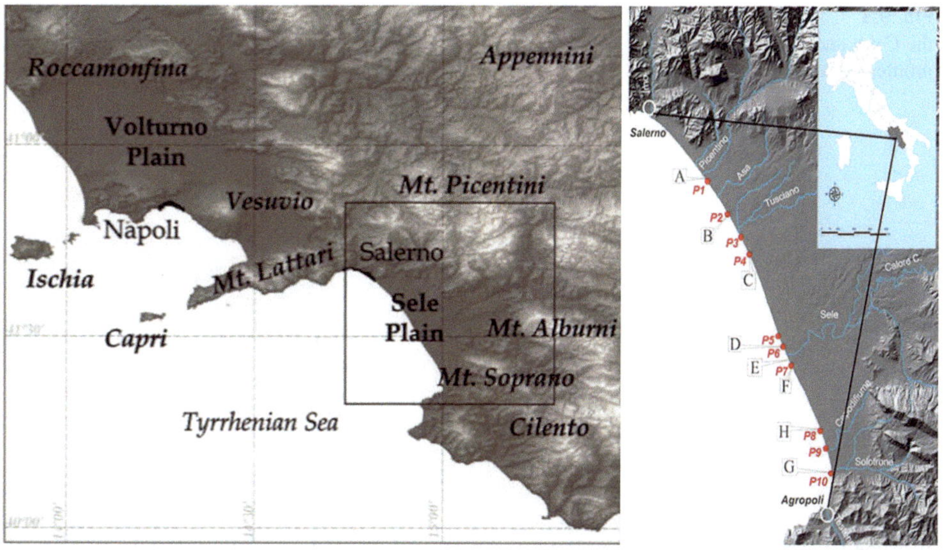

Fig. 9.1 Map of the test site (**a**) and of the profiles used for test results (**b**)

Table 9.2 Beach profile variations for 2003–2009 for the profiles considered

Profiles	P1, P2	P3, P4	P5–P7	P8–P10
Coastal surface 2003 (m^2)	38,805	88,548	83,980	105,352
Coastal surface 2009 (m^2)	26,938	73,200	56,841	1,06,352
Beach length (m)	2,500	2,500	2,500	2,000
Surface variation (m^2)	−1,867	−5,348	−27,139	1,000
Annual surface variation (m^2/y)	−1,695	−3,070	−5,428	167
Beach width variation (m)	−5	−6	−11	1
Max beach width variation (m)	−16	−14	−24	25
Annual beach width variation (m/y)	−0.71	−1.20	−2.17	0.08
Variation percentage (%)	4.60	3.39	6.46	0.16

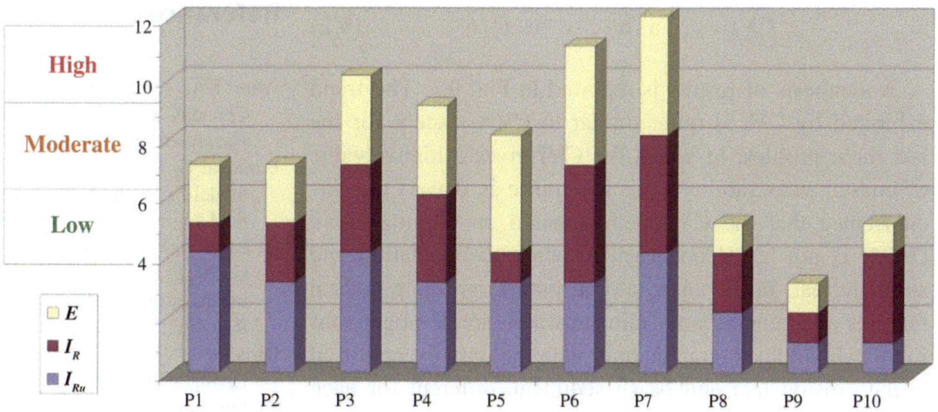

Fig. 9.2 Experimental values of the CVA index on the profiles considered

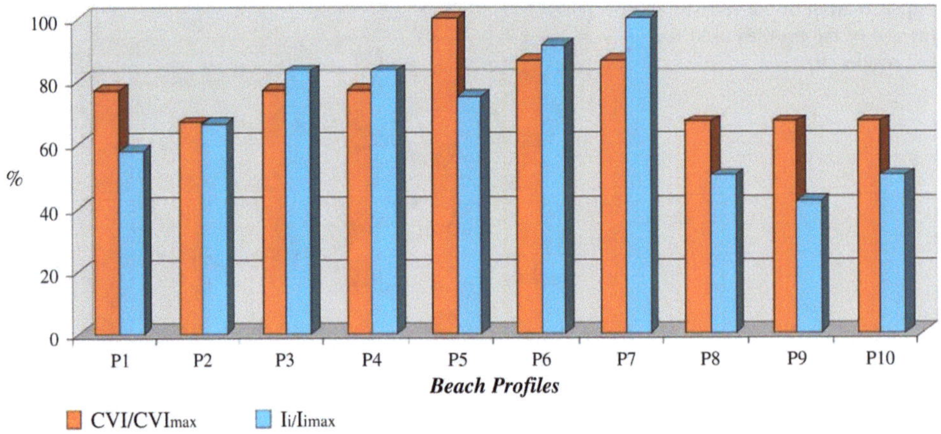

Fig. 9.3 Comparison between the CVA and CVI results for the profiles considered

profiles P1, P5 and P9 exhibit the lowest values; with regard to the component E of the impact index, its values closely follow the annual shoreline erosion rate, which is maximum for profiles from P3 to P7, intermediate for P1 and P2, and minimum for profiles from P8 to P10.

9.4 Validation of the Proposed Method

In order to validate the proposed method, we did a comparison between CVA and CVI (Gornitz et al. 1994). First, the CVI was calculated on the basis of the geomorphology, wave climate and beach erosion rate parameters already available on the different profiles. The aforementioned variables were ranked on a linear scale from 1 to 5 in order of increasing vulnerability. Following Gornitz et al. (1994), square root of the product mean of the six chosen variables is used to calculate the CVI of the coastal region studied. That is,

$$\mathrm{CVI} = \sqrt{(\mathrm{a} \cdot \mathrm{b} \cdot \mathrm{c} \cdot \mathrm{d} \cdot \mathrm{e} \cdot \mathrm{f})/6} \tag{9.2}$$

A synthesis of results is reported in Fig. 9.3. The trend exhibited by CVI is quite similar to CVA, except for the last three profiles, in which the CVI gives a higher value of coastal vulnerability. This difference is due to the circumstance that the CVI method based on Gornitz et al. (1994) is not based on beach response to inundation and storm retreat, but it takes into account large scale coastal features and general wave climate which are homogeneous in the whole physiographic unit, with the only exception of beach erosion rate and beach sediment, different for each profile.

On the other hand, the comparison between the two methods for P1 and P5 profiles shows that CVI method is more cautious for coastal vulnerability of gravel profiles, because it gives a warning meaning to the gravel sediment size, associating it to higher energy impact. On the contrary, CVA gives lower values of retreat to the same beach profiles, because the Kriebel and Dean (1993) retreat equation is inversely proportional to the mean diameter (Benassai et al. 2013). However, CVI seems to be less suitable, due to its high level of generality, to give proper stress to different vulnerability conditions of coastline stretches belonging to the same physiographic unit, in comparison to CVA method, which through a better computation of the specific geomorphological and hydraulic parameters, gives a more precise vulnerability assessment (i.e. closer to the degree of damage experienced on different stretches of coastline).

References

Arun KA, Kunte PD (2012) Coastal vulnerability assessment for Chennai, east coast of India using geospatial techniques. Nat Hazards 64:853–872. doi:10.1007/s11069-012-0276-4

Benassai G, Montuori A, Migliaccio M, Nunziata F (2013) Sea wave modeling with X-band COSMO-SkyMed© SAR-derived wind field forcing. Ocean Sci 9:325–341. doi:10.5194/os-9-325-2013

Di Paola G, Igcesias j, Rosriguer G, Benassai G, Aucecci P., Pappoye G (2011) Estimating coastal vulnerability in a meso-tidal beach by means of quantitative and semi-quantitative methodologies. J Coast Res SI61:303–308

Directorate General Environment European Commission (2004) Living with coastal erosion in Europe: Sediment and Space for Sustainability. PART I—Major findings and Policy Recommendations of the EUROSION project

Gaki-Papanastassiou K et al (2010) Coastal vulnerability assessment to sea-level rise based on geomorphological and oceanographical parameters: the case of Argolikos Gulf, Peloponnese, Greece. Hellenic J Geosci 45:109–122

Gornitz VM et al. (1994) The development of a coastal risk assessment database: vulnerability to sea-level rise in the US Southeast. J Coast Res SI12:327–338

Kriebel DL, Dean RG (1993) Convolution method for time-dependent beach-profile response. J Waterway Port Coast Ocean Eng 119(2):204–207

Pappone G et al (2012) Relative sea-level rise and marine erosion and inundation in the Sele river coastal plain (Southern Italy): scenarios for the next century. Rend Fis Acc Lincei 23:121–129. doi:10.1007/s12210-012-0166-4

Stockdon HF, Holman RA, Howd PA, Sallenger AH (2006) Empirical parametrization of setup, swash, and run-up. Coast Eng 53(7):573–588

Behaviour of a Thin Compressible Clay Horizon Under Geogrid Reinforced Sand with a Wide Platform Load

10

B. R. Jones and J. L. Van Rooy

Abstract

Basal reinforcement, where high tensile strength geogrids are employed beneath structures, is becoming an increasingly accepted construction technique along the eastern coast of southern Africa. The presence of compressible, soft, thin and shallow clay horizons usually associated with complex estuarine or lagoonal deposits are a major consideration when using basal reinforcement as a founding technique. In order to define the deformation behaviour under a wide platform load, three centrifuge tests were undertaken. These tests compared the behaviour of the clay horizon in three scenarios, namely (1) without reinforcement, (2) with a single layer of reinforcement, and (3) with a double layer of reinforcement. It was found that the inclusion of geogrid reinforcement increased the vertical load-bearing capacity of the system and the inclusion of an additional geogrid layer further increases the load bearing capacity. Additionally, the inclusion of geogrid reinforcement also changed the deformation mechanisms of the materials. Without reinforcement, deformation is localised to the portion of clay directly beneath the applied load from the platform. Whilst the inclusion of geogrid reinforcement allows the applied load to be spread to the sand beyond the platform, which contributes to an increase in load-bearing capacity, it comes at the disadvantage of deforming a wider portion of the clay horizon.

Keywords

Centrifuge modelling • Compressible clay • Geogrid • Reinforced sand

10.1 Introduction

Harbours along the eastern coastline of southern Africa have been established on river estuaries where several rivers give rise to an interlinked lagoonal system, comprising complex unconsolidated deposits of sands, silts and clays. These sediments are typically characterised by extremely compressible, soft materials. The poor geotechnical nature of these deposits has lead to limitations on the type of foundation treatment that can be applied to various engineering structures. Accordingly, specialised foundation designs and construction techniques are required when constructing on these unfavourable materials. Initial development for structures constructed on these deposits primarily involved the construction of costly piling solutions. However, with the recent advancement of geosynthetics it has been possible to provide an alternative to traditional piling techniques using basal reinforcement, whereby suitable high tensile strength geogrids are employed. Moore et al. (2012) and Zannoni et al. (2012) have presented case studies where basal reinforcement techniques have been employed. The basal reinforcement utilized in both studies was specifically for warehouse structures whereby large platform loads developed as a result of product stockpiling. The major concern on site for

B. R. Jones (✉) · J. L. Van Rooy
Department of Geology, Faculty of Natural and Agricultural Sciences, University of Pretoria, Pretoria, South Africa
e-mail: bren.r.jones@gmail.com

G. Lollino et al. (eds.), *Engineering Geology for Society and Territory – Volume 4*,
DOI: 10.1007/978-3-319-08660-6_10, © Springer International Publishing Switzerland 2014

both of the studies was the presence of compressible, soft, thin, shallow, clay horizons below the foundation footprint.

Understanding failure mechanisms are an important component of stability analysis of earth structures. It is therefore essential to have a thorough understanding of how the subsoil's beneath the structure will behave, upon foundation loading. If reinforcement is used, the mechanisms become altered, and choosing the right mechanism for stability analysis may not always be straightforward (Michalowski and Shi 2003). Geotechnical centrifuge modelling provides a means to replicate a situation, which may exist in a prototype. Fundamentally, a geotechnical centrifuge allows reproduction of the same stress levels in a small-scale model as in a full-scale prototype, which is useful for analysing complex soil-structure interaction problems.

The present study focused on defining the behaviour of a compressible, soft, thin, shallow clay horizon, under a wide platform load. Accordingly, modelling of this situation in a geotechnical centrifuge allows for the mechanisms of deformation to be defined.

10.2 Geotechnical Centrifuge Model

Three geotechnical centrifuge model tests were performed at a 1:50 scale using the 3 m radius balanced beam centrifuge housed in the Department of Civil Engineering at the University of Pretoria. The first model was tested without any reinforcement. The second and third models had a single and double layer respectively, of geogrid reinforcement present in the sand layer overlying the soft clay layer. Each model was constructed identically with the model set-up illustrated in Fig. 10.1. Initially, sand was pluviated to create a dense basal sand horizon of 100 mm thickness which was then carefully saturated from the base by introducing pore water via a header tank. Following which, the dry kaolin clay powder was prepared by mixing under partial vacuum with deionized water, to twice its liquid limit. The slurry was poured over the basal sand horizon, and allowed to consolidate in the centrifuge for 14 h at 50 G, under a load of approximately 25 kPa. This ensured the clay was normally consolidated and had a very soft consistency, with an undrained shear strength of approximately 6 kPa for each test. The clay horizon had a thickness of 60 mm for each centrifuge test. The geotechnical properties of the sand and clay are presented in Table 10.1. Finally, the reinforced sand was then pluviated over the clay to a thickness of 30 mm. For the second and third tests with reinforcing included, the sand pluviation was ceased temporarily to allow for placement of the geogrids within the reinforcing sand. The geogrids were placed at equally spaced vertical intervals within the sand layer, and extended the entire length of the centrifuge model box. The geogrids employed were polypropylene biaxial geogrids, which had an ultimate tensile strength of 20 kN/m, at 13 % strain, with a stiffness of approximately 154 kN/m.

During loading of the model, an aluminium platform of 200 mm width was attached to a jack and lowered at a constant rate of 0.081 mm/s. A load cell recorded the load achieved against the platform, while an LVDT measured the vertical displacement.

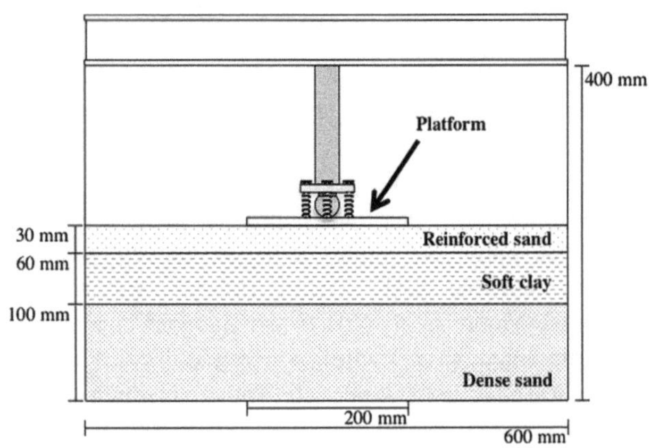

Fig. 10.1 Centrifuge model test set-up. The geogrids were placed in the *upper* sand layer

10.3 Analyses of Centrifuge Tests

Analyses of the centrifuge tests comprised the visual assessment of the material deformation, as well as evaluation of the load-displacement data. Camera's installed on the centrifuge basket recorded the events in-flight. This was undertaken to compare the deformation at specific displacement intervals. An HBM Quantum system acquired data from the load cell and LVDT placed on the jack system.

10.3.1 Deformation Behaviour

The deformation of the centrifuge model, at 15 mm displacement is presented in Fig. 10.2. In the unreinforced test, a shear zone develops through the upper sand horizon at the

Table 10.1 Geotechnical characteristics of the materials used in the centrifuge models

Sand	Angle of internal friction, φ	37°
	Maximum dry density, $\gamma_{d(\max)}$	16.69 kN/m^3
Clay	Liquid limit	47 %
	Plastic limit	22 %
	Undrained shear strength	6.09 kPa
Geogrid	Ultimate tensile strength (UTS)	20 kN/m
	Strain at UTS	13 %

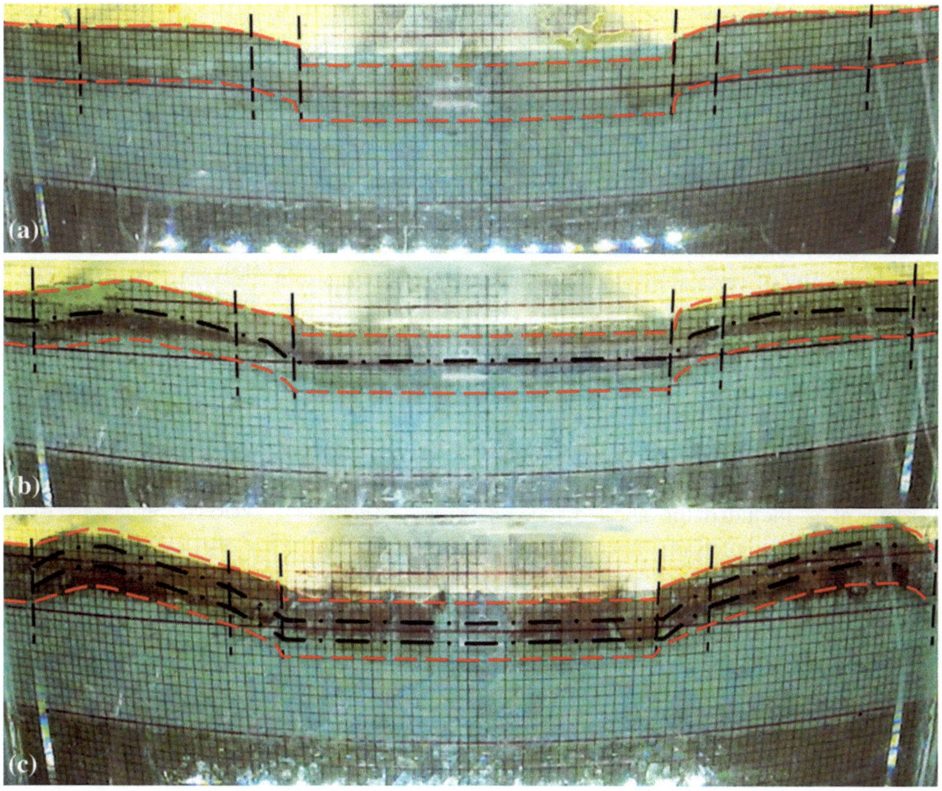

Fig. 10.2 Deformation patterns for each centrifuge tests. **a** Unreinforced, **b** one geogrid, and **c** two geogrids

edge of the platform. As such, the sand beneath the platform essentially behaves as a much deeper rigid platform, to the top of the clay horizon. The deformation of the clay horizon is confined to the portion below the overlying sand directly beneath the applied load from the platform. The clay is squeezed laterally, with little disturbance to the surface of the model, as the 'bulging' on the surface is limited.

With the inclusion of geogrid reinforcement, the mechanism of failure changes with the failure having a wider lateral impact. Huang and Tatsuoka (1990) termed this the 'wide-slab' effect, for their studies of reinforced sand loaded by footings. By including reinforcement beyond the platform footprint, the applied load is spread, mobilising the sand beyond the platform, and subsequently a wider portion of the clay deforms. This is visible by the surface deformation and 'bulging' of the clay horizon. When two layers of reinforcement are included, the effect is enhanced, as the sand is increasingly stiffened.

10.3.2 Load-Displacement

The load-displacement curves are presented in Fig. 10.3. At small displacements all three load-displacement gradients are almost identical, but with increased loading the reinforced curves differ in their relationship with the unreinforced curve. Both unreinforced and reinforced curves reach

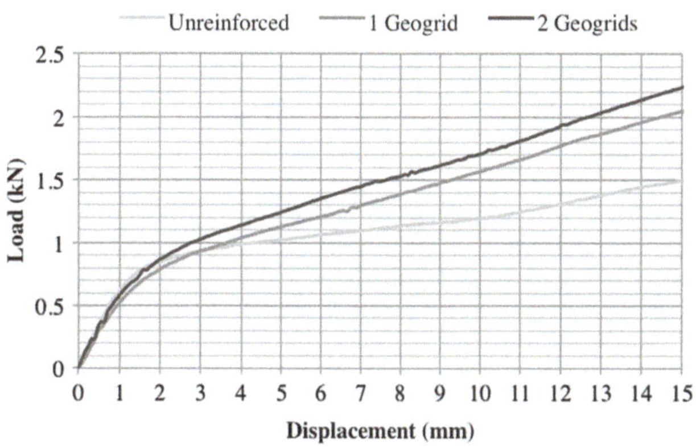

Fig. 10.3 Load-displacement *curves* for each centrifuge model test

a load of approximately 0.8–0.9 kN, at approximately 1.5 mm displacement. Thereafter the reinforced gradients are much steeper than the unreinforced gradient. Both reinforced curves (1 and 2 geogrids) exceed a 2 kN load and stress of 40 kPa, when 15 mm displacement is achieved. At the same displacement the unreinforced test only reaches a load of 1.5 kN and a stress of 30 kPa. Similar load-displacement relationships exist for both the single and the double reinforcement conditions. However, the test with two geogrids installed, achieved an increase in vertical load-bearing capacity compared to the test with a single layer of geogrid reinforcement.

10.4 Conclusion

The addition of the geogrid reinforcement to sand under a wide platform load increases the vertical load-bearing capacity of the system. Furthermore, the addition of multiple layers of reinforcement further contributes to this increase in load-bearing capacity.

However, this resultant increase is only evident after a certain amount of vertical displacement, which is likely due to the reinforcement requiring a certain amount of strain in order to contribute to an increase in bearing capacity. The inclusion of geogrids changes the failure deformation of the clay layer. With no reinforcing present deformation is localised to the portion of clay directly beneath the footprint of the applied load from the platform. When reinforcement is included the applied load is spread and the sand beyond the platform contributes to the increase in load-bearing capacity, resulting in a wider portion of clay deforming.

References

Huang CC, Tatsuoka F (1990) Bearing capacity tests of reinforced horizontal sandy ground. Geotext Geomembr 9(1):51–81

Michalowski RL, Shi L (2003) Deformation patterns of reinforced foundation sand at failure. J Geotechn Geoenviron Eng 129(6): 439–449

Moore L, Richter MV, Gilli A (2012). The use of wire mesh steel geogrids in South Africa. In: Proceedings of the 5th European geosynthetics congress, Valencia, Spain

Zannoni E, Freeman T, Richter MV (2012) Basal reinforcement over soft soil using high strength bonded geogrid at Agriport Maydon Wharf. Proceedings of the 5th European geosynthetics congress, Valencia, Spain

11 Numerical Methods for Deformation Analysis of Liquefiable Soils

Miao Yu, Yu Huang, and Qiang Xu

Abstract

Deformation induced by liquefaction during an earthquake is one of the leading causes for damage to soils and foundations. Because of the significant reduction of the shear strength and shear modulus of post-liquefaction soils, lateral spread can develop from small deformation to the flow failure stage where the phase of the liquefied soils changes from solid to fluid. In this paper, two numerical methods are introduced: the finite element method (FEM) and the smoothed particle hydrodynamics (SPH) method. For the initial stage of small deformation of liquefaction, the FEM based on solid mechanics can accurately simulate the soil behavior. For the flow stage of liquefaction deformation, the SPH method in the framework of fluid dynamics is more suitable because the mechanical and deformation properties in the flow stage are similar to that of a viscous fluid. In this paper, recent advances in the field of liquefaction deformation analysis using the FEM and SPH methods are introduced, as a basis for future research.

Keywords

Liquefaction • Numerical simulation • Finite element method • Smoothed particle hydrodynamics

11.1 Introduction

When subjected to ground motion, the pore water pressure of soils increases sharply and the effective mean stress decreases accordingly until it reaches zero, leading to liquefaction. Significant reduction of shear strength and shear modulus of post-liquefaction soils can cause large lateral deformation and subsidence, which can result in serious damage to buildings and infrastructures on the surface. For example, in the 2010 and 2011 Christchurch (New Zealand) earthquakes, liquefaction caused up to 2–3 m of permanent lateral spread within 200–250 m of the city's river (Cubrinovski et al. 2012). In the 1999 Kocaeli (Turkey) earthquake, lateral deformation induced by seismic liquefaction was found mainly along the shore and creeks with maximum displacement greater than 2 m (Kanıbir et al. 2006).

Early research into liquefaction paid more attention to the influencing factors and the likelihood of liquefaction occurring than to deformation prediction of post-liquefaction soils. The destruction induced by liquefaction deformation has highlighted the problem of post-liquefaction soil behavior and has recently attracted the attention of scholars. The finite element (FEM) method based on solid mechanics can

M. Yu · Y. Huang (✉)
Department of Geotechnical Engineering, College of Civil Engineering, Tongji University, 200092 Shanghai, China
e-mail: yhuang@tongji.edu.cn

Y. Huang
Key Laboratory of Geotechnical and Underground Engineering of the Ministry of Education, Tongji University, 200092 Shanghai, China

Q. Xu
State Key Laboratory of Geo-Hazard Prevention and Geo-Environment Protection, Chengdu University of Technology, 610059 Chengdu, China

G. Lollino et al. (eds.), *Engineering Geology for Society and Territory – Volume 4*,
DOI: 10.1007/978-3-319-08660-6_11, © Springer International Publishing Switzerland 2014

accurately simulate the initial small deformation of liquefaction, whereas for the flow stage of liquefaction deformation, the smoothed particle hydrodynamics (SPH) method in the framework of fluid dynamics would be more suitable. This paper presents recent advancements in liquefaction deformation analysis using the FEM and SPH methods.

11.2 The Finite Element Method

FEM is a simple and useful tool for solving problems of continuous media. It has been applied widely in geotechnical engineering analysis, such as in the fields of deformation prediction, strength analysis, and stability estimation.

Huang et al. (2008a, 2009, 2012) have carried out many finite element simulations for small deformation of embankments on liquefiable soils. Because the liquefied soil is a saturated fluid–solid coupling medium, Biot's dynamic coupled theory for saturated porous media simulation and a cyclic elastoplastic constitutive model were used. The liquefaction resistance and deformation characteristics of embankments under earthquakes were studied, and FEM was validated as a useful tool for seismic assessment and design of embankments.

The dynamic interaction of pile–soil–structure in a liquefiable site was also studied by FEM (Huang et al. 2005). A beam-column element, which takes into account the effect of volume and axial force, was selected to represent the dynamic behavior of piles. The vulnerable parts and the deformation laws of the whole system were calculated, providing a theoretical basis for the seismic design of piles in liquefiable sites.

The authors also proved the effectiveness of liquefaction mitigation measures that use the FEM method. A 2D-FEM program was used to verify the effect of reinforcement measures, revealing that the FEM method can provide useful results for seismic design of embankments (Huang et al. 2006). An effective stress based finite element-finite differences (FE-FD) method was used in the study of Huang et al. (2008b). The pore water pressure was calculated by the FD, while the displacement was obtained by the FE method. The simulation results illustrated that cement grouting cannot prevent soils from liquefying but it can substantially reduce the deformation induced by liquefaction.

11.3 The Smoothed Particle Hydrodynamics Method

The state of liquefied soils changes from solid to liquid phase with the full development of shear strain, and the mechanical and deformation properties become similar to those of a viscous fluid. Hence, the traditional method based on solid mechanics cannot simulate precisely the complex hydrodynamic behavior of post-liquefied soils. In recent years, new numerical methods in the framework of fluid dynamics have been proposed to overcome this limitation. In this paper, the SPH method for flow behavior of post-liquefaction soils is introduced.

SPH is a pure Lagrangian, meshless hydrodynamics method that has advanced rapidly in recent years and has been applied in many engineering fields. The continuous fluid is described by a serious of arbitrarily distributed particles in SPH, with each particle being assigned various physical quantities. Then, the mechanical properties of the whole system are obtained by calculating the kinetic equation of the particle group and tracking the movement of each particle. The most prominent advantage of this method is that it can successfully simulate free surfaces, flow deformation, and deformation boundaries (Liu and Liu 2003).

Huang et al. (2011, 2013) conducted extensive research on numerical simulation of large liquefaction-induced deformation using the SPH method. The authors used Bingham fluids and the Verlet neighbor list method to simulate the flow deformation and search neighboring particles, respectively (Huang et al. 2011). The method's accurate simulation of the flow process of liquefied soils was validated by comparing with the test results. Huang et al. (2013) further proposed a soil–water-coupled SPH numerical modeling that treated soil and water as different layers. The application of this method was verified by test data and it was used to simulate the failure of an embankment on liquefied foundations.

11.4 Summary

Because historical cases have manifested the serious damage and destruction induced by liquefaction deformation, there is a need to establish effective numerical methods to simulate the whole process of liquefaction deformation. The FEM can be used to simulate the initial small deformation of liquefaction. For the subsequent flow deformation, the SPH method in the framework of fluid dynamics can produce a more accurate simulation compared to the traditional numerical methods based on soil mechanics, as it models the mechanism of liquefaction process on a grain scale level. Both FEM and SPH are important tools for seismic liquefaction assessment and 3D simulation should be conducted in further study.

Acknowledgments This work was supported by the National Basic Research Program of China (973 Program, Grant No. 2012CB719803), the National Natural Science Foundation of China (Grant Nos. 41372355, 41225011) and the Chang Jiang Scholars Program of China.

References

Cubrinovski M, Robinson K, Taylor M, Hughes M, Orense R (2012) Lateral spreading and its impacts in urban areas in the 2010–2011 Christchurch earthquakes. NZ J Geol Geophys 55(3):255–269

Huang Y, Yashima A, Zhang F (2005) Finite element analysis of pile-soil-structure dynamic interaction in liquefiable site. Chin J Geotech Eng 27(6):646–651 (in Chinese)

Huang Y, Sawada K, Moriguchi M, Yashima A, Zhang F (2006) Numerical assessment of the effect of reinforcement on the performance of reinforced soil dikes. Geotext Geomembr 24(3):169–174

Huang Y, Yashima A, Sawada K, Zhang F (2008a) Numerical assessment of the seismic response of an earth embankment on liquefiable soils. Bull Eng Geol Environ 67(1):31–39

Huang Y, Zhang F, Yashima A, Ye WM (2008b) Numerical simulation of mitigation for liquefaction-induced soil deformation in a sandy ground improved by cement grouting. Environ Geol 55(6):1247–1252

Huang Y, Yashima A, Sawada K, Zhang F (2009) A case study of seismic response of earth embankment foundation on liquefiable soils. J Cent S Univ Technol 16(6):994–1000

Huang Y, Zhang WJ, Mao WW, Jin C (2011) Flow analysis of liquefied soils based on smoothed particle hydrodynamics. Nat Hazards 59(3):1547–1560

Huang Y, Zheng H, Zhuang ZJ (2012) Seismic liquefaction analysis of a reservoir dam foundation in the South-North Water Diversion Project in China. Part II: seismic response simulation. Nat Hazards 60(3):1313–1324

Huang Y, Zhang WJ, Dai ZL, Xu Q (2013) Numerical simulation of flow processes in liquefied soils using a soil–water-coupled smoothed particle hydrodynamics method. Nat Hazards 69(1): 809–827

Kanıbir A, Ulusay R, Aydan Ö (2006) Assessment of liquefaction and lateral spreading on the shore of Lake Sapanca during the Kocaeli (Turkey) earthquake. Eng Geol 83(4):307–331

Liu GR, Liu MB (2003) Smoothed particle hydrodynamics: a mesh-free particle method. World Scientific Press, Singapore

12 Late Glacial and Holocene Sedimentation and Investigation of Fjord Tsunami Potential in Lower Howe Sound, British Columbia

L. E. Jackson Jr., A. Blais-Stevens, R. L. Hermanns, and C. E. Jermyn

Abstract

Preliminary investigation of the history of sedimentation in lower Howe Sound fjord near Vancouver, B.C. indicates that no rock slide capable of generating a destructive displacement wave has occurred during the latter half of the Holocene. Possible rock slide run-out deposits were detected along the western margin of Bowen Island. These could date from any time during the Holocene. Investigation of the adjacent slope suggests it is creeping and is not an imminent large rock slide failure.

Keywords

Fjord tsunami • Fjord geomorphology • Paraglacial sedimentation

12.1 Introduction

Howe Sound (HS) is the southernmost fjord on the mainland coast of British Columbia (Fig. 12.1). Study of swath multibeam bathymetry (SMB) imagery generated in 2007 revealed apparent submarine rock slide run-out deposits on the sea bottom west of Bowen Island in Collingwood Channel (Fig. 12.2D) (Jackson et al. 2008). This caused immediate concern with regard to the likelihood of future landslide activity in this area that could cause displacement waves (landslide-triggered tsunamis) that could potentially affect nearby communities and marine traffic.

This report summarizes the results of the evaluation of slope stability along Bowen Island with preliminary exploration of these sediments beneath the ocean bottom of HS. A compete account is presented in Jackson et al. (in press).

12.2 Setting

HS is the southernmost fjord on the mainland of western Canada (Fig. 2.1). It is located immediately north of the City of Vancouver. It measures 38 km along its long (north-south) axis and has an average width of 6 km. Average water depths within Howe Sound are in the 200–250 m range. HS and surrounding parts of the rugged and glacially sculpted Coast Mountains are underlain by the Jurassic to Eocene crystalline rocks of the Coast Plutonic complex (Woodsworth et al. 1991, p. 515). Mountain peaks rise 2,000 m above sea level within 3 km of the shore. Relief contrasts of 1,000–1,500 m are common. Subaerial landslide and debris flow activity is common around HS (Blais-Stevens and Septer 2008).

L. E. Jackson Jr. (✉)
Geological Survey of Canada, 625 Robson Street, Vancouver, BC V6B5J3, Canada
e-mail: lionelj@sfu.ca

A. Blais-Stevens
Geological Survey of Canada, 601 Booth Street, Ottawa, ON, Canada
e-mail: andree.blais-stevens@nrcan.gc.ca

R. L. Hermanns
Geological Survery of Norway, Leif Eiricssons Vei 39, 7491 Trondheim, Norway
e-mail: reginald.hermanns@ngu.no

C. E. Jermyn
Vlaardingen, Netherlands
e-mail: courtney.jermyn@gmail.com

G. Lollino et al. (eds.), *Engineering Geology for Society and Territory – Volume 4*,
DOI: 10.1007/978-3-319-08660-6_12, © Springer International Publishing Switzerland 2014

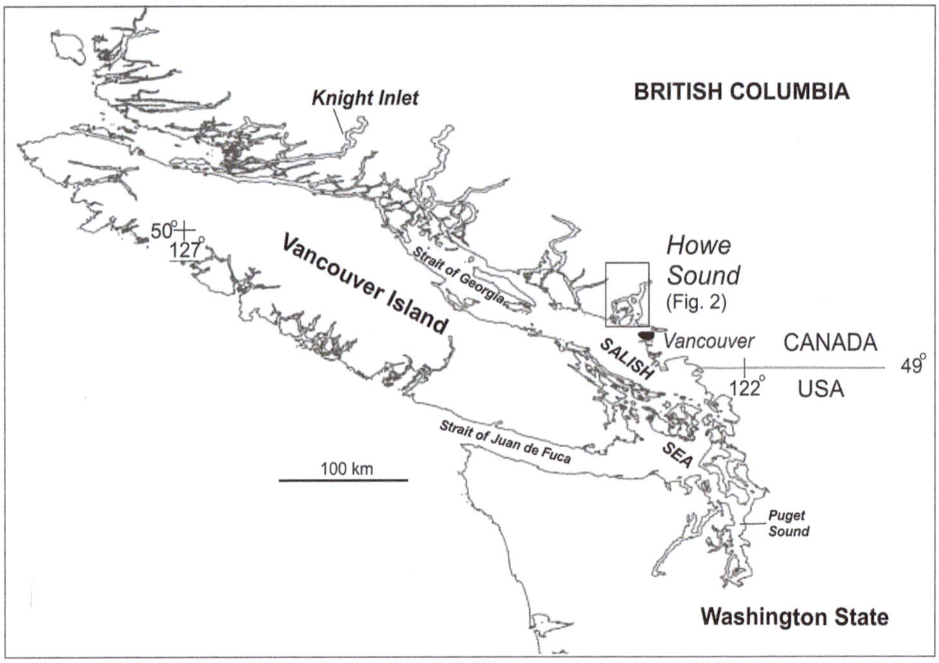

Fig. 12.1 Howe Sound region

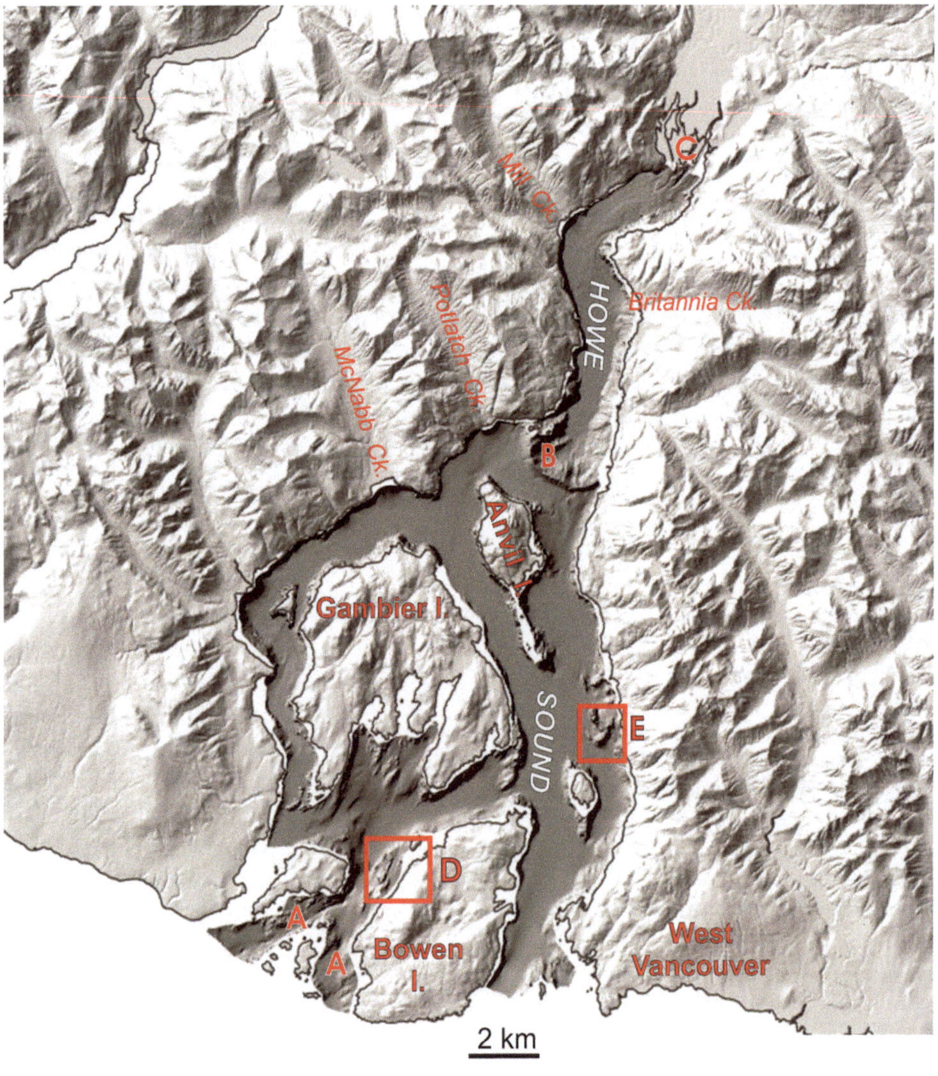

Fig. 12.2 Digital elevation model of Howe Sound: *A* outer sill of ice stagnation glacial sediments; *B* moraine which creates the inner sill that divides Howe Sound into an upper and lower basin; *D* west slope of Mt. Gardner on Bowen Island which has failed as a small rock slide or debris flow. Sediment cores 42–44 were collected from the adjacent sea bottom; *E* Lions Bay area: location sediment cores 40 and 40

The HS area was inundated by the Cordilleran Ice Sheet during the last glaciation. Three radiocarbon ages document deglaciation of Bowen Island by as early as ca. 13,000 radiocarbon years before present.

The bottom of HS is divisible into a lower basin bounded by a sill of ice stagnation sediments and a moraine (Fig. 12.2A and B, respectively) and an upper basin between the moraine and the delta of the Squamish River (Fig. 12.2C).

12.3 Methods

This study consisted of terrestrial and marine components. The western slopes of Bowen Island, where open fissure fissures were noted locally, were traversed to determine the nature of extant slope deformation. Structural geology mapping was carried out along with a stability analysis as a part a M.Sc. thesis (van Zeyl 2009). The coastline of the island immediately across Collingwood Channel was traversed to look for evidence of prehistoric run-up of displacement waves. The marine geology component consisted of sub-bottom seismic reflection profiling, piston coring, and radiocarbon dating of organic materials from the cores to determine the sea bottom stratigraphy and sedimentation rates.

12.4 Findings

van Zeyl (2009) found that the northwest slope of Mount Gardner appears to have deformed gravitationally due to adverse structure and stratigraphy and erosion at the toe of the slope. This large-scale deformation apparently has ceased or is taking place at a rate that does not generally manifest itself in active surface deformation features. He concluded that the slope does not appear to be actively deforming toward a failure. However, he recommended precautionary long term monitoring.

Three sediment cores (42–44) ranging from 10.5 to 1.5 m in length were collected from piston coring of the sea bottom within about 200 m west of the west coast of Bowen Island and two other sediment cores (40 and 41, ~7 m in length) were taken from the Lions Bay area (D and E respectively, Fig. 12.2). Textural, stratigraphic, biostratigraphic investigations and radiocarbon dating of the shells of marine invertebrates within the cores have yielded insights into the sedimentological history of HS (Jackson et al. in press).

Rainout of gravel-size clasts from icebergs ceased by ca. 9,400 ^{14}C years before present (core 42). This apparently indicates the cessation of glacier margin calving and iceberg production presumably at the head of HS. Holocene sedimentation has been highly non-uniform: it has been almost zero in Collingwood Channel along the west coast of Bowen Island (cores 42–44). There has been significant Holocene sedimentation in the Lions Bay area (E, Fig. 12.2). However, almost all took place during the early half of the Holocene. Radiocarbon ages determined on invertebrates within core 41 are commonly inverted with respect to one another. The disturbance that caused disruption of stratigraphy is the either the result of bioturbation or the geyser-like eruption of methane from bottom sediments. The floor of HS is marked by craters formed by the latter process. These gas-escape craters (or pockmarks) are common features on the floors of other fjords and the continental shelf of British Columbia (Barrie et al. 2005) and Chile (Dowdeswell and Vásquez 2013).

12.5 Discussion

The site of sediment core 41 (in widest reach of lower HS) is separated from the complex of submarine fans along the eastern coast of HS by submarine topography. Holocene sedimentation has been entirely from suspended sediment. The Squamish and Fraser rivers are the sources of suspended sediment within HS. Although accurate estimation of sedimentation rates within sediment core 41 is not possible due to inversion of some ^{14}C ages, the aggregate range of ^{14}C ages suggests that sedimentation has been in the order of more than a metre per thousand years until the mid Holocene (ca. 5,000–4,000 ^{14}C years BP). Less than 1 m of sediment was deposited since that time. The site of sediment core 40 is located within the submarine fan below the mouth of Lone Tree Creek, one of many mountain torrents that have built an apron of coalescent submarine fans along the eastern margin of HS (Jackson et al. 2008). Radiocarbon ages are not sequential which would be expected in a submarine fan environment where scour and reworking of the fan can take place. However, there is less than 100 cm of sediment in the core that appears to be related to the last 4,000 ^{14}C years. This suggests that the rate of sedimentation on the submarine fan below the mouth of Lone Tree Creek has decreased dramatically in the area of core 40 since the mid-Holocene.

The decrease in the rates of sedimentation apparent in cores 40 and 41 during the latter half of the Holocene is in agreement with the findings of Brooks (1994) who investigated the fluvial geomorphology of the Squamish River basin with respect to the of transport of residual glacial sediment from tributaries to the main stem of the Squamish River during the Holocene. He found that this fluvial system incised through glacial fill to bedrock thousands of years before the present so that present sediment transport of the Squamish River to HS is small compared to sediment transport earlier in the Holocene. This pattern, whereby erosion and transport of sediment relict from a recent glaciation was most intense early in the post glacial period and progressively approaches new rates adjusted to non-glacial conditions, is referred to as the 'paraglacial effect' (Church and Ryder 1972). This has been documented on alluvial fans and river systems throughout North American Cordillera (Jackson et al. 1982; Ritter and Ten Brink 1986), and recently deglaciated mountains elsewhere (Iturrizaga 2008). Likewise, sediment core 40 indicates that most of the submarine fan below Lone Tree Creek was deposited during the first half of the Holocene.

12.6 Conclusion

One of the prime objectives of this study was to determine if subsequent sedimentation would conceal past rock slides from SMB imagery in HS. Holocene sedimentation ranged from almost negligible along the west margin of Bowen Island to more than 11 m in the Lion's Bay. The run-out deposits along west side of Bowen Island originated from small rock slides or debris flows that may or may not have been coeval during the Holocene. In the Lion's Bay area, the lack of similar deposits, as recorded by SMB imagery (Jackson et al. 2008), indicate that no rock slide events that produced relief in run-out areas of more than 1 m has occurred since at least mid Holocene time. The lack of such run-out features in SMB imagery of lower HS suggests that large rapidly moving rock slides capable of generating displacement waves have not occurred during the latter half of the Holocene.

References

Barrie JV, Hill PR, Conway KW, Iwanowska K, Picard K (2005) Environmental marine geoscience 4: Georgia basin: seabed features and marine geohazards. Geosci Can 32:145–152

Blais-Stevens A, Septer D (2008) Historical accounts of landslides and flooding events along the sea to sky corridor, British Columbia, from 1855–2007. Geological survey of Canada, open file 5741, 1 CD-ROM, p 117

Brooks GR (1994) The fluvial reworking of Late Pleistocene drift, Squamish River drainage basin, southwestern British Colombia. Géographie physique et Quaternaire 48(no 1):51–68. (8 fig., 4 tables)

Church M, Ryder JM (1972) Paraglacial sedimentation: consideration of fluvial processes conditioned by glaciation. Geol Soc Am Bull 83:3059–3072

Dowdeswell JA, Vásquez M (2013) Submarine landforms in the fjords of southern Chile: implications for glacimarine processes and sedimentation in a mild glacier-influenced environment. Quatern Sci Rev 64:1–19

Iturrizaga L (2008) Paraglacial landform assemblages in the Hindukush and Karakoram Mountains. Geomorphology 95(1–2):27–47

Jackson LE, MacDonald GM, Wilson MC (1982) Paraglacial origin for terraced river sediments in Bow Valley, Alberta. Can J Earth Sci 19:2219–2231

Jackson LE Jr, Hermanns RL, Jermyn CE, Conway K, Kung R (2008) Annotated images of submarine landslides and related features generated from swath multibeam bathymetry, Howe Sound, British Columbia. Geological survey of Canada, open file 5662, 1 CD-ROM

Jackson LE Jr, Blais-Stevens A, Hermanns RL, van Zeyl DP, Stead D, Jermyn CE, Barrie V, Conway K, Hetherington R (in press) Late glacial and Holocene sedimentation and investigation of fjord tsunami potential in lower Howe Sound, British Columbia. Geological survey of Canada open file

Ritter DF, Ten Brink NW (1986) Alluvial fan development and the glacial-glaciofluvial cycle. Nenana Valley, Alaska. J Geol 94: 613–615

van Zeyl DP (2009) Evaluation of subaerial landslide hazards in Knight Inlet and Howe Sound, British Columbia. Unpublish M.Sc. thesis, Department of Earth Sciences, Simon Fraser University, p 184

Woodsworth GJ, Anderson RG, Armstrong RL (1991) Plutonic regimes, chapter 15 in 'Geology of Cordilleran Orogen in Canada'. In: Gabrielse H, Yorath CJ (ed) Geological survey of Canada, geology of Canada, issue number 4, pp 491–531. (also Geological Society of America, The geology of North America, v. G-2)

Catalogue of Historical Displacement Waves and Landslide-Triggered Tsunamis in Norway

13

Reginald L. Hermanns, Thierry Oppikofer, Nicholas J. Roberts, and Gro Sandøy

Abstract

Due to Norway's more than 100,000 km long coast line and tens of thousands of lakes, Norwegian society is strongly exposed to displacement waves caused by landslides impacting into water bodies and to tsunamis caused by submarine landslides. In a database that goes back to the fourteenth century 270 such events are reported, related to rock falls, rock avalanches, debris flows, quick clay slides, snow avalanches and submarine landslides. By far the most severe effects are related to rock avalanches impacting into water bodies setting off waves with near-field heights of several tens of meters and far-field heights locally exceeding 10 m. Consequently, the highest numbers of fatalities (283) are related to those events reaching up to 73 fatalities in a single event. Our database has the potential to aid prediction of future landslide-generated displacement waves and tsunamis in Norway. However, it became evident that detailed data comprising landslide type, landslide volume, wave heights around the water body, only exist for few events and it is needed to study in detail future events in order to build better data sets that might be used for empirical prediction of those secondary effects.

Keywords

Landslide-generated displacement wave • Landslide-generated tsunami • Rock avalanche • Database • Fjordland

13.1 Introduction and Geological Setting

Landslides are common in Norway due to the occurrence of steep-sided valleys and adverse climate. Deep erosion during multiple glacial cycles together with Holocene isostatic rebound produced an extensive fjord system in Western Norway comprising more than hundred thousands of kilometers of either high, steep coastline or sensitive marine sediments (Vorren and Mangerud 2008; Vorren et al. 2008). Most valleys and fjords are settled, with populations concentrated at low elevation close to lakes, marine shorelines and rivers. Consequently, Norway's population is heavily exposed to displacement waves generated by impact of subaerial landslides and to tsunamis generated by movement of submarine landslides (following the definition given in Hermanns et al. 2013).

The three displacement wave events that were caused by the impact of rock avalanches into the water bodies at Loen (in 1905, 1936) and in Tafjord (in 1934) are events from the past century and each caused loss of several tens of lives; they are well reported in the scientific literature (e.g. Jørstad 1968; Grimstad and Nesdal 1990; Blikra et al. 2006; Hermanns et al. 2006). However, many more events have gone undocumented or received minimal attention. Although the loss of life due to submarine landslides generating tsunamis is by far smaller, detailed studies on selected sites have evaluated the causative relations between the landslide process and the

R. L. Hermanns (✉) · T. Oppikofer · G. Sandøy
Geological Survey of Norway, P.O. Box 6315, 7491 Trondheim, Sluppen, Norway
e-mail: Reginald.hermanns@ngu.no

N. J. Roberts
Centre for Natural Hazard Research, Simon Fraser University, 8888 University Drive, Burnaby, BC V5A 1S6, Canada

G. Lollino et al. (eds.), *Engineering Geology for Society and Territory – Volume 4*,
DOI: 10.1007/978-3-319-08660-6_13,

generation of the wave (L'Heureux et al. 2010). Unlike Aysén fjord in southern Chile (Lastras et al. 2013), no Norwegian sites have been found to experience a combination of subaerial and submarine slope failures. We are developing a catalogue of landslide-generated displacement waves and tsunamis in Norway to improve understanding of the phenomena, thereby aiding hazard and risk assessment on both regional and local scales.

13.2 Methods

The catalogue is derived from a subset of a the Norwegian national landslide inventory database—an ongoing project cataloguing gravitational mass movements of soil, rock, snow and ice in Norway since the fourteenth century (Hermanns et al. 2012)—supplemented by scientific publications and by accounts of landslides summarized from church records. The national landslide inventory database was compiled by the Geological Survey of Norway (NGU) and includes contributions from the Norwegian Public Roads Administration, the Norwegian agency for railway services, the Norwegian Geotechnical Institute (NGI), and the Norwegian Water Resources and Energy Directorate (NVE), which is now tasked with managing the national inventory database and making it publicly available (Nordahl et al. 2013).

The road and railway authorities collect information about events that have an impact on their infrastructure. NGU has contributed historic information on events resulting in injuries, fatalities or major damages to infrastructure (see also Furseth 2006). NGI has contributed

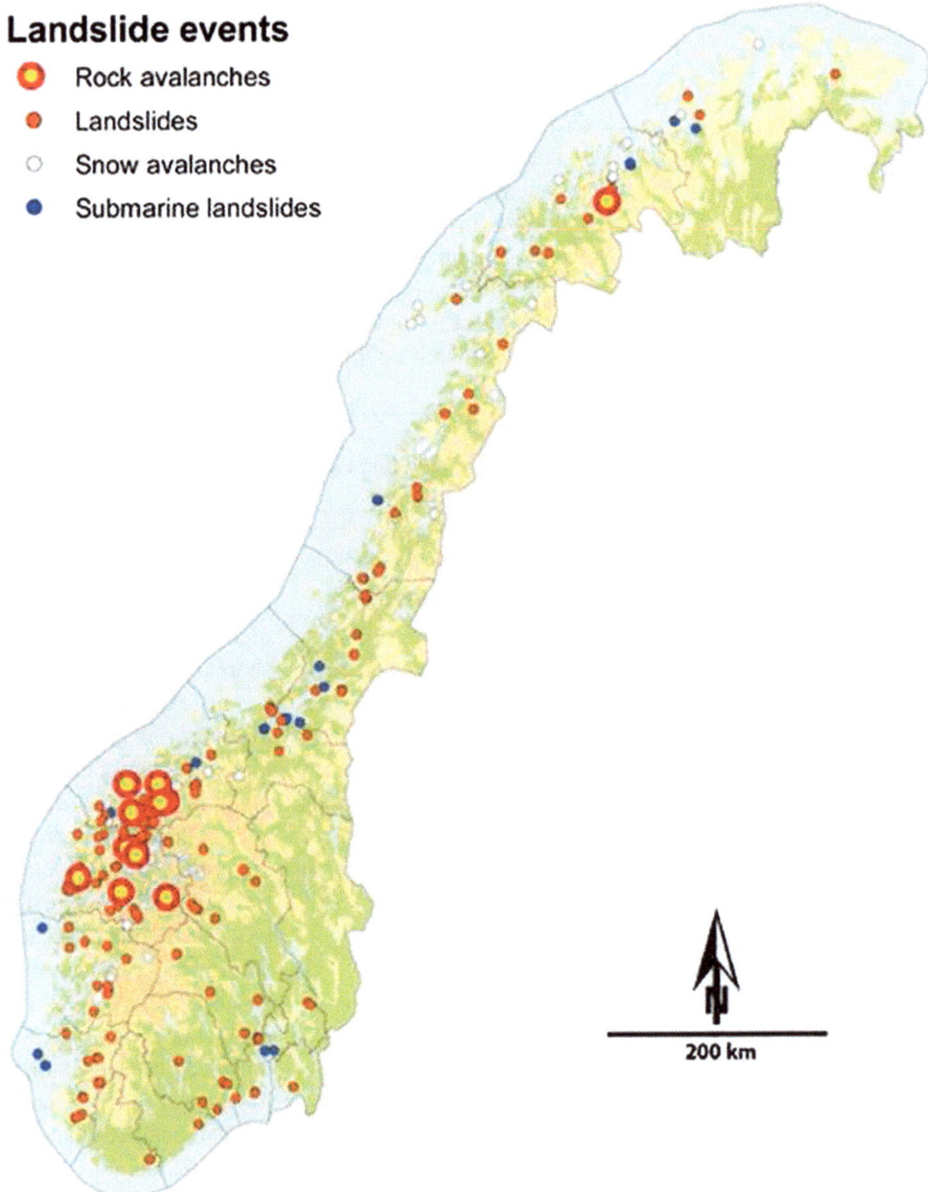

Fig. 13.1 Distribution of landslide-generated displacement waves and landslide-generated tsunamis in Norway classified by landslide type

information mainly on snow avalanches and quick clay landslides. Our subset database on landslide-triggered displacement waves and landslide-triggered tsunamis includes details on mass movement events, the water bodies they impact, consequent displacement waves or tsunamis, and wave and tsunami impacts (Fig. 13.1).

13.3 Results

Of the ~42 500 events in the Norwegian national landslide inventory, 241 are indicated to have generated displacement waves (212 events) or tsunamis (29 events). All submarine landslides in the database were recognized because they generated tsunamis. Fifty-nine displacement waves are related to snow avalanches, and 12 to rock slope failures that exceed the threshold of forming rock avalanches (500,000 m^3); five additional rock slope failures with poorly constrained volumes might have exceeded the threshold of forming rock avalanches. The remaining 136 sub-aerial landslides have been mainly rock falls (92 entries), but also quick clay slides (25 entries) and debris flows (14 entries) or could not have been identified (5 entries).

In total 339 fatalities are reported due to either landslide-generated displacement waves and tsunamis. This is 7.5 % of the total 4,475 fatalities reported in relation with landslides in Norway so far (Hermanns et al. 2012). Displacement waves from rock avalanches are the most devastating events, accounting for 283 of all fatalities, but including only 7 events. The fatalities related to displacement waves are related to high (tens of meters) near-field and moderate (~10 m) far-field (exceeding 10 km distance) displacement wave heights.

Detailed reports of multiple run-up heights around a water body exist only for the three infamous twentieth century events in Loen Lake and Tafjord (Jørstad 1968) and the Rissa quickclay slide (Townson and Kaya 1988). For the other events run-up is reported for 40 events of which 6 events have two run-up measurements. These are related to rock fall events (21 events with maximum run-up heights ranging between 1 and 20 m), rock avalanche events (12 events with maximum run-up heights ranging between 14 and 70 m), quick clay slides (5 events with maximum run-up heights ranging between 1 and 6.8 m), and debris flows (3 events with maximum run-up heights ranging between 1 and 5 m). In most cases, however, run-up is not reported at all, as records do not exist for most of the fourteen century to nineteen century events (79). Ongoing updating of the catalogue will include addition of new events and updating of existing events through collection of new background material and translation of pre-twentieth century reports into modern Norwegian and English.

13.4 Discussion and Conclusions

Displacement waves caused by subaerial landslides impacting into a water body and landslide-triggered tsunamis caused by submarine landslides represent a serious threat to Norwegian society. Among these rock avalanches generating displacement waves have caused the most fatalities, due to the typically great heights of both near-field and far-field waves. Other types of landslides generated either displacement waves or tsunamis have resulted in fatalities, although those occurred close to the area of the landslide due to lower wave heights. Snow avalanches are the only wave-producing gravitational process that has not caused any documented fatalities. Hence in order to reduce the negative impact of these secondary effects of landslides, at all sites where landslide hazard assessment is carried out and landslides can impact a water body that could lead to loss of life, injury or loss of property also the secondary effect should be assessed in future with care. Empirical relations of landslide type, volume, and distance from the impact area provided by this database might be used for a first-level assessment of possible effects of displacement waves and tsunamis generated by future landslides. Our compilation, although likely one of the most detailed of any country, indicates that geological records are poor. Therefore, effort should be given in future to better establish the relations between landslide type, landslide volume, resulting wave height and distance from the source in order to produce empirical tools that much better help to forecast future landslide scenarios/wave scenarios.

References

Blikra LH, Longva O, Braathen A, Anda E, Dehls JF, Stalsberg K (2006) Rock slope failures in Norwegian fjord areas: examples, spatial distribution and temporal pattern. In: Evans SG, Scarascia Mugnozza G, Strom A, Hermanns RL (eds) Landslides from massive rock slope failure; NATO science series, IV. Earth and environmental sciences, vol 49. Springer, Dodrecht, pp 475–496

Furseth A (2006) Skredulykker i Norge. Tun Forlag, Oslo

Grimstad E, Nesdal S (1990) The Loen rockslides—a historical review. In: Barton M, Stephanson W (eds) Rock joints. Balkema, Rotterdam, pp 1–6

Hermanns RL, Blikra LH, Naumann M, Nilsen B, Panthi KK, Stromeyer D, Longva O (2006) Examples of multiple rock-slope collapses from Köfels (Ötz valley, Austria) and western Norway. Eng Geol 83:94–108

Hermanns RL, Hansen L, Sletten K, Böhme M, Bunkholt H, Dehls JF, Eilertsen R, Fischer L, ĹHeureux J-S, Høgaas F, Nordahl B, Oppikofer T, Rubensdotter L, Solberg I-L, Stalsberg K, Yugsi Molina FX (2012) Systematic geological mapping for landslide understanding in the Norwegian context. In: Eberhardt E, Froese C, Turner AK, Leroueil S (eds) Landslide and engineered slopes: protecting society through improved understanding. Taylor & Francis Group, London, pp 265–271

Hermanns RL, L'Heureux J-S, Blikra LH (2013) Landslide-triggered tsunami, displacement wave. In: Bobrowsky PT (ed) Encyclopedia of natural hazards, Springer Science + Business Media B.V., Dordrecht, pp 611–615, doi 10.1007/978-1-4020-4399-4

Jørstad FA (1968) Waves generated by landslides in Norwegian fjords and lakes. Norwegian Geotechnical Institute, Oslo, 32p

Lastras G, Amblas D, Calafat AM, Canals M, Frigola J, Hermanns RL, Lafuerza S, Longva O, Micallef A, Sepúlveda S, Vargas G, De Batista M, Van Daele M, Azpiros M, Bascuñan I, Duhart P, Iglesias O, Kempf P, Rayo X (2013) Landslide cause tsunami waves: insights from Aysén fjord, Chile. EOS 4(34):297–298

L'Heureux J-S, Hansen L, Longva O, Emdal A, Grande L (2010) A multidisciplinary study of submarine landslides at the Nidelva fjord delta, Norway—implications for geohazards assessments. Norw J Geol 90:1–20

Nordahl B, Sletten K, Grøtan BO (2013) Nå kan DU registrere skred. Posisjon, Magasinet for geomatikk 2:22–24

Townson JM, Kaya Y (1988) Simulations of the waves in Lake Botnen created by Rissa landslide. In: Proceedings of the institution of civil engineers part 2—Research and theory 85, pp 145–160

Vorren TO, Mangerud J (2008) Glaciations come and go. In: Ramberg IB, Bryhni I, Nøttvedt A, Rangnes K (eds) The making of land: geology of Norway. Geological Society of Norway, Trondheim, pp 481–533

Vorren TO, Mangerud J, Blikra LH, Sveian H (2008) The emergence of modern Norway. In: Ramberg IB, Bryhni I, Nøttvedt A, Rangnes K (eds) The making of land: geology of Norway. Geological Society of Norway, Trondheim, pp 534–559

14 Earthquake-Triggered Subaerial Landslides that Caused Large Scale Fjord Sediment Deformation: Combined Subaerial and Submarine Studies of the 2007 Aysén Fjord Event, Chile

Reginald L. Hermanns, Sergio A. Sepúlveda, Galderic Lastras, David Amblas, Miquel Canals, María Azpiroz, Ignacio Bascuñán, Antonio M. Calafat, Paul Duhart, Jaime Frigola, Olaia Iglesias, Philipp Kempf, Sara Lafuerza, Oddvar Longva, Aaron Micallef, Thierry Oppikofer, Xavier Rayo, Gabriel Vargas, and Freddy Yugsi Molina

Abstract

On 21 April 2007 (M_w 6.2) an earthquake triggered more than 500 landslides near the epicenter along the Aysén fjord, Chile. One of the major failures occurred at the Punta Cola Valley involving a volume of 20.9 million cubic meters of rock. The main rockslide was followed by a rock/debris avalanche involving talus and glacio-fluvial deposits in the slope toe and valley floor that added a volume of 7.3 million cubic meters as entrained material. About half of the material involved in the rockslide-debris avalanche reached the shoreline and entered the fjord pushing deltaic deposits offshore while inducing a shoreline retreat of 100 m. The impact of the debris avalanche deformed an area of 7.6 km^2 of the otherwise featureless and smooth sedimented fjord floor. The central part of the deformed area is currently deeper with respect to the undeformed floor, which suggests that between 1 and 10 m of sediment were eroded from an area of 1.85 km^2 due to the direct impact of the avalanche. The combination of debris avalanche impact of this and other landslides, subaqueous failures and fjord floor deformation generated a series of displacement waves within the fjord with several meters to tens of meters high run-up along the shoreline.

R. L. Hermanns (✉) · O. Longva · T. Oppikofer · F. Y. Molina
Geological Survey of Norway, P.O. Box 6315, 7491 Sluppen, Trondheim, Norway
e-mail: Reginald.hermanns@ngu.no

S. A. Sepúlveda · G. Vargas
Universidad de Chile, Plaza Ercilla 803, Santiago, Chile

G. Lastras · D. Amblas · M. Canals · M. Azpiroz · A. M. Calafat
J. Frigola · O. Iglesias · S. Lafuerza · X. Rayo
GRC Geociències Marines, Universitat de Barcelona, Carrer Martí i Franquès s/n, 08028 Barcelona, Spain

I. Bascuñán · P. Duhart
Servicio Nacional de Geología y Minería, La Paz 406, Puerto Varas, Chile

P. Kempf
Renard Centre of Marine Geology, Ghent University, Krijgslaan 281/S8, 9000 Ghent, Belgium

A. Micallef
University of Malta, Msida, MSD 2080, Malta

G. Lollino et al. (eds.), *Engineering Geology for Society and Territory – Volume 4*,
DOI: 10.1007/978-3-319-08660-6_14, © Springer International Publishing Switzerland 2014

Keywords
Rock avalanche • Earthquake trigger • Displacement wave • Fjord sediment deformation

14.1 Introduction and Geological Setting

A number of events show that devastating displacement waves have occurred from subaerial landslides (e.g. Hendron and Patton 1987; Blikra et al. 2006). In some regions of the world, such as the fjords of western Norway, displacement waves generated by subaerial rock avalanches have taken place several times during the last century often resulting in large loss of life (e.g., Blikra et al. 2006; Hermanns et al. 2012). In addition, submarine landslides have also been identified as tsunami sources (e.g., Tappin et al. 2008). However, subaerial rock avalanches impacting a water body with synchronic submarine landslides and soft-sediment deformation have not been investigated as detailed topographic and bathymetric data from the same event are not usually available. In this contribution we present the first detailed combined high resolution data set of topographic and bathymetric data of the Punta Cola rockslide debris avalanche that entered the Aysén fjord on April 21st, 2007.

In early 2007, after an unusual four-month period of moderate seismicity in the Aysén Fjord (45.5° S, Chilean Patagonia), a Mw 6.2 earthquake struck the region. This seismic episode was related to tectonic activity along the regional, strike-slip Liquiñe-Ofqui Fault Zone (LOFZ), which accommodates the oblique convergence of the Nazca and South American plates. The mainshock (21st April 2007) was localized in one of the several branches of the LOFZ that crosses the fjord, and triggered hundreds of subaerial landslides in the epicentral area along the fjord and surroundings. Slope failure occurred on the glacially oversteepened slopes composed of granitic rocks of the North Patagonian Batholith and within the volcanic soil layer (Sepúlveda et al. 2010). The largest rock slides and rock avalanches that entered the fjord water mass generated a series of displacement waves that together with debris flows caused ten fatalities and severe damage to the salmon industry, which is the most important economic activity of the area (Naranjo et al. 2009; Sepúlveda and Serey 2009).

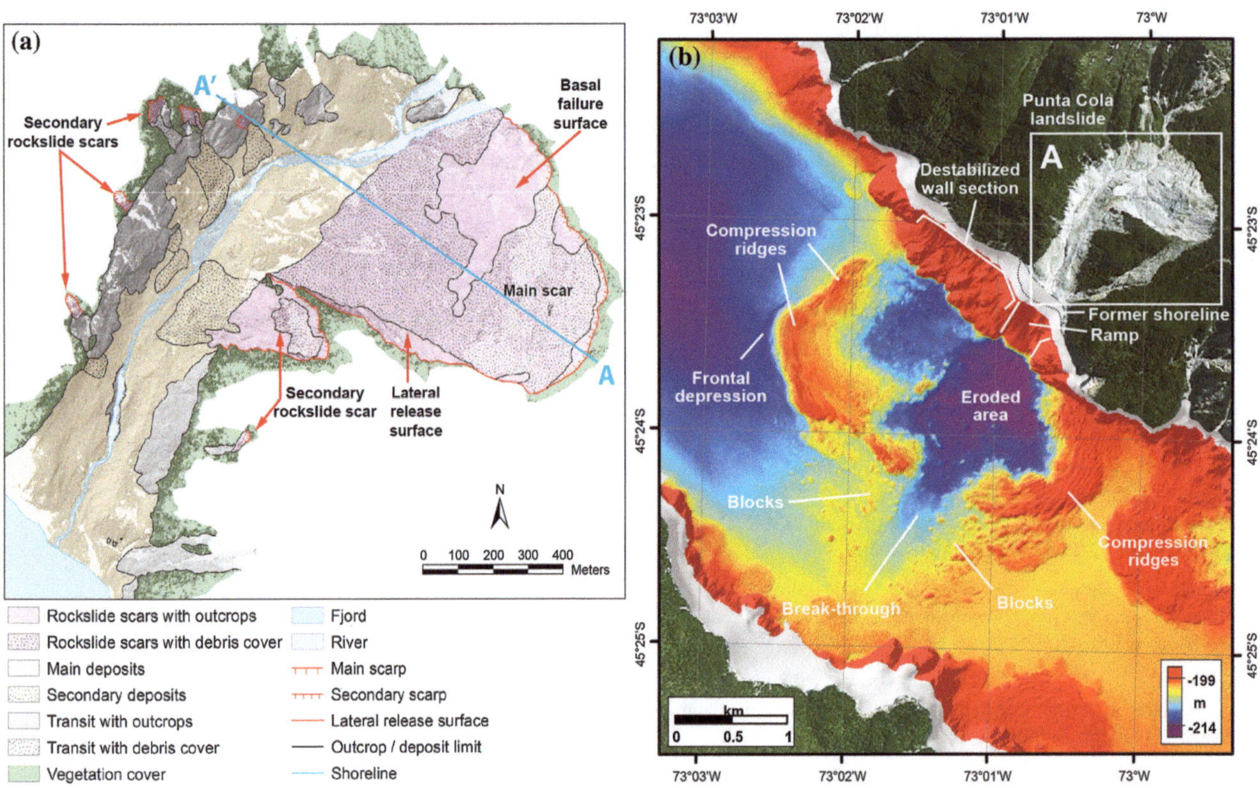

Fig. 14.1 The Punta Cola subaerial and submarine landslide. **a** TLS mapping of the Punta Cola rockslide. *A–A′* marks the profile location of Fig. 14.2. **b** Combined satellite imagery and bathymetric map of the Punta Cola landslide area, showing its imprint in the fjord floor

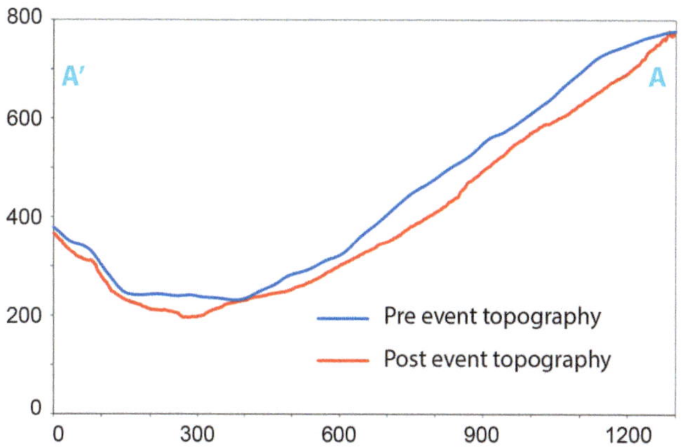

Fig. 14.2 Topographical profiles along main rock slope failure into the Punta Cola valley (location shown in Fig. 14.1). The pre event topography is based upon the photogrammetric DEM with 7 m pixel resolution and the post event topography is based upon the DEM obtained by TLS with an average point density of 35 points/m^2. Note that topography got reduced along the slope by the rock slope failure and in the valley by entrainment of debris into the debris avalanche (see text for details)

The Punta Cola rockslide failure sourced from a NW facing 220–750 m high slope within the Punta Cola valley 1–1.75 km away from the water front. It descended along the Punta Cola valley down to the waterline and entered the Aysén fjord (Sepúlveda and Serey 2009) (Fig. 14.1a). The rockslide resulted in the deformation of the fjord floor sediments in a large area in front of the Punta Cola valley (Van Daele et al. 2013; Lastras et al. 2013). Modeling of the displacement wave has been carried out, reproducing near source run-up heights observed on islands in the fjord, but the modeled and observed run-up in more distal areas differ by a factor 2–3 (Scheele et al. 2013).

14.2 Methods

Detailed Terrestrial Laser Scanning (TLS) was carried out in the Punta Cola valley in January 2010 in order to document the precise post failure morphology of the valley and obtain a volume estimate of the rockslide (Oppikofer et al. 2012). The post failure DEM obtained by TLS was built from 58 scans and covered an total area of 1.4 km^2 with an average point density of 35 points/m^2. The pre failure DEM reconstructed by photogrammetry based upon aerial photography and had a pixel resolution of 7 m (Yugsi Molina et al. 2012) (Fig. 14.2).

In order to study the impact of this subaerial landslide on the ocean sediments, swath bathymetry data was obtained during the DETSUFA cruise onboard BIO Hespérides in March 2013 with a Simrad EM-1002 multibeam echosounder. Data were tide-corrected and processed with Caris HIPS and SIPS, yielding a final 4 m cell-size DEM.

14.3 Results

Based upon the TLS digital elevation model, an structural analysis using Coltop3D software revealed that the basal failure surface of the rockslide is complex, with a 30° NW dipping fault plane being the most critical (Redfield et al. 2011). The post failure digital elevation model (DEM) obtained by TLS was compared to a pre failure DEM reconstructed by photogrammetry, indicating that the rockslide volume that failed from the main scar was 20.9 Mm^3. In addition 7.3 Mm^3 sediments were eroded out of the valley and entrained into the avalanche (Yugsi Molina et al. 2012).

The combination of both data was used to establish the rockslide-debris avalanche on land suggesting that failure occurred in two events. The first and main event was composed of a block up to 135 m in thickness that broke in smaller fragments, slid into the valley and run-up the opposing slope up to 180 m. Further close to the shoreline, run-up was 45 m, however flow direction here had to be nearly valley-parallel. Valley fill entrainment occurred mostly along the valley axis, while along the SE valley slope the rockslide had a net loss of volume (Redfield et al. 2011; Oppikofer et al. 2012). The debris avalanche entered the fjord causing the failure of the delta and a shoreline retreat of 100 m. Following the main event a second compartment of highly fractured rock failed upslope from the main failure and deposited at the slope foot. This deposit necessarily dammed the valley, however the dam did not persist long as no lake sediments have been found. Furthermore the entrainment of debris in the valley undercut its slopes were secondary landslides occurred that deposited on top of the debris avalanche.

This sub-aerial landslide entered the water mass, generating a displacement wave and leaving its imprint on the fjord floor. The debris avalanche descended and accelerated from the sea surface down to 203 m water depth through a 320 m wide, 580 m long, 10–25° ramp in the submerged part of the fjord wall (Fig. 14.1b). In turn, a 1.25 km long section of the northern wall west of the ramp also destabilized. The combined avalanche produced an area of 7.6 km^2 of deformation in the basin-plain sediment of the otherwise featureless and smooth fjord floor. The central part of the deformed area is currently deeper with respect to the undeformed floor, which suggests that between 1 and 10 m of sediment were eroded from an area of 1.85 km^2 due to the direct impact of the avalanche (Fig. 14.1b). This depressed area is most evident at the foot of the ramp following three main directions, whereas erosion is more modest further west, thus indicating the area of higher flow energy. The eroded sediment from the central part was folded and piled forming a deformation ring that attains a positive relief of up to 12 m with respect to the undeformed fjord floor, characterized by a series of concentric compressional ridges and a faint frontal depression. In front of the ramp, the area impacted with the strongest energy, the deformation ring has no continuity. This suggests that the sediment was there over-deformed by the arrival of large rock fragments from the sub-aerial landslide, probably pulling it off and forming dozens of up to 70 m large sediment chunks, which can be observed at the sides of the break-through. These blocks could be formed by rock fragments coated by sediment. The imprint of the Punta Cola landslide can be observed up to 3.5 km away from the coastline, at the foot of the opposite fjord wall (Lastras et al. 2013).

14.4 Discussion and Conclusion

The Aysén event exemplifies the enormous imprint on submarine sediments that a subaerial landslide can leave. The subaerial landslide at Punta Cola caused the failure of a 100-m-wide part of the delta along the shoreline and slope failure on the submarine slope and in addition strong deformation of 7.6 km^2 of flat ocean bottom sediments. The subaerial landslide represents only a minor fraction of the total area and volume of the event including its submarine part. In general, several parameters have been taken into account to model the magnitude of displacement waves generated by a landslide impact into a water body including: (i) volume of moving material, and especially the morphology of the front area, (ii) water depth of the water body, (iii) velocity of the slide (e.g. Tinti et al. 1999; Løvholt et al. 2008). However those parameters might not entirely explain the water surface elevation, since the synchronic submarine deformation is not taken into account. In the Aysén fjord event, omitting such deformation might help explaining the discrepancy of the measured run-up along the shorelines in the fjord and the modeled surface elevation (Scheele et al. 2013). Displacement wave modelling should therefore not only take into account the entering of a large rock avalanche in the fjord water mass, but also the sudden upwards movement of the marine sediments.

Acknowledgments This research was supported by Acción Complementaria DETSUFA (CTM2010-09891-E MAR) of the Spanish RTD programme, and a Generalitat de Catalunya "Grup de Recerca Consolidat" grant (2009-SGR 1305).

References

Blikra LH, Longva O, Braathen A, Anda E, Dehls JF, Stalsberg K (2006) Rock slope failures in Norwegian Fjord Areas: examples, spatial distribution and temporal pattern. In: Evans SG, Scarascia Mugnozza G, Strom A, Hermanns RL (eds) Landslides from massive rock slope failure. NATO science series, IV. Earth and environmental sciences, vol 49. Springer, Dodrecht, pp 475–496

Hendron AJ Jr, Patton FD (1987) The Vaiont slide; a geotechnical analysis based on new geologic observations of the failure surface. Eng Geol 24:475–491

Hermanns RL, Hansen L, Sletten K, Böhme M, Bunkholt H, Dehls JF, Eilertsen R, Fischer L, ĹHeureux J-S, Høgaas F, Nordahl B, Oppikofer T, Rubensdotter L, Solberg I-L, Stalsberg K, Yugsi Molina FX (2012) Systematic geological mapping for landslide understanding in the Norwegian context. In: Eberhardt E, Froese C, Turner AK, Leroueil S (eds) Landslide and engineered slopes: protecting society through improved understanding. Taylor & Francis Group, London, pp 265–271

Lastras G, Amblas D, Calafat AM, Canals M, Frigola J, Hermanns RL, Lafuerza S, Longva O, Micallef A, Sepúlveda S, Vargas G, De Batist M, Van Daele M, Azpiroz M, Bascuñán I, Duhart P, Iglesias O, Kempf P, Rayo X (2013) Landslide cause tsunami waves: insights from Aysén fjord, Chile. EOS Trans Am Geophys Union 4(34):297–298

Løvholt F, Pedersen G, Gisler G (2008) Oceanic propagation of a potential tsunami from the La Palma Island. J Geophys Res 113:C09026

Naranjo JA, Arenas M, Clavero, Muñoz O (2009) Mass movement-induced tsunamis: main effects during the Patagonian. Fjordland seismic crisis in Aisén (45°25′S), Chile. Andean Geol 36:137–145

Oppikofer T, Hermanns RL, Redfield T, Sepúlveda SA, Duhart P, Bascuñan I (2012) Morphologic description of the Punta Cola rock avalanche and associated minor rockslides caused by the 21 April 2007 Aysén earthquake (Patagonia, southern Chile). Revista Asociación Geológica Argentina 69:339–353

Redfield T, Hermanns R, Oppikofer T, Duhart P, Mella M, Derch P, Bascuñán I, Arenas M, Fernández J, Sepúlveda S, Rebolledo S, Loew S, Yugsi Molina F, Abächerli A, Henderson IHC, Jaboyedoff M, Kveldsvik V (2011) Analysis of the 2007 earthquake-induced Punta Cola rockslide and tsunami, Aysén fjord, Patagonia, Chile (45.3° S, 73.0° W). In: 5th international conference on earthquake geotechnical engineering, Santiago, paper 12.8

Scheele L, Yugsi Molina F, Oppikofer T, Hermanns RL, Kveldsvik V, Glimsdal S (2013) ICG P10 2006, tsunami modeling and prediction: modeling of the Aysén fjord rock slide tsunami. ICG report 3013-10-1, NGI report 20061073-02-R, p 19

Sepúlveda SA, Serey A (2009) Tsunamigenic, earth-quake-triggered rock slope failures during the 21st of April 2007 Aisén earthquake, Southern Chile (45.5° S). Andean Geol 36(1):131–136

Sepúlveda SA, Serey A, Lara M, Pavez A, Rebolledo S (2010) Landslides induced by the 2007 Aysen Fjord earthquake, Chilean Patagonia. Landslides 7:483–492

Tappin DR, Watts P, Grilli ST (2008) The Papua New Guinea tsunami of 17 July 1998: anatomy of a catastrophic event. Nat Hazards Earth Syst Sci 8:1–24

Tinti S, Bortolucci E, Armigliato A (1999) Numerical simulation of the landslide-induced tsunami of 1988 on Vulcano Island, Italy. Bull Volc 61:121–137

Van Daele M, Versteeg W, Pino M, Urrutia R, De Batist M (2013) Widespread deformation of basin-plain sediments in Aysén fjord (Chile) due to impact by earthquake-triggered, onshore-generated mass movements. Mar Geol 337:67–79

Yugsi Molina F, Oppikofer T, Hermanns RL, Red-Field TF, Bascuñán I, Loew S, Sepúlveda SA (2012) Mechanism and volume estimation of the 2007 Punta Cola rockslide-debris avalanche using terrestrial laser scanning and aerial photogrammetry. In: Eberhardt E, Froese C, Turner AK, Leroueil S (eds) Landslides and engineered slopes: protecting society through improved understanding, vol 1. Taylor & Francis Group, London, pp 553–559

Pre-shear Effect on Liquefaction Resistance of Sand

15

Jiafeng Lu and Bin Ye

Abstract

Pre-shear history has been proven to be a critical factor in the liquefaction resistance of sand. In contrast to prior experimental studies in which triaxial shear tests were used to examine the effect of pre-shear on the liquefaction resistance of sand, hollow cylinder torsional shear tests were used in this study, to avoid the influence of inherent anisotropy that is inevitably produced during the sample preparation process due to gravitational deposition. A series of cyclic undrained shear tests were carried out on sand samples that had experienced different degrees of pre-shear loading. The test results showed that the liquefaction resistance of sand is greatly reduced by its pre-shear history, and a small pre-shear strain can cause sand to be much more prone to liquefaction. During the cyclic shear tests, the samples that had experienced pre-shear loading exhibited different behaviors when cyclic shear loading started in different directions, i.e. the clockwise direction and the counterclockwise direction.

Keywords

Pre-shear effect • Liquefaction • Hollow cylinder shear test • Anisotropy

15.1 Introduction

Liquefaction is a major threat to the safety of civil engineering structures constructed in sandy grounds when earthquakes occur. The liquefaction resistance of sand is governed by many factors, such as relative density, particle size gradation, particle shape etc. These normal factors have been studied extensively in the past few decades. In addition to these factors, the "pre-shear history" of sand is another factor that has a significant effect on its liquefaction resistance, but this factor has not received much attention. Finn et al. (1970) were the first researchers to note that the pre-shear history of sand can greatly influence its liquefaction resistance. Finn et al. (1970) demonstrated that a small pre-shearing can cause an increase in liquefaction resistance, and a large pre-shearing can cause a drastic reduction in liquefaction resistance. Ishihara and Okata (1982) also found that the stress-induced anisotropy developed by pre-shearing can greatly influence the rate of increase of pore pressure during liquefaction. Recently, Yamada et al. (2010) and Ye et al. (2012) showed experimentally and theoretically that the pre-shearing induced by liquefaction can cause continuous, orderly and rapid changes in anisotropy, which influences the reliquefaction resistance of sand.

Most previous studies on the effect of pre-shear history on the liquefaction resistance of sand are based on experimental results obtained from triaxial shear tests. During triaxial compression or extension tests, stress-induced anisotropy develops in the same direction as the inherent anisotropy that is inevitably produced during the sample preparation process. However, this deficiency of the triaxial shear test apparatus can be overcome by another type of shear test, i.e., the hollow cylinder torsional shear test. In a hollow cylinder torsional shear test, the shear load can be applied along the horizontal plane, as shown in Fig. 15.1.

J. Lu · B. Ye (✉)
Tongji University, Siping Road 1239, Shanghai 200092, China
e-mail: yebinmail1977@gmail.com

G. Lollino et al. (eds.), *Engineering Geology for Society and Territory – Volume 4*,
DOI: 10.1007/978-3-319-08660-6_15,

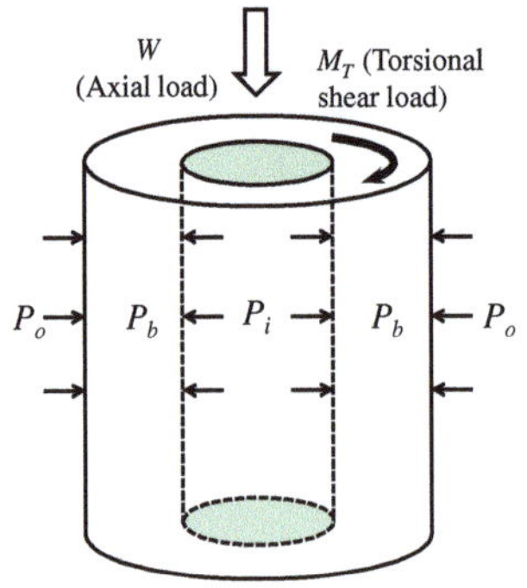

Fig. 15.1 Torsional shear load applied to a hollow cylindrical sample

In this study, a series of hollow cylinder torsional shear tests were conducted to investigate how the liquefaction resistance of sand changes after pre-shear loading. The test results proved that pre-shearing can greatly reduce the liquefaction resistance of sand.

15.2 Description of the Hollow Cylinder Torsional Shear Test

The hollow cylinder testing apparatus can simultaneously impose axial load W, torque M_T, outer cell pressure P_o, inner cell pressure P_i and back pressure P_b (pore pressure) on the sample and control all five of them individually.

15.2.1 Sample Preparation

The tested sand in this study was Fujian standard sand, which is widely used in geotechnical experiments in China. The cumulative grain size distribution of the tested sand and physical properties are shown in Fig. 15.2 and Table 15.1, respectively.

The outer and inner diameters of the hollow cylindrical samples were 100 and 60 mm, respectively. The heights of the samples were 200 mm. Each sample was prepared by the water pluviation method. The Skempton's B value was 0.96 or higher after the saturation procedure during which the mean effective stress was kept constant at 110 kPa.

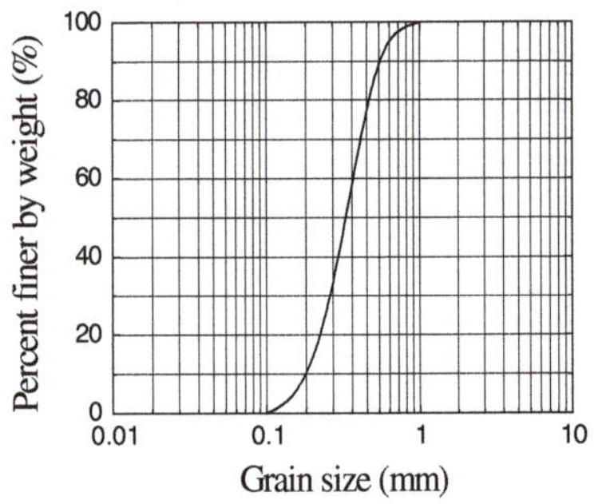

Fig. 15.2 Cumulative grain size distribution of the tested sand

Table 15.1 Physical properties of the tested sand

Density of soil particles ρ_s(g/cm^3)	2.64
Maximum void ratio $e_{\max}$	0.78
Minimum void ratio $e_{\min}$	0.50

15.2.2 Test Scheme

To investigate the pre-shear effect, the following test scheme was designed:

Stage 1: The samples were isotropically consolidated under equal outer and inner cell pressures of 200 kPa and a back pressure (pore pressure) of 90 kPa.

Stage 2: The consolidated samples were subjected to strain-controlled torsional shear loading under drained conditions. Table 15.2 lists the different degrees of pre-shear strain that were applied to the samples.

Stage 3: After the pre-shear strain reached the targeted value, the shear stress was unloaded to zero under drained conditions at the same rate that it was applied.

Stage 4: The samples were subjected to a cyclic torsional shear load under undrained conditions until liquefaction occurred.

The cyclic torsional loads were applied in sinusoidal mode, as shown in Fig. 15.3.

15.3 Test Results and Discussions

The stress–strain relations for Samples [b-1]–[g-1] in the pre-shear and subsequent unloading processes (Stages 2 and 3) are shown in Fig. 15.4 During the pre-shear tests, the mean effective stress of the sand was held constant under drained conditions. The changes in the relative densities of all samples were very small during the pre-shear tests.

15.3.1 Cyclic Shear Test on the Sample Without Pre-shear History

Figure 15.5 displays the stress path from the cyclic shear test of Sample [a], which had no pre-shear before cyclic shear loading. The parameter N_{re}, which is defined as the cycle number at which the mean effective stress of sand first reached the zero stress state, was used to indicate the liquefaction resistance of the sand.

Table 15.2 List of the tested samples

Sample no.	Pre-shear strain, γ_{pr} (%)	D_r before pre-shear (%)	D_r after unloading (%)	Starting direction of cyclic shear load
[a]	0	79.5	–	Clockwise
[b-1]	0.5	79.5	80.0	Clockwise
[b-2]			79.8	Counterclockwise
[c-1]	1.0	79.5	79.9	Clockwise
[c-2]			79.8	Counterclockwise
[d-1]	2.0	79.5	79.8	Clockwise
[d-2]			79.8	Counterclockwise
[e-1]	3.0	79.5	79.5	Clockwise
[e-2]			79.7	Counterclockwise
[f-1]	4.0	79.5	78.3	Clockwise
[f-2]			79.0	Counterclockwise
[g-1]	5.0	79.5	78.5	Clockwise
[g-2]			78.2	Counterclockwise

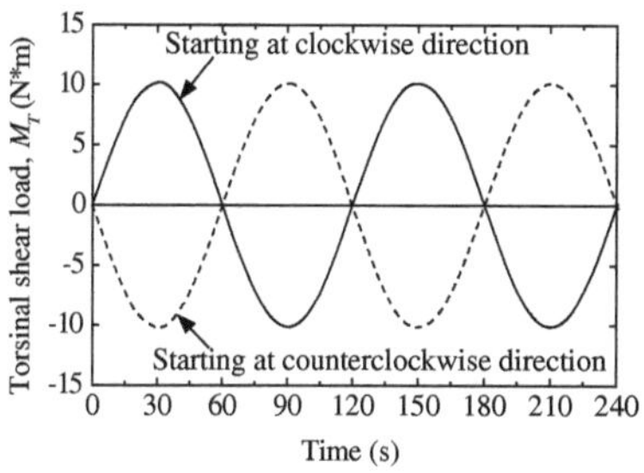

Fig. 15.3 Time history of the cyclic shear loads in the pre-shear tests

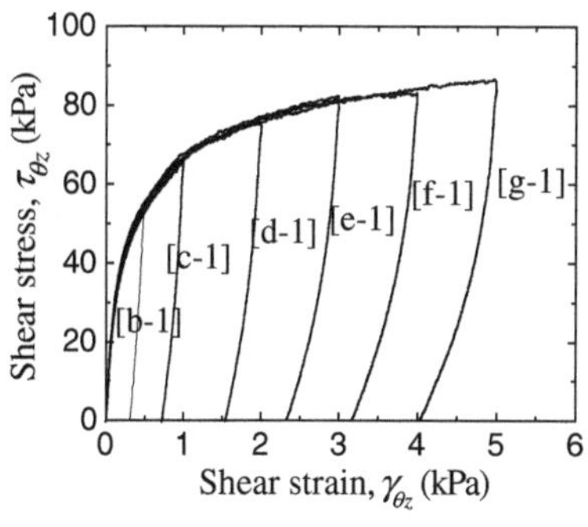

Fig. 15.4 Stress path and stress–strain relation

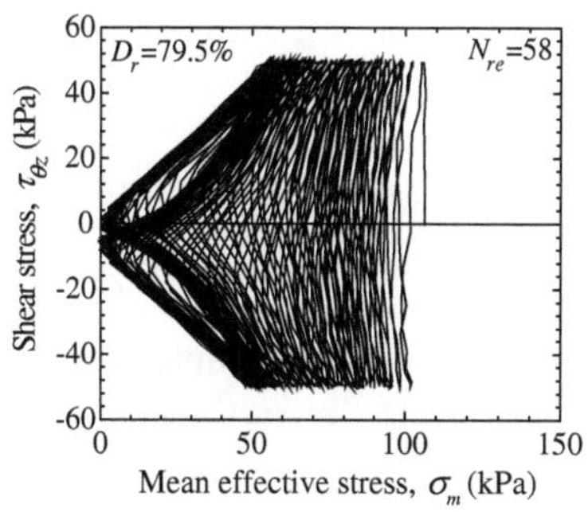

Fig. 15.5 Cyclic shear behavior of sample [a]

15.3.2 Cyclic Shear Tests on the Samples with Pre-shear History

Figure 15.6 displays the cyclic shear behaviors of Samples [b-1, 2]–[d-1, 2], which experienced pre-shear strains from 0.5 to 2 %. Comparing the stress paths in Figs. 15.5 and 15.6, it can be observed that he mean effective stresses of the samples with pre-shear history decreased much faster than that of Sample [a]. Many fewer loading cycles were required for Samples [c-1, 2], [d-1, 2] to reach the zero stress state.

From Fig. 15.6, it can also be observed that the shear tests begun in the clockwise direction and those begun in the counterclockwise direction resulted in almost the same values of N_{re}. However, the stress paths exhibit obvious differences, especially during the first half loading cycle. This difference can be theoretically explained by the development of stress-induced anisotropy, using rotational hardening theory (Zhang et al. 2007). During the pre-shear tests, the stress-induced anisotropy developed in the same direction as the pre-shear loading. The anisotropy can be represented by the upward rotation of the yield surface in the stress space, as shown in Figs. 15.7 and 15.8. After unloading, the anisotropy remained in the soil, and the yield surface still inclined upward. If the subsequent cyclic loading started in the clockwise direction (Fig. 15.7), the stress increment vector, dτ, pointed toward the interior of the yield surface, and an unloading process began. Thus, the mean effective stress did not change under undrained conditions. Alternatively, if shear loading started in the counterclockwise direction (Fig. 15.8), the stress increment vector, dτ, pointed outward from the yield surface, and the loading process started, causing the mean effective stress to decrease more substantially. In a more detailed comparison of the cyclic shear tests started in the clockwise and

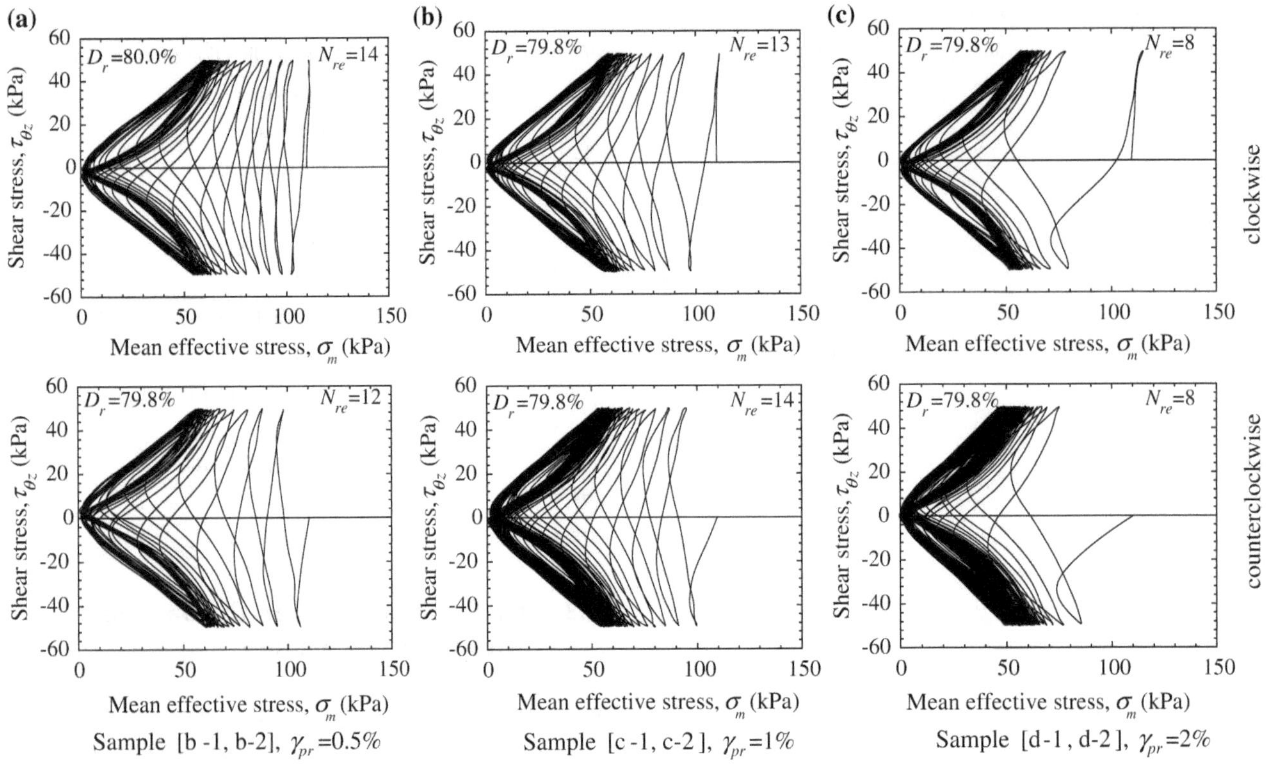

Fig. 15.6 Cyclic shear behavior of samples [b]–[h] (with pre-shear history)

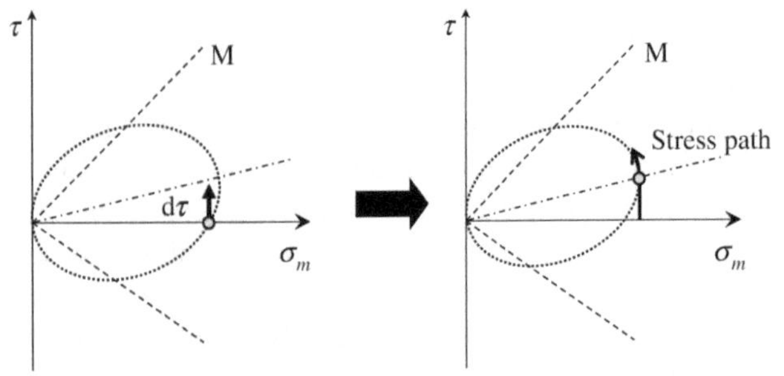

Fig. 15.7 Cyclic shear loading starting in the clockwise direction

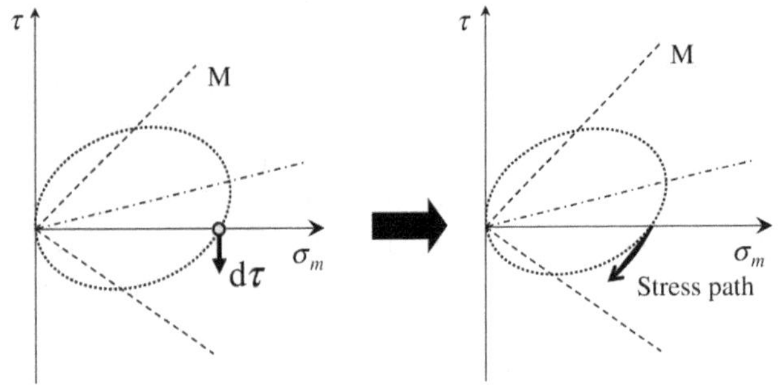

Fig. 15.8 Cyclic shear loading starting in the counterclockwise direction

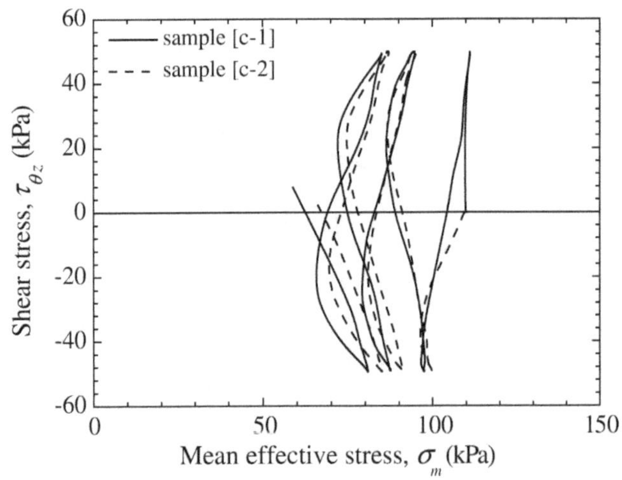

Fig. 15.9 Stress path of the first three cycles of cyclic loading of Samples [c-1] and [c-2]

counterclockwise directions, Fig. 15.9 illustrates the stress paths of Samples [c-1] and [c-2] during the first three loading cycles.

In addition, stress path of Sample [c-1] for the second half of the loading cycle was very similar to that of Sample [c-2] from the beginning. This implies that the anisotropic state of Sample [c-1] after the first half of the loading cycle was close to the initial state of Sample [c-2]. This means that the influence of the anisotropy induced by pre-shearing nearly disappeared after the first loading cycle.

The above discussion suggests that the influence of pre-shear loading could lead to a substantial decrease in the liquefaction resistance of sand and that this influence runs through the whole liquefaction process.

15.4 Conclusions

In this study, hollow cylinder shear tests were carried out to examine the effect of pre-shear on the liquefaction resistance of sand. Cyclic undrained shear loadings were applied to samples that had experienced different degrees of pre-shear loading. The main conclusions drawn from the results of this study are summarized below.

(1) The liquefaction resistance of sand is greatly influenced by its pre-shear history. Sand is more prone to liquefaction after experiencing pre-shear, even though its relative density is almost unchanged. The larger the pre-shear strain is, the lower the liquefaction resistance will be.
(2) After pre-shear loading, sand exhibits anisotropic behavior during subsequent cyclic shear loading, especially during the first loading cycle. From the second half of the loading cycle onward, the stress paths of the tests started in the clockwise direction were similar to those of the tests started in the counterclockwise direction.
(3) The anisotropy induced by pre-shear loading disappeared after the first loading cycle. From the second loading cycle on, the anisotropy induced by the following cyclic loading played a more significant role in the liquefaction process.

References

Finn WDL, Bransby PL, Pickering DJ (1970) Effects of strain history on liquefaction of sands. Soil Mech Found Div Proc ASCE 96(6):1917–1934

Ishihara K, Okada S (1982) Effects of large preshearing on cyclic behavior of sand. Soils Found 22(3):109–125

Yamada S, Takamori T, Sato K (2010) Effects on reliquefaction resistance produced by changes in anisotropy during liquefaction. Soils Found 50(1):9–25

Ye B, GL Y, Zhang F (2012) Numerical modeling of changes in anisotropy during liquefaction using a generalized constitutive model. Comput Geotech 42:62–72

Zhang F, Ye B, Noda T, Nakano M, Nakai K (2007) Explanation of cyclic mobility of soils: approach by stress-induced anisotropy. Soils Found 47(4):635–648

Engineering Geology and Geohazards of Sefidrud Delta, South Caspian Coast, North Iran

16

M. Hashemi, M. R. Nikoudel, M. Khamehchiyan, and N. Hafezi Moghadas

Abstract

The engineering geological conditions and geohazards of Sefidrud Delta were assessed using a stepwise approach. The approach included the geological, sedimentological and geotechnical studies of the area. First of all the sedimentary environments of Delta and their associated deposits were identified. Investigation and assessment of engineering geological characteristics of these deposits revealed that the most Holocene deltaic deposits have poor geotechnical conditions. Most of these deposits are associated with geohazards including liquefaction, erosion and land subsidence.

Keywords

Sefidrud delta • Engineering geology • Geohazards

16.1 Introduction

Sefidrud delta (Fig. 16.1) is a large, wave-dominated clastic sediment prism associated with wide beach ridges, coastal dunes, lagoonal and related sediments in the southern Caspian Sea, Iran. It has been formed by the River Sefidrud which has a drainage area of 61,300 km^2. At present, surface of subaerial delta plain and coastal length of the delta are 1,405 km^2 and 91 km respectively. Due to rapid Caspian sea-level fluctuations, the Sefidrud delta has been developed during repetitive transgressive and regressive coastal conditions (Kazanci et al. 2010; Kazanci and Gulbabazadeh 2013).

The delta is an area of great socioeconomic importance to northern part of Iran because of its high population density, valuable natural resources, tourism potentials and etc. This important area has experienced geological problems in the past decades like liquefaction in a vast area of delta during 1990 Manjil earthquake that caused the destruction of many houses and commercial and public buildings (Ishihara et al. 1992; Yegian et al. 1995). In this research, the engineering geological conditions and geohazards of the Sefidrud delta were assessed using a stepwise approach. This approach includes geological, sedimentological and geotechnical studies of the study area.

16.2 Methodology

As mentioned previously, this research was done using a stepwise approach. First of all, based on geological and sedimentological studies, the sedimentary sub-environment of the delta and associated sediments were identified. Then the engineering geological characteristics of the sediments

M. Hashemi (✉)
Department of geology, Faculty of sciences,
University of Isfahan, Isfahan, Iran
e-mail: M-hashemi@sci.ui.ac.ir

M. R. Nikoudel · M. Khamehchiyan
Department of engineering geology, Faculty of sciences,
Tarbiat Modares University, Tehran, Iran

N. Hafezi Moghadas
Department of geology, Faculty of s ciences ,
Ferdowsi University of Mashhad, Mashhad, Iran

G. Lollino et al. (eds.), *Engineering Geology for Society and Territory – Volume 4*,
DOI: 10.1007/978-3-319-08660-6_16, © Springer International Publishing Switzerland 2014

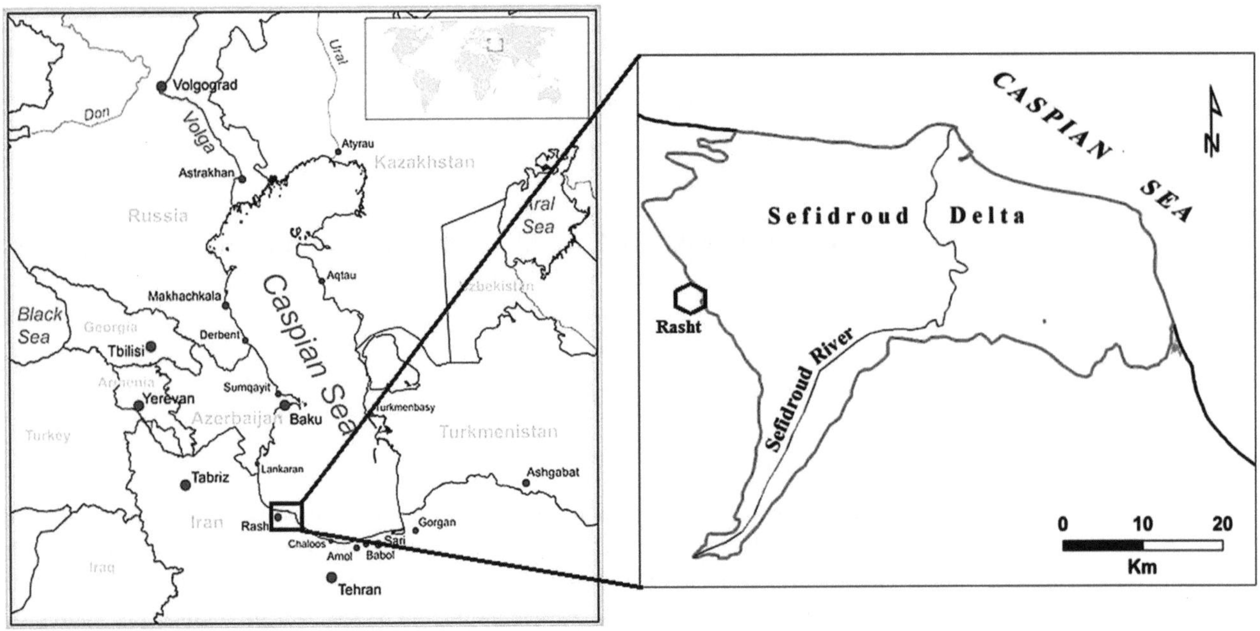

Fig. 16.1 Sefidrud delta

were assessed and finally the associated geo-hazards including liquefaction, subsidence and erodibility of sediments were identified. For geotechnical data, various borehole information from both the private and public projects across the study area were used. In addition to the available geotechnical data, 20 Dynamic Cone Penetrometer (DCP) tests were carried out to investigate the characteristics of surficial soils.

For liquefaction potential assessments, safety factor of liquefiable soils against liquefaction was determined based on the SPT and DCP data using the approach introduced by Youd and Idriss (2001). Finally, Liquefaction Potential Index (LPI) of soils was determined for SPT and DCP test locations using Sonmez (2003) and Luna and Frost (1998), respectively.

Erodibility defines the resistance of the soil to both detachment and transport. Erodibility varies with various factors including soil texture, aggregate stability, shear strength, infiltration capacity and organic and chemical content (Knapen et al. 2006; Morgan 2005). In this research, two main factors including texture and shear strength of soil were considered for evaluation of soil erodibility. The role of soil texture is that large particles are resistant to transport because of the greater force required to entrain them and that fine particles are resistant to detachment because of their cohesiveness. The least resistant particles are silts and fine sands (Morgan 2005). Considering these facts, we developed Table 16.1 for classification of erodibility of sediments based on soil texture factor. As this table shows, silty and fine sandy sediments that are not resistant to detachment and transport have very high erodibility, so they were ranked 1 (first rank). On the other hand, gravelly sediments that are resistant to transport and detachment, have very low erodibility, so they were ranked 5 (last rank). Other sediments were ranked between these two categories.

The detachability of soil depends not only on its texture but also on soil shear strength (Cruse and Larson 1977). The shear strength of the soil is a measure of its cohesiveness and resistance to shearing forces exerted by gravity, moving fluids and mechanical loads (Morgan 2005). This factor that can be easily measured is the most suitable parameter for characterizing soil erosion resistance (Knapen et al. 2006). So, in this research for a more accurate evaluation of soil erodibility, shear strength factor were considered in addition to the soil texture factor. In this regard, Tables 16.2 and

Table 16.1 Classification of soil erodibility based on texture of soil

Soil texture	Erodibility classification	Rank of erodibility
Gravely soils	Very low erodible	5
Clayey soils	Low erodible	4
Coarse sandy soils	Medium erodible	3
Clayey silty soils	High erodible	2
Silty and fine sandy soils	Very high erodible	1

Table 16.2 Classification of soil erodibility (cohesive soils) based on shear strength

SPT	Cu (kPa)	Shear strength	Erodibility	Rank of erodibility
0–2	<12.5	Very low	Very high erodible	1
3–5	12.5–25	Low	High erodible	2
6–9	25–50	Medium	Medium erodible	3
10–16	50–100	High	Low erodible	4
>16	>100	Very high	Very low erodible	5

Table 16.3 Classification of soil erodibility (granular soils) based on shear strength

SPT	DPI (mm/blow)	Shear strength	Erodibility	Rank of erodibility
2–3	>45	Very low	Very high erodible	1
4–7	25–45	Low	High erodible	2
8–20	15–25	Medium	Medium erodible	3
21–40	5–15	High	Low erodible	4
>40	<5	Very high	Very low erodible	5

Table 16.4 Classification of settlement susceptibility of soil based on Cc and LI

Cc	LI	Settlement susceptibility
<0.25	<0.5	Low
0.25–0.5	0.5–1	Medium
>0.5	>1	High

16.3 were developed for evaluation of soil erodibility based on shear strength of cohesive and granular soils, respectively. The shear strength of various sediments was determined based on results of the SPT and DCP tests.

Compaction of Holocene strata (Tornqvist et al. 2008; Tosi et al. 2009; Stiros 2001) was considered as the main cause of subsidence of deltaic deposits. In this regard, to assess the subsidence hazard, the compressibility characteristics especially the compression coefficient (Cc) and liquidity index (LI) of sediments were considered to classify the settlement susceptibility as shown in Table 16.4.

16.3 Geology and Engineering Geology

For engineering geological assessment and geo-hazards evaluation of the Sefidrud Delta, at first the sedimentary environments (or sub-environments) of the delta and their associated deposits were identified based on previous sedimentological studies (Kazanci and Gulbabazadeh 2013; Kazanci et al. 2010; Nogol-e-Sadat 1991; Annells et al. 1975). Figure 16.2 shows the distribution of various deposits of the delta. As shown in this figure, main deposits of the delta are undifferentiated deltaic deposits including Early Holocene Deltaic Deposits (DD1), Middle Holocene Deltaic Deposits (DD2) and Late Holocene Deltaic Deposits (DD3). The latest deposits form the present active lobe of the delta. Other deposits include Coastal Deposits (CD), Natural Levee Deposits (NLD), Lagoonal Deposits (LD), Marine Deposits (MD), fine grained Alluvial Deposits (ADF) and coarse grained Alluvial Deposits (ADC). Main Geotechnical properties of deltaic deposits are given in Table 16.5.

16.4 Geohazards of the Delta

After investigation of the engineering geological characteristics of the sediments, their associated geo-hazards including liquefaction, subsidence and erodibility were identified as follow:

16.4.1 Liquefaction Potential

To assess the liquefaction potential, first of all the most liquefiable deposits were identified based on geological criteria (Youd and Perkins 1978). Then the liquefaction potential (LPI) of liquefiable deposits was determined based on SPT and DCP data. The most liquefiable deposits of the delta are DD1, DD2, DD3, ADC, ADF, CD and NLD. The LPI of DD1, DD2 and ADC are less than 5, so these deposits have low to medium liquefaction potential. The LPI of CD is

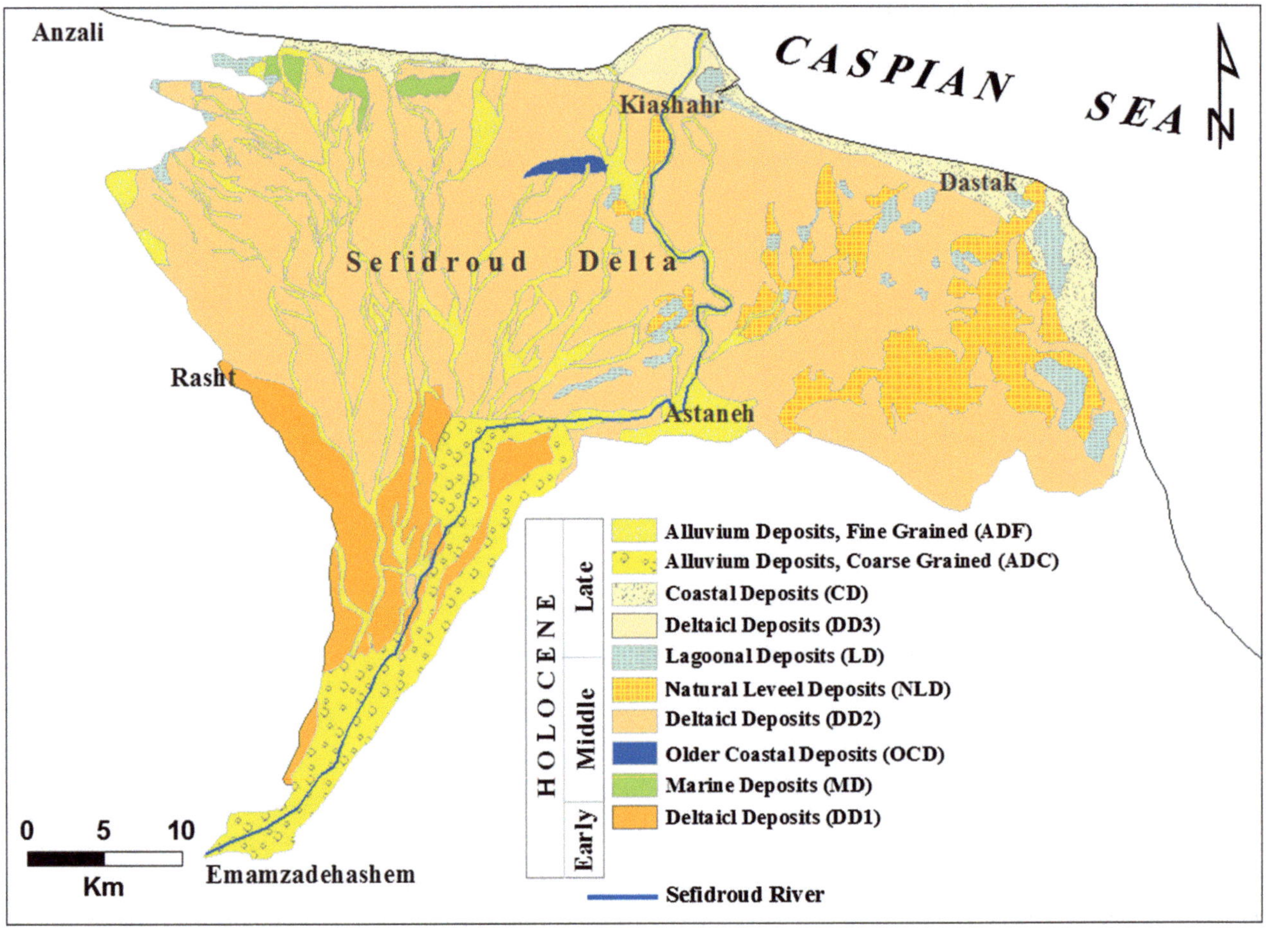

Fig. 16.2 Distribution of various deposits of Sefidrud delta

Table 16.5 Main geotechnical properties of deltaic deposits

Soil parameters	DD1	DD2	DD3
Gravel (%)	47–78	0–33	0–6
Sand (%)	14–46	10–49	20–65
Silt and Clay (%)	3–30	51–90	30–80
Natural moisture content (%)	7–56	25–86	28–79
Liquid limit (%)	–	27–85	18–69
Plasticity index (%)	–	1–49	0–38
Soil density (kN/m^3)	18.81–20.2	17.4–20.03	15.76–18.7
Void ratio	0.22–0.86	0.68–2.1	0.71–1.9
SPT blow count (N)	26 to >50	1 to >50	1–31

between 10 and 15, so these deposits have high liquefaction potential and the LPI of the NLD, DD3 and ADF is more than 15, so these deposits have very high liquefaction potential based on Sonmez (2003). Figure 16.3 shows the liquefaction potential map of the Sefidrud Delta.

16.4.2 Erodibility

Considering the texture of deposits (Table 16.1) DD1 and ADC are very low erodible; MD and LD are low erodible; OCD, CD and ADF are medium erodible; DD2 is high

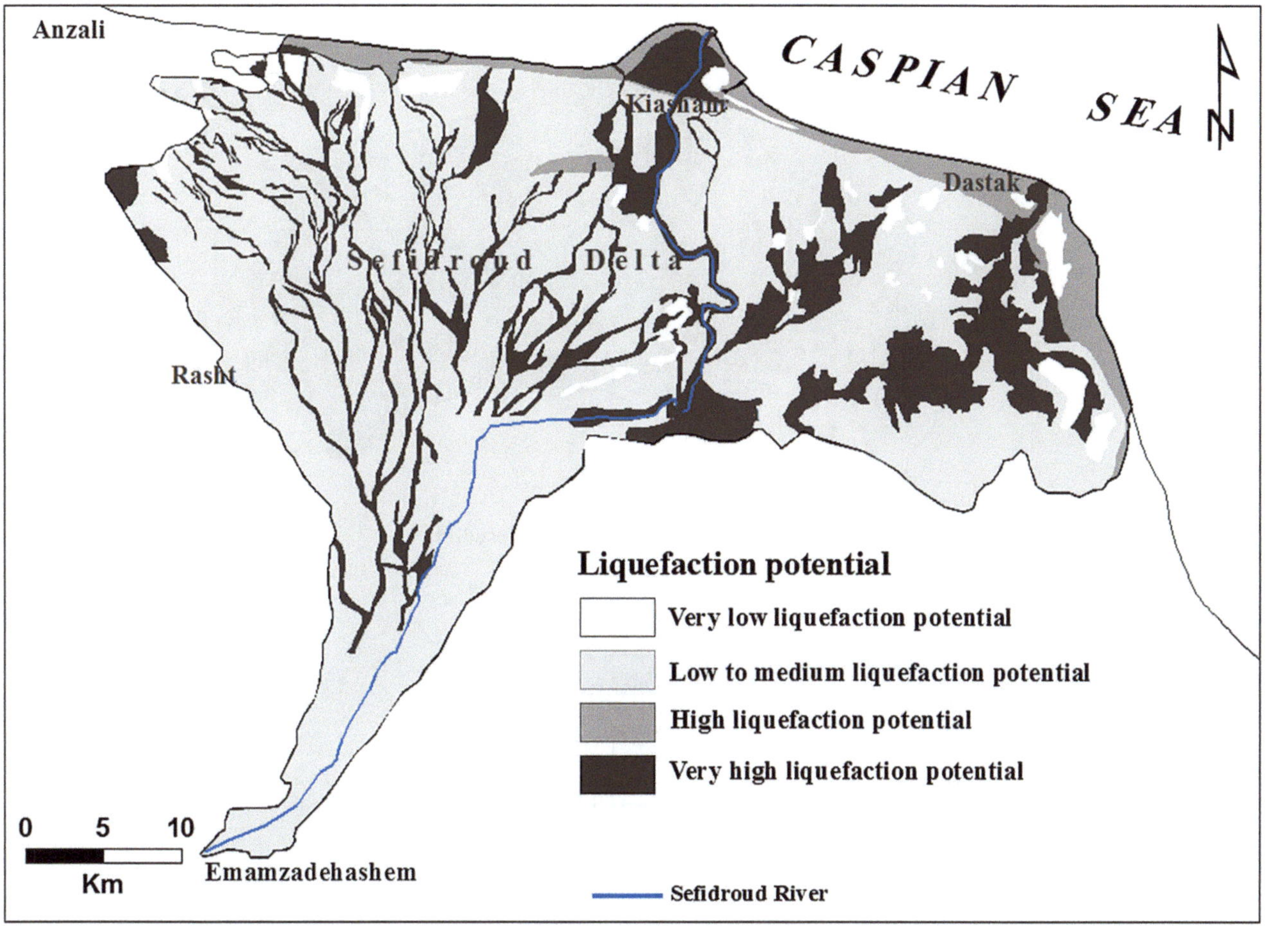

Fig. 16.3 Liquefaction potential map of Sefidrud delta

erodible and DD3 and NLD are very high erodible. On the other hand, based on shear strength of deposits (Tables 16.2 and 16.3) DD1 and ADC are very low to low erodible; MD and LD are low to medium erodible; OCD and CD are medium erodible; DD2 is high erodible and ADF, DD3 and NLD are very high erodible. By combining the factors of texture and shear strength of deposits, DD1 and ADC are very low erodible; MD and LD are low erodible; OCD and CD are medium erodible; DD2 and ADF are high erodible and finally DD3 and NLD are very high erodible. Figure 16.4 shows the erodibility map of the Sefidrud delta.

16.4.3 Subsidence

As previously mentioned compaction of Holocene strata was considered as the main cause of subsidence of deltaic deposits. DD2, DD3, LD and MD contain susceptible strata to compaction that causes subsidence of the area. In most cases of LD, the LI is more than 1 and the Cc is more than 0.5, so based on Table 16.4, the LD has high settlement susceptibility. In most cases of DD2, the LI is between 0.5 and 1 and the Cc is between 0.25 and 0.5, so the DD2 has medium settlement susceptibility. In most cases of MD, the LI is less than 0.5 and Cc is less than 0.25, so the DD2 has low settlement susceptibility. Other deposits like DD3, NLD and ADF have medium to high settlement susceptibility due to their loose nature. Figure 16.5 shows the settlement susceptibility map of the Sefidrud delta.

16.5 Conclusion

The results of the research show that the most Holocene deposits of Sefidrud Delta have poor engineering geological conditions. These poor conditions will cause (or have caused) rising geological problems including liquefaction potential, erosion and subsidence. DD1 in comparison with other deltaic deposits has better condition and poses few problems. But DD2 is associated with land erosion and land subsidence problems. The youngest deltaic deposits DD3 is associated with all problems including liquefaction potential, soil erosion and land subsidence. Other problematic deposits are NLD, ADF and LD that pose geological

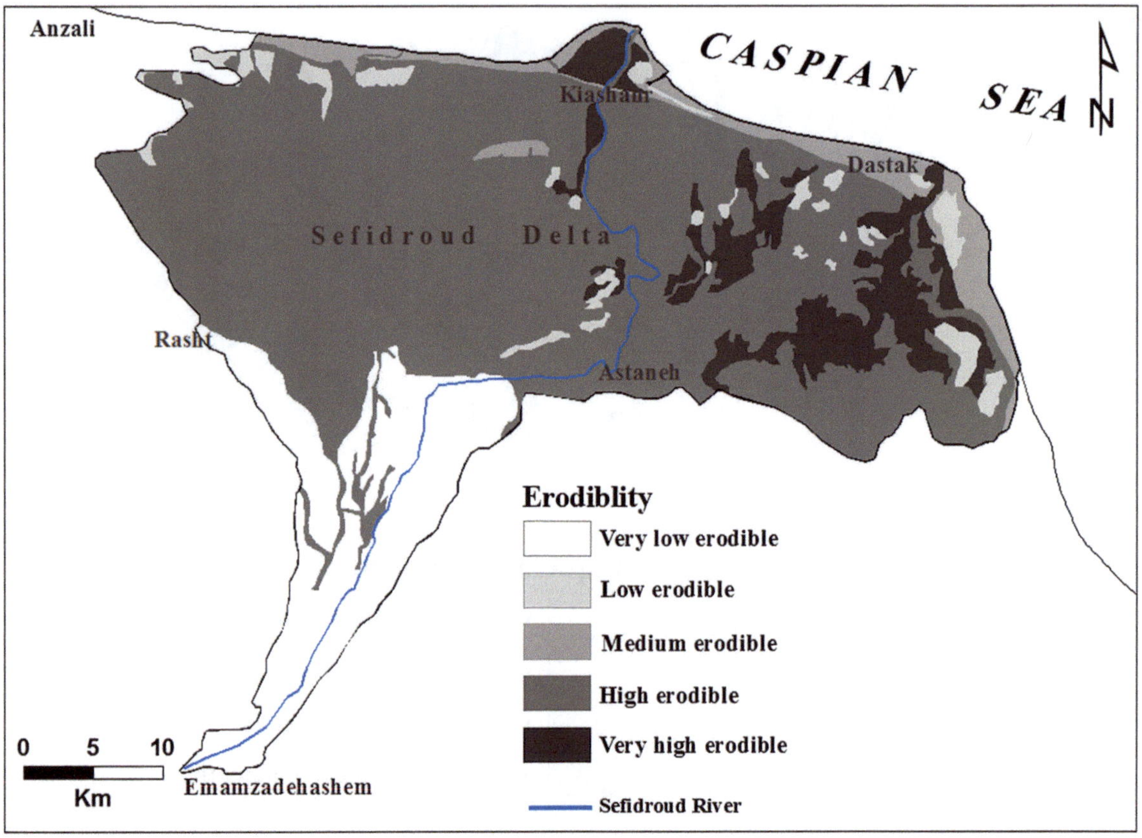

Fig. 16.4 Erodibility map of Sefidrud delta

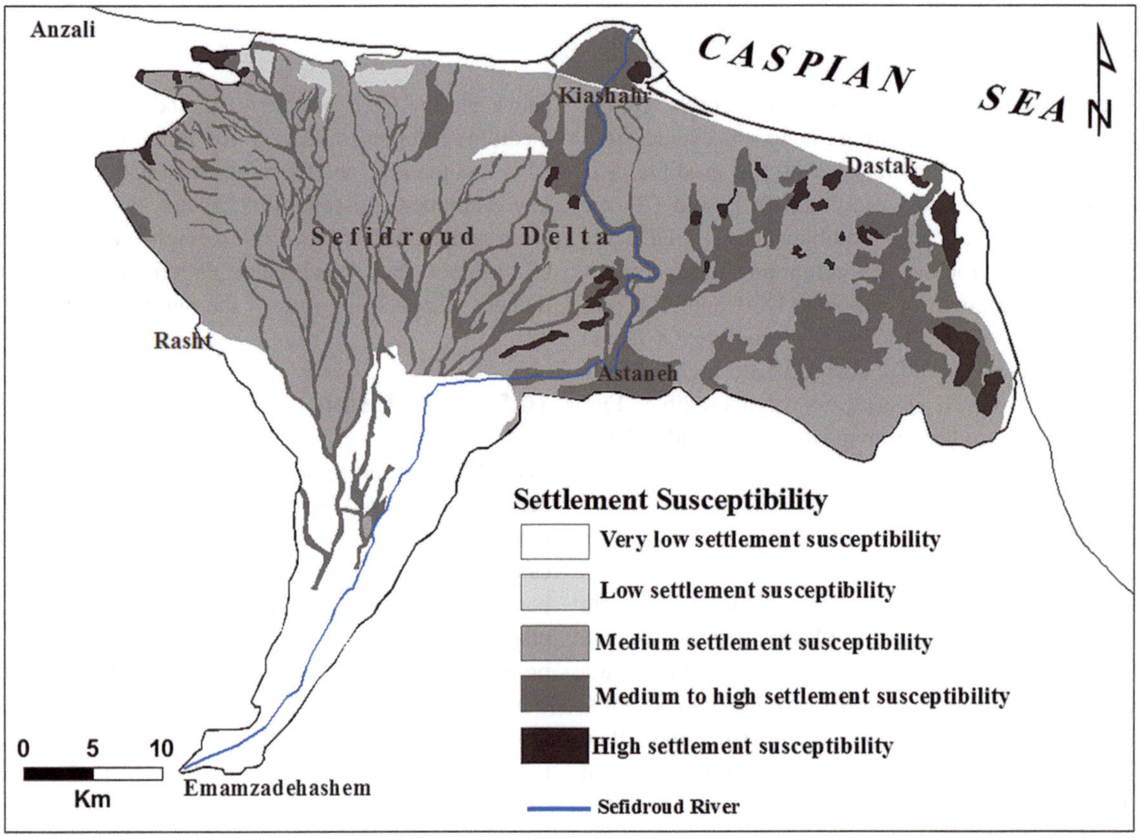

Fig. 16.5 Settlement susceptibility map of Sefidrud delta

problems to the surrounding area. The results of this research could be used as a useful tool for urban planning of the investigated area to a sustain development of the area.

References

Annells RN, Arthurton RS, Bazley RA, Davies RG (1975) Explanatory text of the Qazvin and Rasht quadrangles map 1:250000. Geological Survey of Iran, Tehran

Cruse RM, Larson WE (1977) Effect of soil shear strength on soil detachment due to raindrop impact. Soil Sci Soc Am J 41:777–781

Ishihara K, Haeri SM, Moinfar AA, Towhata I, Tsunjino S (1992) Geotechnical aspects of the June 20, 1990 Manjil Earthquake in Iran. Soil Found 32(3):61–78

Kazancı N, Gulbabazadeh T (2013) Sefidrud delta and quaternary evolution of the southern Caspian Lowland, Iran. Mar Pet Geol 44:120–139

Kazanci N, Gulbabazadeh T, Suludere Y (2010) Geological evolution of the Sefidrud Delta, Southern Caspian Sea, Iran. In: International conference on tectonic crossroads: evolving orogens of Eurasia-Africa-Arabia, Middle East Technical University, Ankara, Turkey, 4–8 Oct 2010

Knapen A, Poesen J, Govers G, Gyssels G, Nachtergaele J (2006) Resistance of soils to concentrated flow erosion: a review. Earth Sci Rev 80:75–109

Luna R, Frost JD (1998) Spatial liquefaction analysis system. J Comput Civil Eng 12(1):48–56

Morgan RPC (2005) Soil erosion and conservation, 3rd edn. Blackwell Publishing Ltd, Oxford, p 304

Nogol-e-Sadat MAA (1991) Comprehensive geological studies of Guilan Province. Governmental Office of Guilan Province, Rasht

Sonmez H (2003) Modification of the liquefaction potential index and liquefaction susceptibility mapping for a liquefaction-prone area (Inegol, Turkey). Environ Geol 44(7):862–871

Stiros SC (2001) Subsidence of the Thessaloniki (northern Greece) coastal plain, 1966–1990. Eng Geol 61:243–256

Tornqvist TE, Wallace DJ, Storms JEA, Wallinga J, Van Dam RL, Blaauw M, Derksen MS, Klerks CJW, Meijneken C, Snijders EMA (2008) Mississippi Delta subsidence primarily caused by compaction of Holocene strata. Nat Geosci 1:173–176

Tosi L, Teatini P, Carbognin L, Brancolini G (2009) Using high resolution data to reveal depth-dependent mechanisms that drive land subsidence: the Venice coast, Italy. Tectonophysics 474:271–284

Yegian MK, Ghahraman VG, Nogole-Sadat MAA, Daraie H (1995) Liquefaction during the 1990 Manjil, Iran, earthquake, II: case history analysis. Bull Seismol Soc Am 85(1):83–92

Youd TL, Idriss IM (2001) Liquefaction resistance of soils: summary report from the 1996 NCEER and 1998 NCEER/NSF workshop on evaluation of Liquefaction resistance of soils. J Geotech Geoenviron 127(10):817–833

Youd TL, Perkins DM (1978) Mapping of liquefaction induced ground failure potential. J Geotech Engrg Div ASCE 104 (4):433–446

17 Investigations on the Possible Source of the 2002 Landslide Tsunami in Rhodes, Greece, Through Numerical Techniques

Filippo Zaniboni, Gianluca Pagnoni, Alberto Armigliato, Katharina Elsen, and Stefano Tinti

Abstract

The island of Rhodes (Aegean Sea, Greece) has been repeatedly hit by tsunamis in the past due to the numerous tsunamigenic sources present in the area, most of which are seismic. Here an investigation is made on the most recent event that occurred on March 24, 2002 in the northeast of the island: unexpected waves affected a 2 km long coastal segment, overtopping part of the sea-wall (3–4 m high) that runs along the littoral road a few km away from the city of Rhodes. Data on the tsunami are poor. Due to the lack of evidence of seismological or meteorological causes, the hypothesis of a landslide source for the tsunami is here explored by means of numerical codes implementing models both for the landslide dynamics and the tsunami propagation. The reconstructed failing mass is located at about 100 m sea depth with a volume of 30 million m^3. A sensitivity analysis is further run by varying the source volume, to evaluate the effects on the tsunami impact.

Keywords

Landslide tsunami • Numerical modeling • Coastal hazard

17.1 Introduction

During the evening of March 24th, 2002, unusual waves affected a coastal segment of about 2 km in the northeast side of the island of Rhodes (marked in yellow in Fig. 17.1). The waves overtopped the sea wall protecting the littoral road, 3–4 m high, damaging some shops and stores located on the landside of the road.

The island of Rhodes, located in the south-eastern Aegean Sea, was affected in the past by several large earthquake-generated tsunamis, as for example the events of 1303, 1481, 1609 and 1741, that caused many casualties and severe damages also in other Greek islands (Papadopoulos et al. 2007). In the case of the 2002 event, however, the seismic origin for the tsunami can be ruled out, since no shock was recorded by the Greek seismic network in the evening of March 24. Since also storms and other meteorological anomalies can be excluded from the analysis of meteorological data, the only remaining option is that a submarine failure was the cause for the observed waves. This hypothesis was soon advanced and was further supported by the statements of some local fishermen, reporting an increase of the sea depth about 1 km off the affected shoreline (Papadopoulos et al. 2007).

Also with the goal of investigating such possibility, a high-resolution bathymetric survey was carried out after the event off the northeast coast of Rhodes by the Hellenic Center of Marine Research (HCMR) in the search for evidences of underwater failures (Sakellariou et al. 2002). The most studied zone was the steep slope NW of the northern tip of the island of Rhodes (named Cape of the Mills), that was found to be characterized by several scars and headwalls at depths of about 300–400 m, (visible in Fig. 17.1).

F. Zaniboni (✉) · G. Pagnoni · A. Armigliato · K. Elsen · S. Tinti
Dipartimento di Fisica e Astronomia, Università di Bologna, Bologna, Italy
e-mail: filippo.zaniboni@unibo.it

G. Lollino et al. (eds.), *Engineering Geology for Society and Territory – Volume 4*,
DOI: 10.1007/978-3-319-08660-6_17,

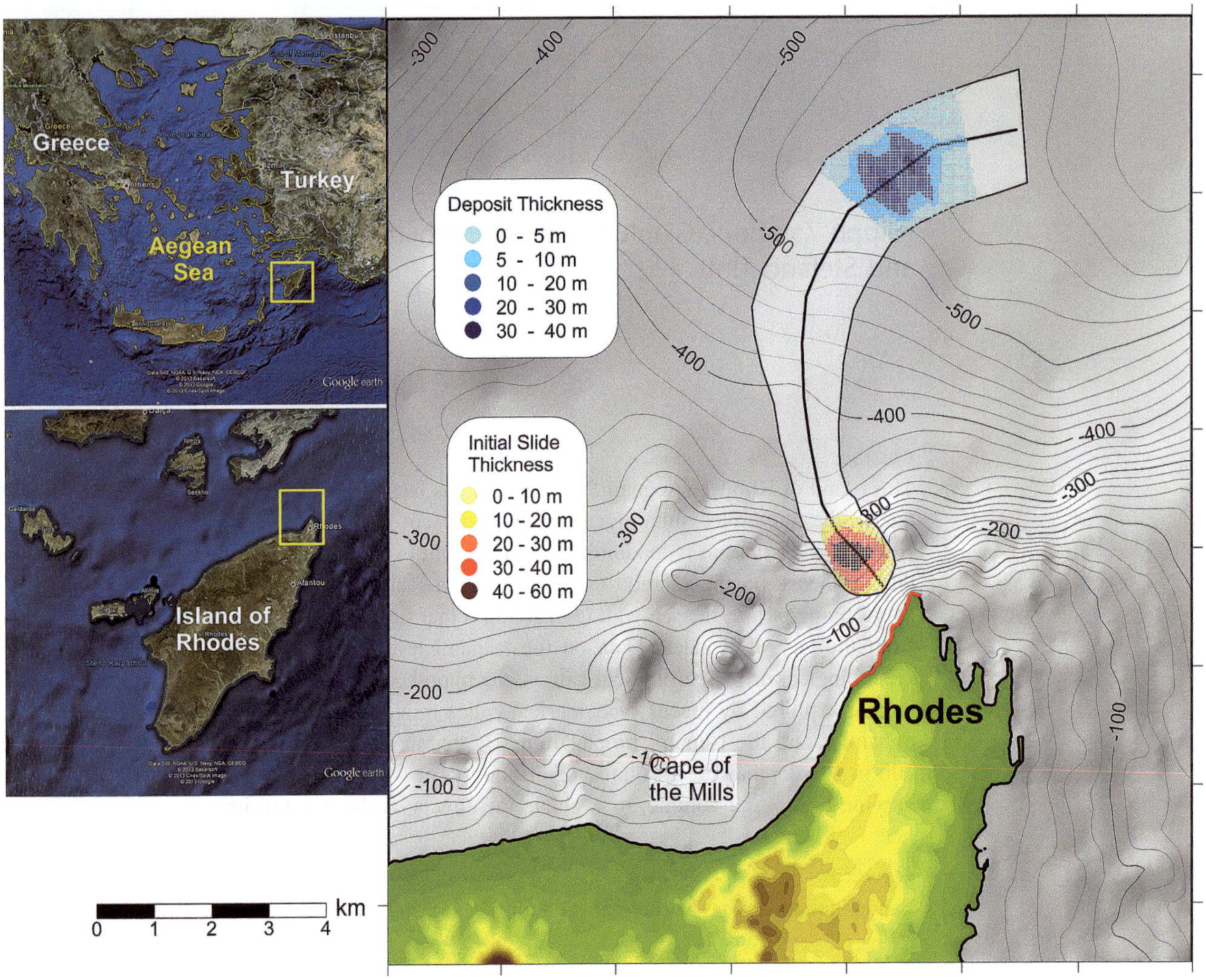

Fig. 17.1 Map of the northern tip of the island of Rhodes, with the 2002 affected coastal stretch evidenced in *red*. The sliding surface is highlighted in *light grey*, the CoM trajectory by the *black line* inside it. The *yellow-red scale area* marks the initial thickness of the landslide, the *cyan-blue* is for the final deposit

This area was seen to be the most prone to mass flows, and since it fits also with the area of sea depth changes reported by fishermen, Sakellariou et al. (2002) suggested it could be the best candidate for the 2002 tsunami generation, though they did not identify any specific scars as the source of the tsunami neither any specific depositional features.

In this work we explore the hypothesis of a landslide generation for the tsunami by studying this occurrence through numerical simulations. First we make assumption on the position of the source and reconstruct the sliding mass, on the basis of the available data. Second, we simulate the landslide motion by means of the numerical code UBO-BLOCK1, that provides the tsunamigenic impulse. Eventually, we calculate the generation and propagation of the tsunami through the numerical code UBO-TSUFD, and compare the results with the observations.

To cope with the uncertainties in the source, we conduct a sensitivity analysis by changing the source volume, more specifically the landslide thickness, which is seen to have the main effect of changing the tsunami size and the impact on the coast.

17.2 Numerical Models

The slide dynamics is simulated through the numerical code UBO-BLOCK1, developed and maintained by the University of Bologna tsunami research team. It implements a

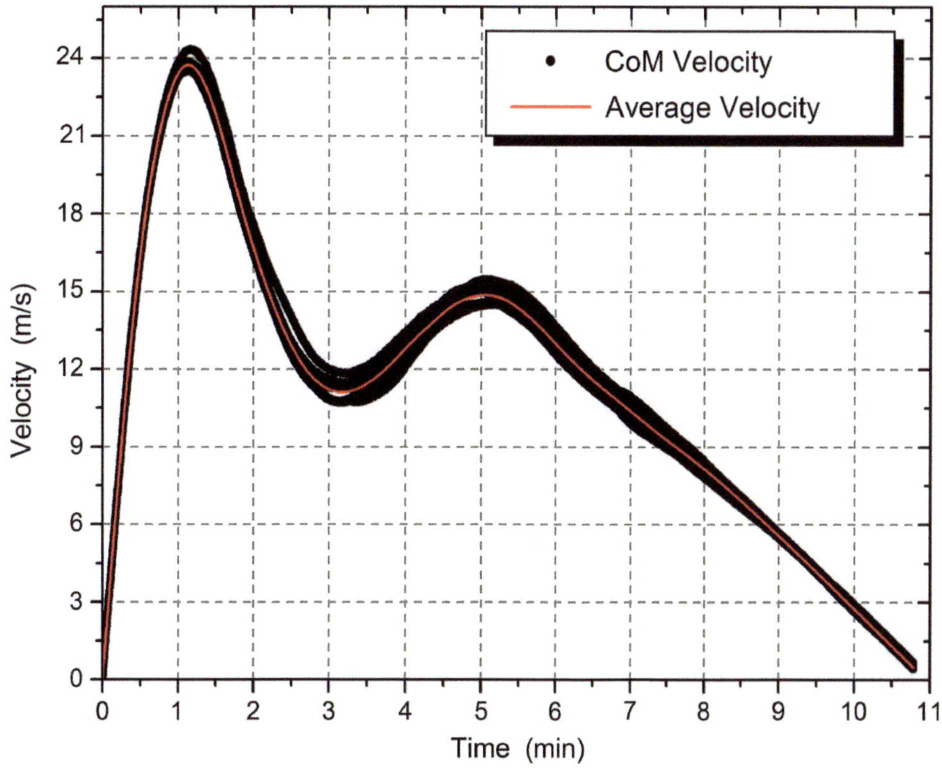

Fig. 17.2 Velocity of the individual blocks (*black dots*) and average velocity (*red line*), obtained as the mean of the CoM velocity at each time step, versus time

Lagrangian block approach, meaning that the mass is partitioned into a series of blocks and that the active forces (gravity, buoyancy, drag, block reciprocal deformation) are referred to the centre of mass (CoM hereafter) of each block. The code needs as input the specification of the sliding surface, the upper surface of the initial mass, the CoM pre-defined trajectory and the lateral boundaries confining the mass spreading during the motion (all reported in Fig. 17.1).

An intermediate software, UBO-TSUIMP, starting from the complete time-history of the slide motion computes the tsunamigenic impulse, which is related to the local changes to sea floor produced by the moving slide attenuated through a filter based on the ratio of the typical slide wavelength over the local sea-depth. It also performs the mapping of such impulse from the landslide grid, Lagrangian, to the tsunami grid, which is Eulerian.

The propagation of the tsunami is computed via the numerical code UBO-TSUFD, developed by the same research team, that solves the shallow water equations with a finite difference technique, where the impulse due to the slide is introduced as a forcing term. The software requires as input a grid with regular steps, with each node representing the local sea depth or topography.

Detailed examples of applications of such models can be found in Tinti et al. (2006, 2011), Tonini et al. (2011), Zaniboni et al. (2013).

17.3 The Simulation of the Submarine Slide

The landslide scenario is determined starting from some considerations in Sakellariou et al. (2002). The mobilized volume is hypothesized in the range of 32 million m^3, and is obtained considering one of the scars along the underwater slope NW of the city of Rhodes, between 120 and 320 m depth, and simply filling it considering the surrounding morphology (Fig. 17.1). The initial mass thickness (in yellow-red scale in Fig. 17.1) that reaches the maximum value of 60 m, and is on average around 27 m, is distributed over an area of 1.2 km^2. The undisturbed sliding surface (in pale grey in Fig. 17.1), is defined on the basis of the bathymetry and is delimited laterally by boundaries that control the mass spreading during the motion; the CoM track (black inner line) is specified following the line of maximum local gradient, and is found to bend north-eastward in deeper sea. The slide motion is then computed by inputting the above elements in the UBO-BLOCK1 code.

Figure 17.1 contains also the location of the final simulated deposit (cyan-blue area). The hypothesized landslide stops at more than 500 m sea depth, at a distance of 8 km from the source. The thickness lowers to an average value of 8 m, the area covered by the slide growing to 4 km^2. As visible in the velocity graph (Fig. 17.2), a strong acceleration phase brings the mass to the speed of 24 m/s and is followed by a fast deceleration when the mass reaches the sub-horizontal foot of

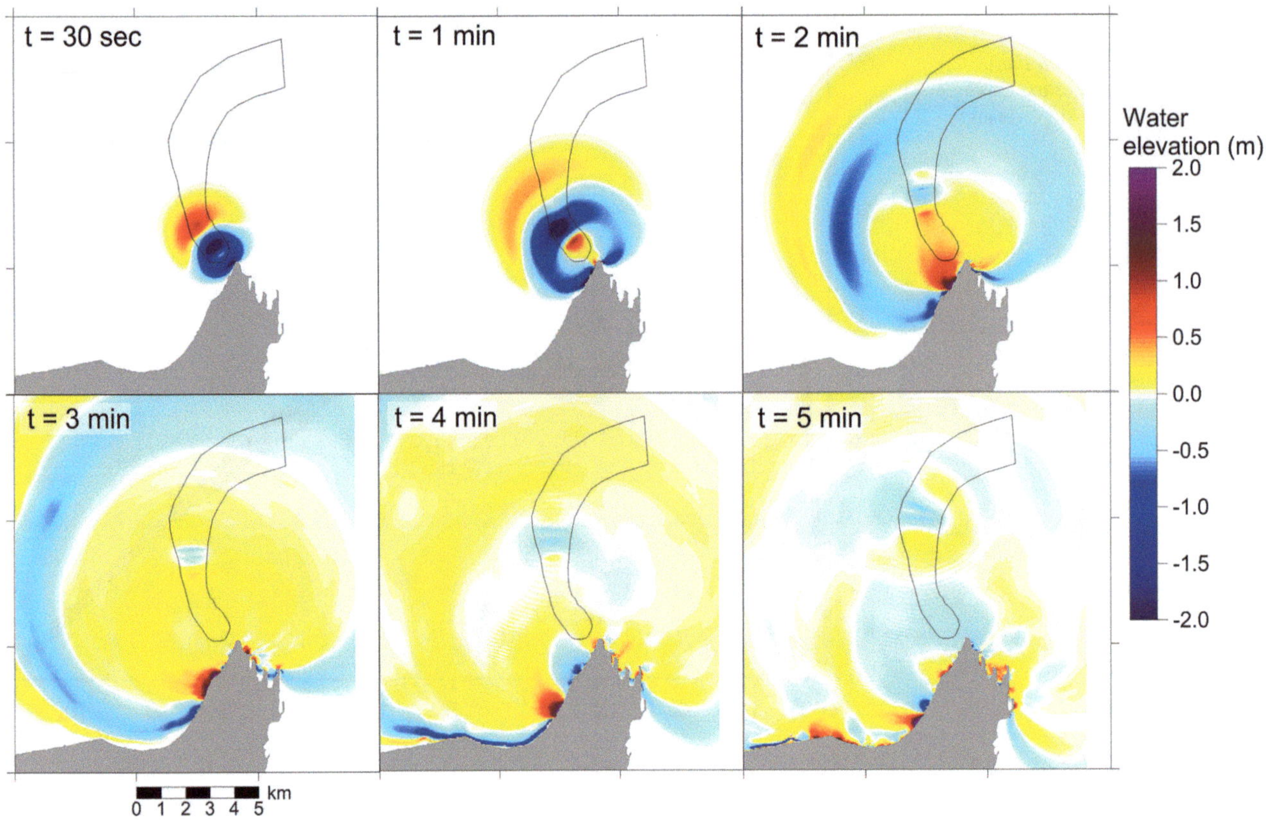

Fig. 17.3 Tsunami propagation over the computational domain, covering the northern part of the island of Rhodes. Positive signal, corresponding to sea level increase, are marked in *yellow-red*, negative (corresponding to water lowering) in *cyan-blue*. The sliding surface boundary is marked in *black* in each sketch

the escarpment, between 300 and 400 m depth. After this, the slope steepens again, and a new acceleration rises the velocity up to values of 15 m/s. The motion stops after about 11 min. Black dots in Fig. 17.2 represent the velocities of the individual CoMs. Since all speeds happen to be quite similar to one another and close to the average value (red line), one can infer that the mass remains rather compact while sliding. The first part of the motion, with the strongest acceleration and highest velocity reached, is the most tsunamigenic, considering also that afterwards the mass moves in deeper water, and therefore its capability of generating perturbation on the sea surface is reduced.

17.4 Tsunami Propagation Modeling

The computational grid of the tsunami model is made of over 160,000 regularly spaced nodes, 40 m apart, covering the northern part of the island of Rhodes (the most affected by the 2002 event) and the bathymetry north of it.

The model output of Fig. 17.3 shows an initial radial propagation for the tsunami, centered in the source zone (southern part of the black contour, Fig. 17.3). The sea surface perturbation exhibits a negative front (marked in cyan-blue) propagating southward, opposite to the landslide motion towards the coast of Rhodes, and a positive signal moving northward, towards the open sea. After about 30 s the tsunami reaches the coast where the 2002 event was observed, first manifesting with a negative signal, i.e. a sea withdrawal, of more than 2 m. The same area is affected by a positive signal, meaning sea inundation, about 2 min after the slide initiation. In all the panels of Fig. 17.3 one can also observe the propagation of a train of waves along the coast, which means that the coastal bathymetry favours the trapping of wave energy along the near-shore zone. The tsunami crosses Cape of the Mills after 2 min, hitting the eastern coast with a front almost parallel to the coast. The same happens along the coast to the west of the source (see the t = 4 min sketch), where before the main sea retreat one can observe a small positive signal, that is part of the fast circular front travelling offshore.

The synthetic marigrams, reported in Fig. 17.4, confirm some features already mentioned above. The first negative arrival, over 2 m deep, in the coastal segment in front of the source area (marigrams 1, 2 and 3), is followed by sea level rise almost equivalent to the withdrawal (in nodes 2 and 3).

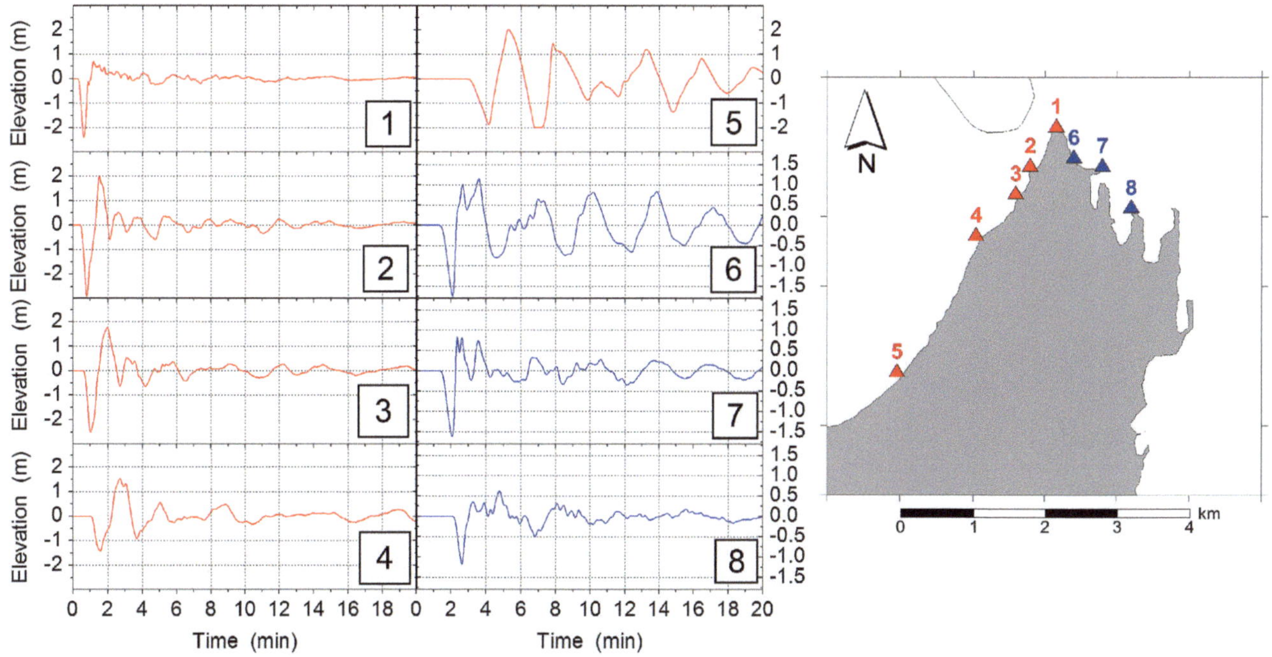

Fig. 17.4 Virtual tide gauges located west (in *red*) and east (in *blue*) of Cape of the Mills

This seems compatible with the observations, reporting that the 3–4 m seawall was overtopped by the waves. The tsunami shows a 2 min dominant period, and affects the coast with a series of crests and troughs with decreasing amplitude. Moving westward the tsunami weakens, but still remains not negligible: node 4 (2 km far from the source) and node 5 (around 4 km away) record a first peak-to-peak perturbation of over 3 m, with sizeable oscillations lasting for more than 20 min. Marigrams 6, 7 and 8 corresponding to nodes beyond Cape of the Mills (reported in blue in Fig. 17.4) exhibit and initial sea retreat and waves of smaller amplitude (−1.5/+1 m) but still capable to produce damages to the harbor structures.

17.5 Sensitivity Analysis and Discussion

The scenario presented here is based on conjectures on the landslide that do not have direct evidence or validation. An accurate geophysical characterization of the initial sliding mass (volume, thickness, shape) does not exist, and likewise also a sound estimate of its initial position is missing. Moreover no observations of the final deposit, that could provide strong constraints on the dynamics modeling, have been made yet. Moreover, as concerns the tsunami only the qualitative effects of the waves on a short segment of coast are known through eyewitnesses accounts (overtopping of the sea wall and watering of the shops on the road), since the tide gauge in the harbor of Rhodes did not work in the evening of the tsunami.

A first attempt to constrain the characteristics of the source and to estimate its influence on the generated tsunami has been made, by reducing the initial thickness of the above illustrated scenario (hereafter denoted as case A) to 75 and 50 %: this produces the same decrease also in the volume of the sliding mass, that turns out to be 24 million (case B) and 16 million m^3 (case C) respectively. Scenarios B and C are computed through the same numerical codes as case A and results are compared in Fig. 17.5, by the respective maximum water elevation produced along the northern coast of the island of Rhodes. A first very interesting observation is that there is an almost linear correlation between the sliding mass thickness and the wave height on the coast, especially along the coast west of Cape of the Mills, where the tsunami propagation is faster and direct. Moving to the east of the Cape, on the other hand, this direct dependence seems to fail, or at least to weaken, due probably to the non-linear effects rising from the tsunami interaction with the small harbors and basins in this zone. Along the coastal segment most affected by the 2002 tsunami, one can notice 2–5 m wave elevation for case A, 1.5–4 m for case B and 1–3 m for case C. If case A seems to slightly overestimate the observations (consisting mainly of the overtopping of the sea-wall, 3–4 m high), case C seems to represent the lower bound for such an occurrence. These two cases can be then interpreted as the maximum and minimum limits for the mass generating a tsunami compatible with the observations.

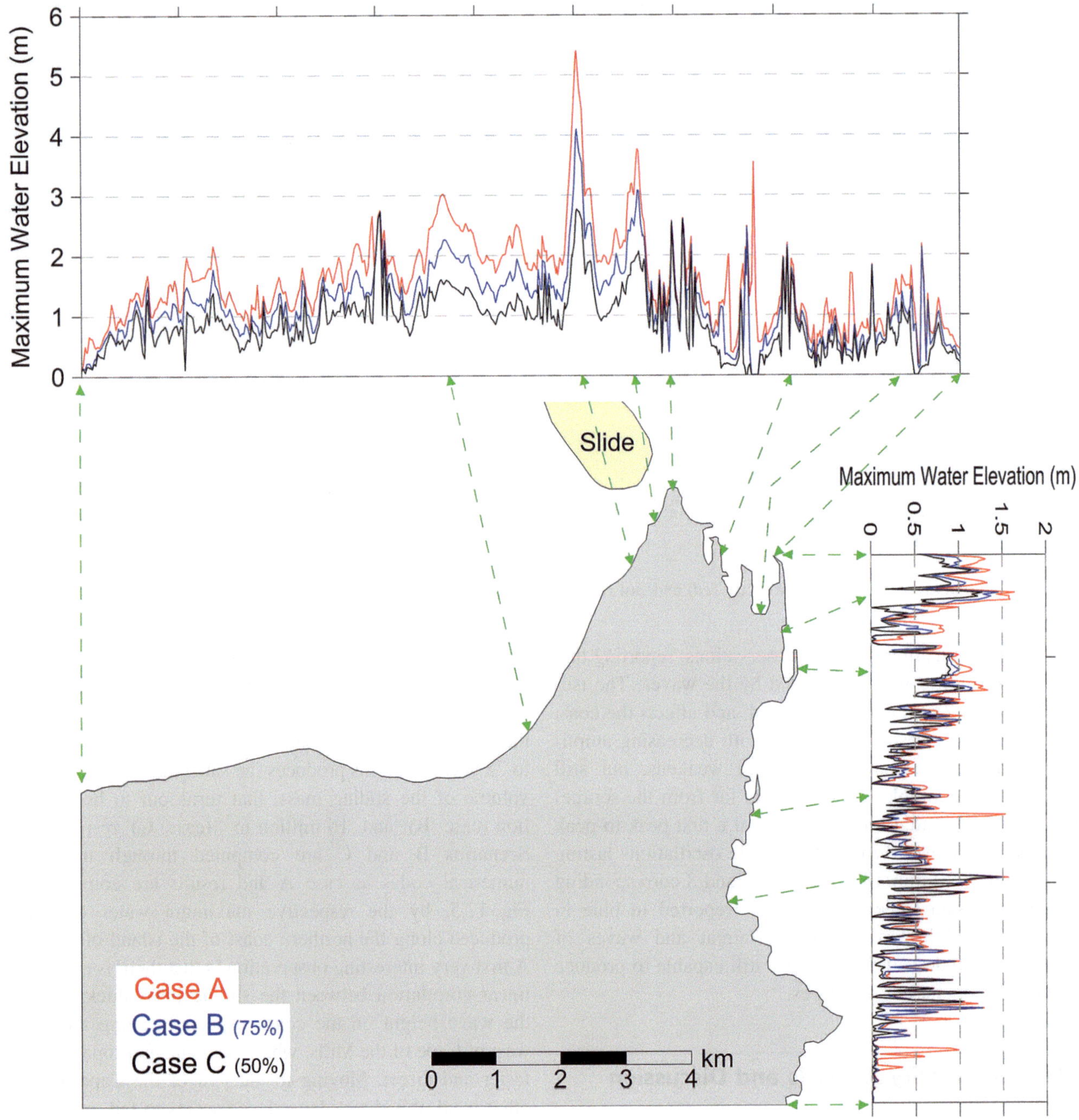

Fig. 17.5 Maximum water elevations along the coast for the three scenarios of slide: the reference case, *A* in *red*; the 75 % reduced thickness case, *B* in *blue*; the halved-thickness case, *C* in *black*

17.6 Conclusions

The 2002 non-seismic tsunami affecting the northern tip of the island of Rhodes is here investigated through numerical techniques. A rough initial mass reconstruction has been performed, basing on some indications coming from bathymetric surveys carried out offshore the NW coast of the city of Rhodes by the HCMR, especially focused along the steep slope located between 100 and 300 m depth. A volume of more than 30 million m^3 has been estimated, detaching from about 120 m sea depth: the slide reached a maximum velocity of about 24 m/s, before slowing down and reaching deeper water, at 500 depth, 8 km far from the source.

The simulated tsunami hits the coast within 1 min, with a leading negative wave, corresponding to a sea withdrawal, followed by an almost equivalent water rise, and continues for tens of minutes to affect the coast with 2 min wave period. Moving east of Cape of the Mills the tsunami loses strength, remaining however in the range between −1 and 1 m oscillations.

Due to the scarce availability of observations, it is difficult to constrain the parameters governing the model. An attempt is here made by reducing the initial mass thickness and evaluating its effects on the generated tsunami, comparing them with the evidences. The main result is that the reference case A and the halved-thickness scenario (case C) represent respectively the upper and lower extremes of the range of possible volumes for the 2002 tsunami source.

As a general comment, it is worth to underline the hazard posed by events like this: small-volume slides can mobilize suddenly along steep slopes, not far from densely populated and industrialized coasts, causing waves that can cause casualties and huge damages on local scale.

Acknowledgments This work has been carried out in the framework of EU-FP7 project ASTARTE, contract no. 603839.

References

Papadopoulos GA, Daskalaki E, Fokaefs A, Giraleas N (2007) Tsunami hazards in the eastern Mediterranean: strong earthquakes and tsunamis in the east Hellenic arc and Trench system. Nat Hazards Earth Syst Sci 7:57–64. doi:10.5194/nhess-7-57-2007

Sakellariou D, Lykousis V, Rousakis G, Georgiou P (2002) Slope failure and associated coastal erosion in tectonically active areas: the coastal zone of Rhodes city (Rhodos island) Greece. In: Yilmaz A (ed) Oceanography of eastern Mediterranean and black sea, Proceedings of the 2nd international conference. TUBITAK Publications, Ankara, 13–16 Oct 2002, pp 978–985

Tinti S, Pagnoni G, Zaniboni F (2006) The landslides and tsunamis of 30th December 2002 in Stromboli analysed through numerical simulations. Bull Volcanol 68:462–479

Tinti S, Chiocci FL, Zaniboni F, Pagnoni G, de Alteriis G (2011) Numerical simulation of the tsunami generated by a past catastrophic landslide on the volcanic island of Ischia, Italy. Mar Geophys Res 32(1):287–297. doi:10.1007/s11001-010-9109-6

Tonini R, Armigliato A, Pagnoni G, Zaniboni F, Tinti S (2011) Tsunami hazard for the city of Catania, eastern Sicily, Italy, assessed by means of worst-case credible tsunami scenario analysis (WCTSA). Nat Hazards Earth Syst Sci 11:1217–1232. doi:10.5194/nhess-11-1217-2011

Zaniboni F, Pagnoni G, Tinti S, Della Seta M, Fredi P, Marotta E, Orsi G (2013) The potential failure of Monte Nuovo at Ischia Island (Southern Italy): numerical assessment of a likely induced tsunami and its effects on a densely inhabited area. Bull Volcanol 75:763. doi:10.1007/s00445-013-0763-9

Storm Hazard Assessment for Urban Areas

18

Luigi Mucerino, Chiara F. Schiaffino, Antonietta Franzé, Marco Firpo, Nicola Corradi, and Marco Ferrari

Abstract

Storm surges are one of the most significant threats for coastal urban areas. In view of finding a way to minimize sea storm damages to anthropic structures, this paper focuses on the analysis of storm hazard investigation methods and on coastal control line definition. Generally, coastal control lines are identified through Run up equations proposed by Mase (1989). Applicability of Van der Meer equations (1998), usually employed for beach profile analysis, will be assessed here as an alternative to Mase equations. Three urban beaches on Genoa coast are considered: Voltri beach, Pegli beach and Vernazzola beach. The analysis takes into account sea and weather condition data collected during three major storm surges in the aforementioned littoral areas. Empirical data gathered in field surveys during storms were subsequently examined in order to analyze seaside area response. Collected data were compared with theoretical computations. Results show a greater significance of data obtained through the application of Van De Meer (1998) equation than Mase equation data (1989). In conclusion, Van De Meer (1998) equation could provide information about coastal control lines on urban gravel beaches. Such information would then be useful to locate high hazard areas and act accordingly.

Keywords

Urban areas • Storm surges • Run up • Hazard

18.1 Introduction

In recent years, the impact of high energy events on coastal urban areas was widely discussed, highlighting the relevance of storm hazard assessment for such zones and the importance of a reliable analysis method. Information concerning the potential consequences of storms could enable the prompt implementation of mitigation or adaptation strategies (see e.g., Godschalk et al. 1989).

Most studies nowadays employ Wave Run up equations (Mase et al. 2004). Wave Run up represents the maximum vertical extent of wave uprush above sea water level on beaches or structures (Sørensen et al. 1997), i.e. the area subjected to the waves' attack. Therefore, wave Run up is an important parameter to determine which areas could be flooded during storms (Briganti et al 2005).

Run up equations (Mase 1989) are here compared to beach profile equations proposed by Van De Meer (1998) in order to define the best possible instrument for coastal control line identification. In Run up equations we have taken into account R_{max} and $R_{1/3}$, the first is calculated with maximum wave height (H_{max}), the second is used the significant wave height ($H_{1/3}$). It would represent the ideal tool to devise opportune interventions and limit damages to anthropic structures.

L. Mucerino (✉) · C. F. Schiaffino · A. Franzé · M. Firpo
N. Corradi · M. Ferrari
DISTAV, University of Genoa, Genoa, Italy
e-mail: luigi.mucerino@edu.unige.it

G. Lollino et al. (eds.), *Engineering Geology for Society and Territory – Volume 4*,
DOI: 10.1007/978-3-319-08660-6_18, © Springer International Publishing Switzerland 2014

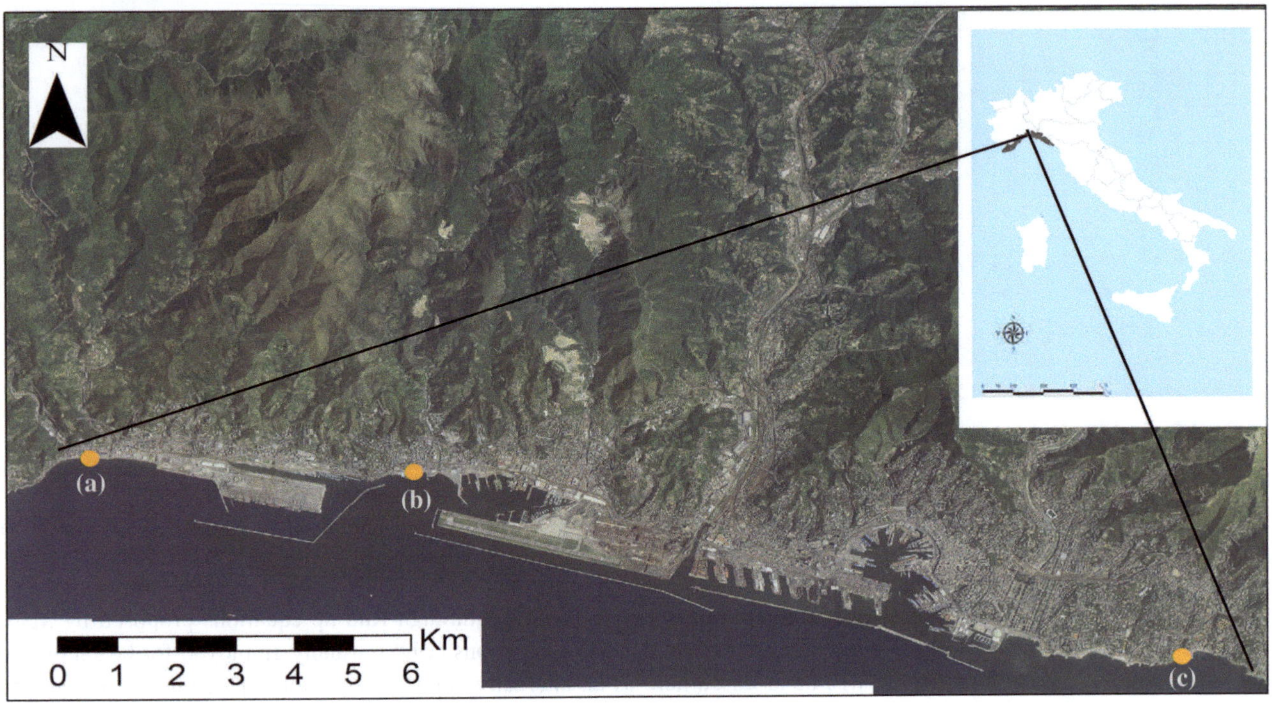

Fig. 18.1 Study area: A Voltri. B Pegli. C Vernazzola (*image* from minambiente.it)

18.2 Study Area

Three urban beaches located along Genoa coast were analyzed: Voltri beach, Pegli beach, Vernazzola beach (Fig. 18.1).

Voltri and Pegli are located in the western part of the city area. Voltri beach, oriented W-E, is 1000 m long; Cerusa and Leira streams flow in this area. Pegli beach, oriented SSW—NNE, is 600 m long and is crossed Varenna stream. Along Voltri and Pegli beaches, 5 coastal hard-structures were built. Beach dynamics are also influenced by Voltri and Genoa ports (Ferrari et al. 2006). Vernazzola beach, located in the eastern part of Genoa, is 130 m long and is oriented SSW—NNE. The beach is flanked by two jetties (Corradi et al. 2008).

Considering the geographical position of Liguria region, the main wave direction of Genoa coast is SW, whereas its secondary wave direction is SE (Cattaneo Vietti et al. 2010). Since the three analyzed beaches are differently oriented, it follows that Voltri beach is highly exposed to wave action, while in Pegli beach wave action is amplified by Voltri port structures. Vernazzola beach, due to its plan and orientation, is protected from main wave action.

18.3 Materials and Methods

This study takes into account the impact of three exceptional storms that hit the area on October 30, 2008, on January 1, 2010 and on November 8, 2011 (Table 18.1).

In particular, due to different geographical orientation, for Voltri beach all three storms are considered, while for Pegli only storms coming from SW are taken into account, since this coastal zone is not prone to SE events. For Vernazzola beach the 2008 storm is left out because a nourishment altered beach profile.

Wave data were recalculated taking into account of the presence of coastal hard-structures.

Table 18.1 Wave characteristics of considered storm events

	30-10-2008	01-01-2010	08-11-2011
H_0 (m)	5.5	4.5	2.6
Wave direction	SW	SW	SE
T (s)	8	12	6
Ttot (h)	18	12	18
Wave setup + Tide	0.46	0.69	0.35

Table 18.2 Mase (1989) and Van De Meer (1998) equation obtained data

	Measured Run up max		Estimated Run up (Mase 1989)				Van De Meer (1998)	
Storm	Height	Width	R max	Width	R 1/3	Width	Height	Width
Voltri								
30.10.2008	4.2	65	5.6	88	3.6	57	4	65
01.01.2010	4.5	70	6.8	102	4.9	74	4.8	71
08.11.2011	1.8	28	2.9	48	1.8	30	2.1	30
Pegli								
30.10.2008	7.2	35	10.7	51	6.7	31	7.5	44
01.01.2010	7.8	43	12.9	67	7.8	41	9.3	50
Vernazzola								
01.01.2010	3.2	25	4	31	2.5	19	3.7	32
08.11.2011	1.5	10	2	13	1.3	8	2.4	13

Backshore field survey was performed with differential GPS, while seabed morphology was detected through a single-beam echo-sounder and a differential GPS. Obtained data were subsequently applied to Mase (1989) and Van De Meer (1998) equations.

18.4 Results

Mase equations (1989) overestimates wave penetration in all three beaches, if compared to collected data. In particular, as Table 18.2 shows, estimated Run up 1/3 is generally lower than real Run up, up to a maximum of 0.6 m lower for Voltri beach, 0.5 m lower for Pegli beach, and 0.7 m lower for Vernazzola beach. On the other hand, estimated maximum Run up is higher than real Run up, up to a maximum of 2.3 m higher for Voltri beach, 5.1 m higher for Pegli beach, and 0.8 m higher for Vernazzola beach.

Results obtained by Van Der Meer equation (1988) are more reliable, and provided data are substantially similar to what was observed during storms (Table 18.2).

18.5 Conclusions

In order to identify a predictive model of coastal storm consequences, equations proposed by Mase (1989) and Van De Meer (1998) were considered.

Obtained results show that Van der Meer equation (1988) is more reliable than Mase equations (1989). Mase's Run up 1/3 underestimates real values, while Run up max values are significantly higher than actual data.

The delta between the results of the two equations might be due to differences in their variables. Mase equations (1989) focus on wave-related data such as wave height, wave period and wave length. Beach morphology plays a minor role, since only beach face slope parameter is considered (Hughes 2004). On the contrary, Van der Meer contains all relevant hydraulic and geometrical parameter: front slope and relative berm height (Burcharth 2013). Therefore the equation considers not only wave characteristics and beach morphology, but also beach sediment grain size and the ratio between solid and liquid substance density (Van De Meer 1998). Beach sediment grain size is particularly important because gravel has higher hydraulic characteristics and water infiltration during swash, and this causes asymmetric uprush and backwash motion with a reduction in backwash transport capacity (Buscombe and Masselink 2006; Clarke et al. 2004; Kulkarni et al. 2004; Lee et al. 2007; Nolan et al. 1999; Osborne 2005; Pedrozo-Acuna et al. 2006). The above mentioned hydraulic and morphodynamic factors generate a different results.

In conclusion, Van De Meer (1998) equation could provide information about coastal control lines for urban gravel beaches. Such information would then be useful to locate high hazard areas and act accordingly.

References

Briganti R, Bellotti G, Franco L, De Rouck J, Geeraerts J (2005) Field measurements of wave overtopping at the rubble mound breakwater of Rome-Ostia yacht harbour. Coastal Eng 52(12):1155–1174

Burcharth HF (2013) On front slope stability of berm breakwaters. Coastal Eng 77:71–76

Buscombe D, Masselink G (2006) Concepts in gravel beach dynamics. Earth-Sci Rev 79(1–2):33–52

Cattaneo Vietti R, Albertelli G, Aliani S, Bava S, Bavestrello G, Benedetti Cecchi L, Bianchi CN, Bozzo E, Capello M, Castellano M, Cerrano C, Chiantore M, Corradi N, Cocito S, Cutroneo L, Diviacco G, Fabiano M, Faimali M, Ferrari M, Gasparini GP, Locritani M, Mangialajo L, Marin V, Moreno M, Morri C, Orsi Relini L, Pane L, Paoli C, Petrillo M, Povero P, Pronzato R, Relini G, Santangelo G, Tucci S, Tunesi L, Vacchi M, Vassallo P, Vezzulli L, Wurtz M (2010) The Ligurian Sea: present status, problems and perspectives. Chem Ecol 26:319–340

Clarke S, Dodd N, Damgaard J (2004) Modeling flow in and above a porous beach. J Waterway Port Coastal Ocean Eng 130(5):223–233

Corradi N, Ferrari M, Schiaffino CF (2008) Evaluation of the effectiveness of a seasonal nourishment programme of the pocket beaches of the city of Genoa. Chem Ecol 24(1):215–223

Ferrari M, Bolens S, Bozzano A, Fierro G, Gentile R (2006) The port of Genoa-Voltri (Liguria, Italy): a case of updrift erosion. Chem Ecol 22(1):361–369

Godschalk DR, Brower DJ, Beatley T (1989) Catastrophic coastal storms, hazard mitigation and development management. Duke University Press, Durham

Hughes SA (2004) Estimation of wave run-up on smooth, impermeable slopes using the wave momentum flux parameter. Coast Eng 51:1085–1104

Kulkarni CD, Levoy F, Monfort O, Miles J (2004) Morphological variations of a mixed sediment beachface (Teignmouth, UK). Cont Shelf Res 24:1203–1218

Lee KH, Mizutania N, Hur DS, Kamiya A (2007) The effect of groundwater on topographic changes in a gravel beach. Ocean Eng 34:605–615

Mase H (1989) Random wave run up height on gentle slope. J Waterway Port Coast Ocean Eng 115(5):649–661

Mase H, Miyahira A, Hedges TS (2004) Randon wave run-up on sea walls near shorelines with and without artificial reefs. Spectal characteristics of random wave run-up. J Coastal Eng, JSCE 46(3):247–248

Nolan TJ, Kirk RM, Shulmeister J (1999) Beach cusp morphology on sand and mixed sand and gravel beaches, South Island, New Zealand. Mar Geol 157(3–4):185–198

Osborne PD (2005) Transport of gravel and cobble on a mixed-sediment inner bank shoreline of a large inlet, Grays Harbor, Washington. Mar Geol 224(1–4):145–156

Pedrozo-Acuña A, Simmonds DJ, Otta AK, Chadwick AJ (2006) On the cross- shore profile change of gravel beaches. Coast Eng 53(4):335–347

Sørensen OR, Madsen PA, Schäffer HA (1997) Surf zone dynamics simulated by a Boussinesq type model. Part I. Model description and cross-shore motion of regular waves. Coast Eng 32(4):255–287

Van De Meer J (1998) Rock slope and gravel beaches under wave attack. Delfttech Univ Pubbl 396

Numerical Simulation of the BIG'95 Debris Flow and of the Generated Tsunami

19

Filippo Zaniboni, Gianluca Pagnoni, Alberto Armigliato, Stefano Tinti, Olaia Iglesias, and Miquel Canals

Abstract

Underwater mass movements along continental margins constitute a relevant threat for coastal communities. In the Mediterranean Sea this issue is particularly critical, both for the concentration of potential sources and for the density of population living along the coast. One of the biggest event recently mapped by submarine geophysical surveys, named BIG'95, is located in the Balearic Sea, between the homonymous islands and the Eastern coast of Spain: it occurred around 11 kyrs BP, involving a volume of around 26 km^3 over an area of 2,200 km^2, from 200 m depth down to 1,600 m and over: the looser fraction of the mass was driven along the Valencia Channel, NE-ward, more than 110 km far from the source area. The BIG'95 event is here simulated through numerical codes, adopting some reasonable simplifications on the slide morphology and dynamics. The tsunamigenic impulse is then evaluated and the propagation of the wave simulated over a computational domain covering the Western Mediterranean Sea. Results show that an event of these characteristics can generate catastrophic tsunami along the coasts of the Balearic Sea, reaching also North Africa, Corsica and Sardinia coasts with relevant waves.

Keywords

Landslide tsunami • Numerical modeling • Coastal hazard

19.1 Introduction

Along the margins of the Mediterranean Sea many evidences exist of mass flows, manifesting as canyons and scars in the source area and mass deposits at and beyond the toe of the margin itself. The steepness of the slopes and the continuous supply of material from rivers increase the possibility of mass mobilization, triggered both by seismic shaking and by simple gravitational instability. The closeness of densely populated and industrialized coasts to these potential tsunamigenic sources are sound reasons to dedicate great attention to such phenomena.

Among the several surveys carried out in the last decades looking for submarine failure occurrences in the Mediterranean Sea, one of the most significant mass movement is undoubtedly the so-called BIG'95 debris flow, affecting the Ebro continental margin, in the Balearic Sea, starting from around 200 m depth between the mainland Spanish coast and the Balearic islands (Fig. 19.1). Bathymetric data show many evidences of submarine canyons cutting the margin and a huge quantity of submarine landslide deposits at its basis, stressing the occurrence of past sliding events (Lastras et al. 2007).

The BIG'95 debris flow, with an estimated date of 11 kyrs BP, covered an area of about 2,200 km^2, and has been widely studied by means of different geophysical and geological techniques, providing a complete characterization of the slide (Lastras et al. 2002, 2004). Its thickness is in the range

F. Zaniboni (✉) · G. Pagnoni · A. Armigliato · S. Tinti
Dipartimento di Fisica e Astronomia, Università di Bologna, Bologna, Italy
e-mail: filippo.zaniboni@unibo.it

O. Iglesias · M. Canals
Grup de Recerca Consolidat Geociències Marines, Universitat de Barcelona, Barcelona, Spain

G. Lollino et al. (eds.), *Engineering Geology for Society and Territory – Volume 4*,
DOI: 10.1007/978-3-319-08660-6_19, © Springer International Publishing Switzerland 2014

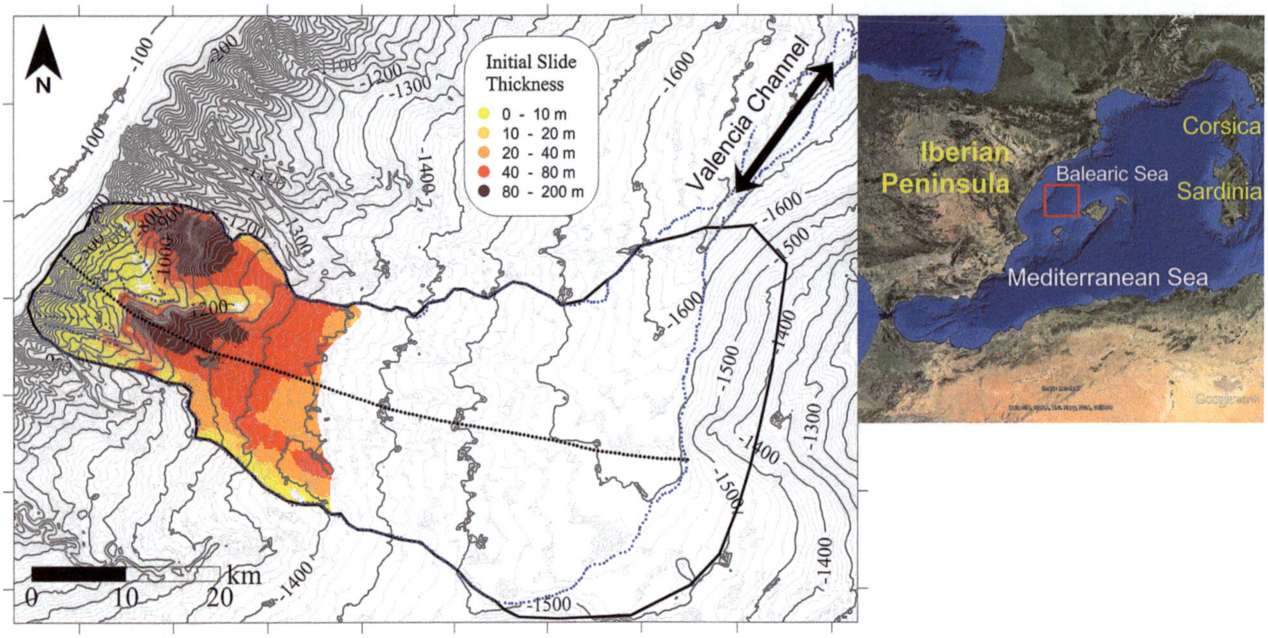

Fig. 19.1 Map of the BIG'95 debris flow area, in the Balearic Sea (water depth is in meters). The initial slide thickness is marked with the *yellow-red-brown scale*, the sliding surface delimited by the *black boundary*. The centre of mass predefined path is represented by the *black dotted line*. The *blue dotted boundary*, partially covered by the black boundary, marks the observed deposit area, the loosest part channeling NE-ward along the Valencia Channel

of 90–150 m in the most proximal area, at the toe of the margin, while at distance of 20–30 km some blocks, 10–35 m high, are found. The loosest part of the slid mass travelled north-east, flowing into the Valencia Channel, extending SE-NW at about 1,800 sea depth and deeper.

The main morphological feature, as concerns the source area, is the main headwall located between 600 and 1,200 m depth, but also minor scars have been identified in shallow water up to the edge of the margin, around 200 m below sea level. The sliding dynamics probably was extremely complex, but in order to evaluate its tsunamigenic potential we have assumed it as a single event, involving approximately 26 km^3 (Lastras et al. 2005; Canals et al. 2004).

In this paper, after a short review of previous works on this issue, we show the results of the BIG'95 scenario. We have applied numerical codes developed by the University of Bologna tsunami research team that include the slide dynamics and the generated tsunami.

19.2 Previous Numerical Simulations

The wave generated by such a huge fall has been already simulated in the past by means of numerical models, with evaluation of the tsunami impact on the surrounding coasts.

The slide dynamics was object of study in Lastras et al. (2005), hereafter mentioned as LAS05, where a conceptual and a numerical model for the sliding motion was developed. The main result was that the looser fraction of the debris flow, moving down to Valencia Channel at more than 110 km distance from the source, reached 50 m/s peak velocity, while the part made of more massive elements, reached a maximum of about 20 m/s.

These features were used in Iglesias et al. (2012), denoted from now on as IGL12, for the evaluation of the generated tsunami, performed through the COMCOT numerical model, implementing a finite difference technique (leap-frog resolution scheme, joined with the scattered grid method) for the resolution of the shallow water equations. The slide dynamics, providing the tsunamigenic impulse, was calculated by considering five different bathymetric configurations, corresponding to as many positions of the sliding mass as resulting from the LAS05 modelling. The application of the numerical code over a numerical grid covering the Balearic Sea, the homonymous archipelago and the Eastern Spanish coast produced interesting results: the island of Ibiza (the closest to BIG'95) is affected by 8 m first positive signal, meaning sea invasion, followed by 4 m trough, with a perturbation period of about 10 min. The island of Mallorca is reached by a 4 m tsunami, with similar features, while the Iberian coast is reached by a first negative wave (sea withdrawal) of about 6 m, after which a 9 m sea rise is observed, with 25 min period. Such results, together with many other described in IGL12, stress the potential hazard connected with events like BIG'95, provoking destructive waves.

19.3 Numerical Techniques

The BIG'95 sliding dynamics is here computed by means of the numerical code UBO-BLOCK1, developed by the University of Bologna tsunami research team. The undisturbed sliding surface has to be provided as input, and the reconstructed sliding mass as well: the latter is divided into a chain of deforming, volume-conserving blocks, that cannot separate from each other. The forces acting on each block contribute to determine the motion equation of the respective centre of mass (CoM), which is computed numerically over a predetermined trajectory. This model does not account for the two different mass phases depicted in LAS05: the sliding body is approached as a deforming but coherent mass, that can be associated with the LAS05 block-like fraction, moving at lower velocity. In this work we have considered that this portion of the slide provides the main contribution to the generation of the tsunami.

The simulation of the tsunami over the computational grid is performed by another in-house developed numerical code, UBO-TSUFE, adopting the finite element technique to solve the shallow water equations. The forcing term, provided by the mass moving over the sea bottom, is computed in dependence of the slide dynamics. An intermediate code, UBO-TSUIMP, deals with mapping data from the landslide reference frame to the tsunami computational grid and with filtering the tsunamigenic impulse on the basis of the local sea depth.

The description of the landslide and tsunami numerical codes can be found in Tinti et al. (1994, 1997) respectively, while some applications to cases of landslide-tsunamis are in Tinti et al. (2006, 2011), Lo Iacono et al. (2012).

19.4 Simulation of the BIG'95 Motion

The application of UBO-BLOCK1 code to the BIG'95 slide requires some simplifications: a unique mass failure is adopted (such as in LAS05); an undisturbed sliding surface is provided, meaning no erosions or mass deposit during the motion; only the block-like dynamics of the sliding body will be the object of our investigation, meaning that the further descent along the Valencia Channel of the looser material is not considered here. We consider this assumption reasonable, because this latter phase probably does not provide relevant tsunamigenic impulse, since the mass moves into deeper water.

The result of the BIG'95 slide reconstruction is visible in Fig. 19.1: the sliding surface is delimited by the black contour, and has been obtained in the final part, at the basis of the Ebro margin, by subtracting the observed deposit (Lastras et al. 2004) from the present bathymetry (provided by the University of Barcelona team). The initial sliding mass has been simplified, as visible in Fig. 19.1 by the blue dotted line marking the observed headwall crown, compared with the simplified slide boundary. The yellow-red-brown scale accounts for the initial thickness, reaching a maximum of 200 m and a mean of about 40 m, for a volume of almost 24 km^3. As mentioned previously, the predefined CoM path has also to be provided: in Fig. 19.1 it is marked by the black-dotted line, following the maximum local slope as a general criterion. Notice that the CoM trajectory ends at about 1,500–1,600 m depth, covering then the proximal (1,200–1,300 m depth, where most of the accumulation is found) and intermediate (over 1,500 m depth, where the 10–35 m blocks are located) depositional areas.

In Fig. 19.2 the velocity time history is reported: the mass reaches a maximum average velocity of 12 m/s, but many blocks (whose single velocity is represented by the black dots) reaches 18 m/s peak values: these are initially located along the steeper slopes, in the upper part of the continental margin. On the contrary, other CoM reaches 6–8 m/s maximum velocity, since they begin their descent downward, where gentler slope are present.

The dynamics shows the typical features of slide motions along the margin, i.e. a strong initial acceleration, here giving the maximum velocity after 6–7 min, and a slow deceleration phase, due to the mass reaching the sub-horizontal area at the base of the slope. The simulation stops after about 38 min, when the mass reaches 1,400–1,500 m sea depth, even if the main part of the deposit is located at the basis of the main headwall, at about 1,200–1,300 m sea depth, well-fitting with the observations.

The Froude number, accounting for the efficiency of tsunami generation by the moving mass, is computed by the ratio between the horizontal velocity of the slide and the wave velocity. In the BIG'95 simulation the critical value, 1, where resonance occurs and the tsunami generation is maximum, is far from being reached: the maximum value, around 0.1, is attained at the same time of the velocity peak, then the mass flows to larger depths and the capability of generate waves lowers consistently.

19.5 Tsunami Propagation Modeling

Provided the tsunamigenic impulse by the intermediate code UBO-TSUIMP, the tsunami simulation is performed over a computational domain that is made of irregular triangles, whose dimension is adapted to the local sea depth. The grid covers the whole Western Mediterranean Sea, providing a general view of the tsunami propagation over a wide domain: the assessment of the tsunami effects on a local scale would require a more detailed bathymetric dataset.

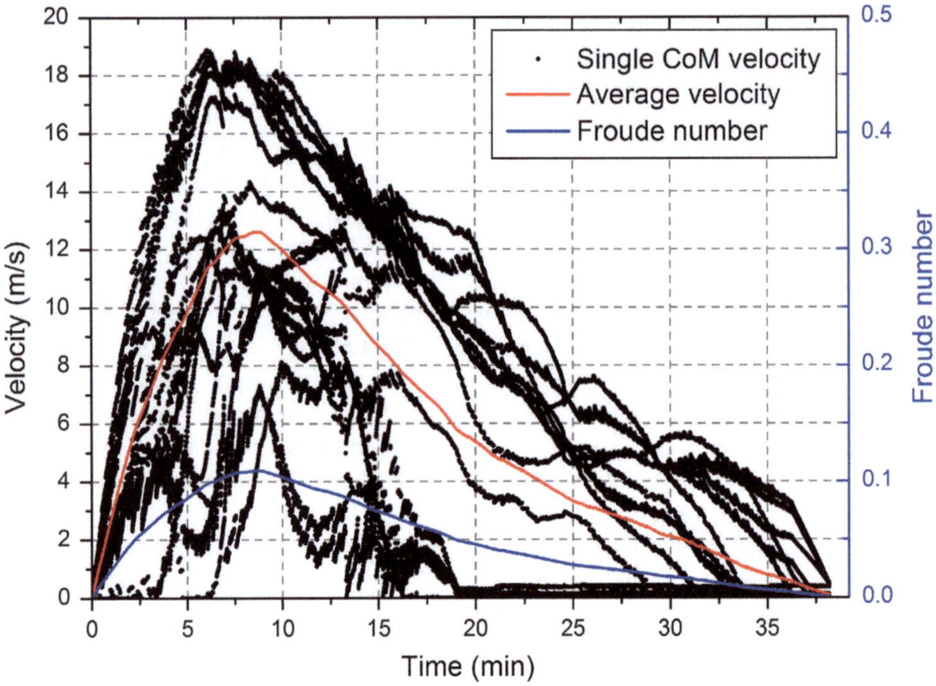

Fig. 19.2 Velocity of the individual centre of mass (*black dots*) and average velocity (*red line*), versus time. The *blue line* marks the Froude number value versus time

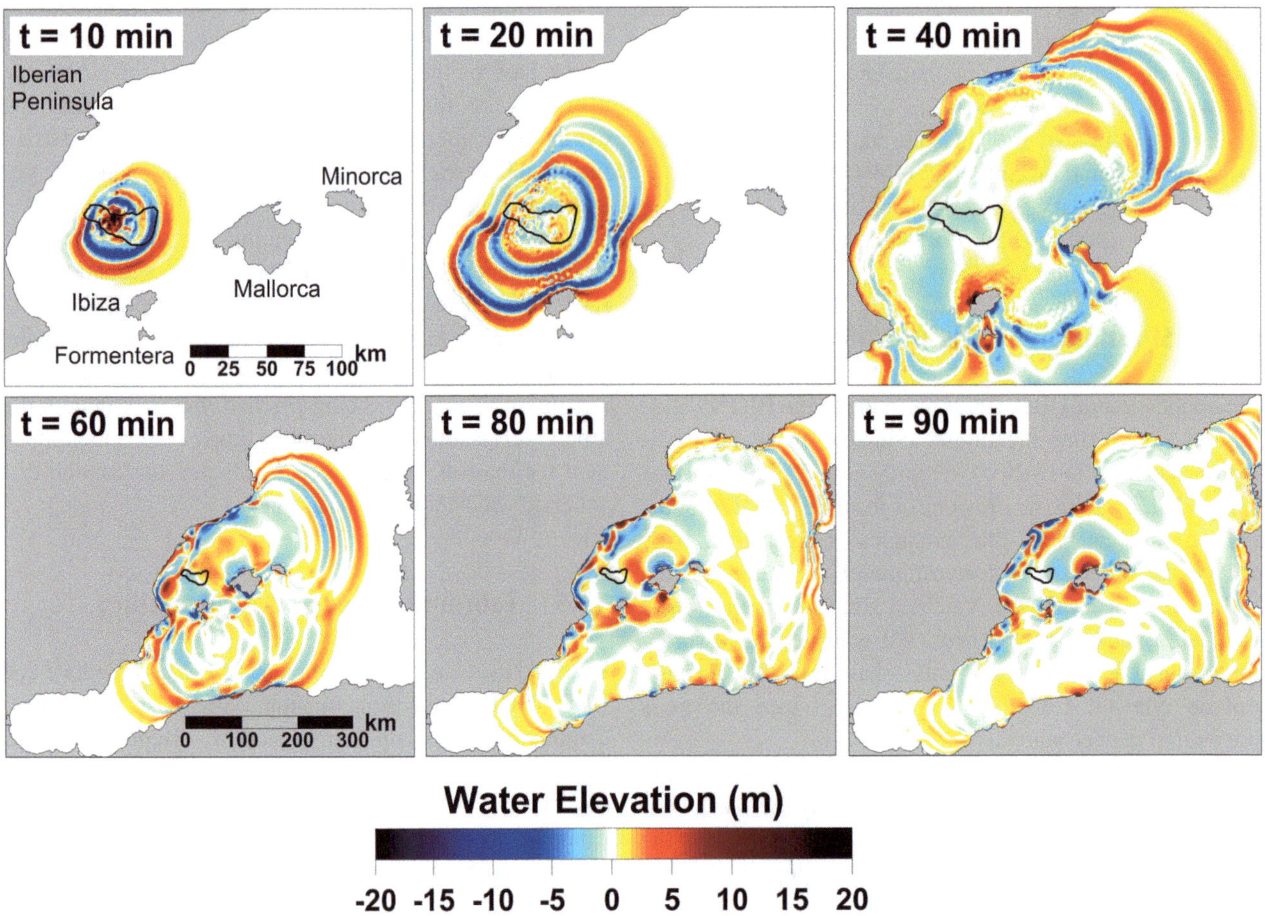

Fig. 19.3 Tsunami propagation. In the three *upper panels* zoom over the Balearic Sea, then on the whole computational domain covering the Western Mediterranean Sea. The *yellow-red scale* marks the sea level increase, the *cyan-blue areas* mark negative signals, i.e. sea lowering. The *black contour* evidences the landslide boundary

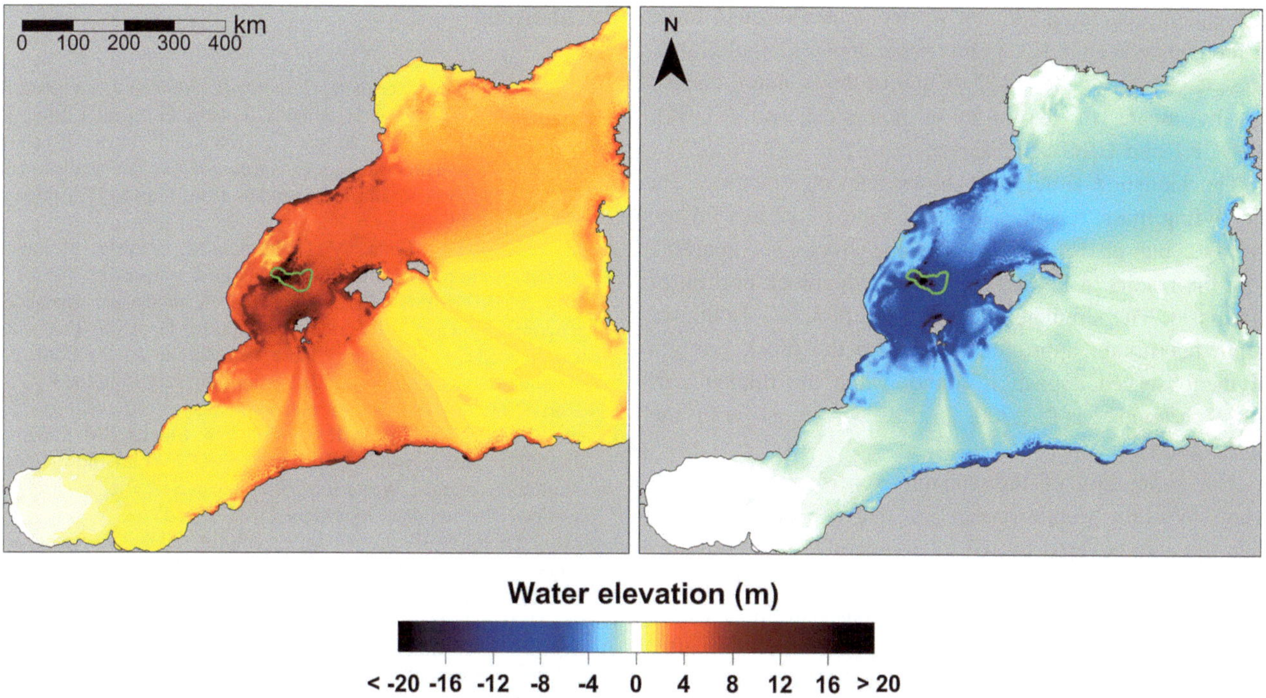

Fig. 19.4 Maximum (*left panel, yellow-red scale*) and minimum (*right panel,* in *cyan-blue*) water elevation over each node of the computational domain. The source area is marked by the *green boundary*

Figure 19.3 reports the wave propagation: initially it radiates almost circularly from the source (see t = 10 min sketch), with a strong positive signal directing towards the Balearic Islands. After 20 min Ibiza has already been affected by a relevant wave (more than 5 m, compatible with IGL12 findings), and also the northern coast of Mallorca is approached by a strong positive front parallel to the coast. The same can be seen NW-ward, close to the coast of Spain. These sketches show also the great number of positive-negative oscillations characterizing the perturbation, reaching the coast of mainland Spain in around 40 min. Enlarging the view, we can notice that the coasts of Africa are affected by the tsunami within 1 h, and the western coasts of Corsica and Sardinia around 80 min after the slide beginning: in all these areas the tsunami manifests with a relevant first positive signal, ranging 5–7 m, and in general one can observe that the whole Western Mediterranean basin is interested by the tsunami within 90 min.

The maximum and minimum water elevation maps provide (Fig. 19.4) further elements for the assessment of the BIG'95 tsunami. The source area, evidenced by the green contour, shows a strong perturbation of the sea surface, up to 20 m. The northern coasts of Ibiza, Mallorca and Minorca are affected by more than 10 m high waves, while the southern coasts seem to be more protected. Some tsunami energy beams are clearly visible to head towards the coast of mainland Spain (especially one very evident NE-ward and another W-ward), and then to deviate to the south, all these features depending on the bathymetry that mainly drives the tsunami propagation. Other tsunami energy directions are found south of Formentera, directed towards the northern coast of Africa, and E-ward towards the coasts of Sardinia. This energy distribution is expected not to change significantly offshore even when tsunami simulations will be run with more refined bathymetry. In the coastal areas however higher waves can be expected since the use of detailed bathymetric data is known to be important to compute the coastal amplification of tsunamis and the associated flooding.

19.6 Conclusions

The BIG'95 debris flow is a major event occurred about 11 kyrs BP that has been extensively studied and characterized by marine geophysical and geological surveys. This abundant data set allows for a reasonable reconstruction of a slide scenario that results to be sufficiently constrained by the knowledge of the final deposit and that in turn can be used for tsunami simulations.

The reconstructed landslide, with volume about 26 km^3, has been hypothesized as a single event in view of the tsunami generation. The slide was simulated in the past through conceptual and numerical models (LAS05), and the consequent generated tsunami was computed through finite difference code (IGL12). In this paper another simulation is presented, both of the landslide and of the tsunami, carried out by means of the in-house numerical codes UBO-BLOCK1 and UBO-TSUFE respectively.

The landslide simulation shows relevant velocities for the sliding mass, reaching the peak value of 12 m/s for the average, but up to 18 m/s for some blocks (comparable with the results in LAS05 as concerns the block portion of the mass). The simulated deposit reaches 1,500–1,600 sea depth maximum, approximately where the block fraction of the observed deposit is found, while the thicker part settles around 1,200–1,300, in full agreement with the observations.

The simulation of the tsunami is performed over the whole Western Mediterranean Sea that is a domain wider than the area covered by IGL12 simulation. Some features are remarkable and comparable with the results in IGL12. A strong tsunami affects the northern coasts of Ibiza (more than 10 m, the first wave approaching the coast 10 min after the slide occurrence), Mallorca (after 20 min, over 10 m high waves) and Minorca (around 30 min travel time). Strong energy beams direct toward the coast of mainland Spain, mainly driven by bathymetry, the northern coasts of Africa and the eastern coast of Sardinia and Corsica, reached by 5–6 m tsunami.

A more detailed grid covering the Balearic Sea could provide a better description of the effects of the BIG'95 generated tsunami on the local scale, accounting for multiple reflections between the coasts of mainland Spain and Balearic Islands, and for the amplification of the tsunami due to resonance effects.

Acknowledgments This work has been carried out in the framework of EU-FP7 project ASTARTE, contract no. 603839

References

Canals M, Lastras G, Urgeles R, Casamor JL, Mienert J, Cattaneo A, De Batist M, Haflidason H, Imbo Y, Laberg JS, Locat J, Long D, Longva O, Masson DG, Sultan N, Trincardi F, Brynk P (2004) Slope failure dynamics and impacts from seafloor and shallow sub-seafloor geophysical data: case studies from the COSTA project. Mar Geol 213:9–72

Iglesias O, Lastras G, Canals M, Olabarrieta M, Gonzalez M, Aniel-Quiroga I, Otero L, Duran R, Amblas D, Casamor JL, Tahchi E, Tinti S, De Mol B (2012) The BIG'95 submarine landslide–generated tsunami: a numerical simulation. J Geol 120(1):31–48

Lastras G, Canals M, Hughes Clarke JE, Moreno A, De Batist M, Masson DG, Cochonat P (2002) Seafloor imagery from the BIG'95 debris flow, western Mediterranean. Geology 30:871–874

Lastras G, Canals M, Urgeles R, De Batist M, Calafat AM, Casamor JL (2004) Characterisation of the recent BIG'95 debris flow deposit on the Ebro margin, Western Mediterranean Sea, after a variety of seismic reflection data. Mar Geol 213:235–255

Lastras G, De Blasio FV, Canals M, Elverhøi A (2005) Conceptual and numerical modeling of the BIG'95 debris flow, Western Mediterranean Sea. J Sediment Res 75:784–797. doi:10.2110/jsr.2005.063

Lastras G, Canals M, Amblas D, Frigola J, Urgeles R, Calafat AM, Acosta J (2007) Slope instability along the northeastern Iberian and Balearic continental margins. Geol Acta 5:35–47

Lo Iacono C, Gràcia E, Zaniboni F, Pagnoni G, Tinti S, Bartolomé R, Masson D, Wynn R, Lourenço N, Pinto de Abreu M, Dañobeitia JJ, Zitellini N (2012) Large and deep slope failures in the Gorringe Bank: evidence for landslide-generated tsunamis along the SW Iberian margin. Geology 40:931–934. doi:10.1130/G33446.1

Tinti S, Gavagni I, Piatanesi A (1994) A finite-element numerical approach for modeling tsunami. Ann Geophys 37:1009–1026

Tinti S, Bortolucci E, Vannini C (1997) A block-based theoretical model suited to gravitational sliding. Nat Hazards 16:1–28

Tinti S, Pagnoni G, Zaniboni F (2006) The landslides and tsunamis of 30th December 2002 in Stromboli analysed through numerical simulations. Bull Volcanol 68:462–479

Tinti S, Chiocci FL, Zaniboni F, Pagnoni G, de Alteriis G (2011) Numerical simulation of the tsunami generated by a past catastrophic landslide on the volcanic island of Ischia, Italy. Mar Geophys Res 32(1):287–297. doi:10.1007/s11001-010-9109-6

Macro Deformation and Micro Displacement Characteristics of Granular Materials

20

Hu Zheng, Zhifang Zhou, and Jinguo Wang

Abstract

Granular material is the most common thing in the nature, including sand, rice and coffee beans et al. Soil and sand are typical granular materials, the deformation of granular materials like soil and sand could induce different kinds of natural hazards, like landslides, debris flow and ground settlement. The macro deformation of granular material is caused by the integral of micro displacement of each particle. However, traditional experimental method could only get the macro deformation rather than micro displacement. Hence, the three typical kinds of granular displacement are explained in this paper to reveal the macro deformation mechanism, firstly. Two latest novel experimental techniques on granular materials deformation studying and corresponding working principles are introduced in detail.

Keywords

Granular materials • Macro deformation • Micro displacement • Experimental techniques

20.1 Introduction

A granular material is a collection of solid particles or grains, such that almost all of particles are in contact with some of their neighboring particles (Liu and Nagel 1998; Jop et al. 2006). The deformation of granular materials such as sand, coal, and catalyst particles is a common occurrence in natural and industrial setting. Unfortunately, the deformation characteristics of these materials are not well understood (Jaeger et al. 1996; Rao et al. 2008; Zhou et al. 2013).

H. Zheng (✉) · Z. Zhou · J. Wang
School of Earth Science and Engineering, Hohai University, Nanjing, 210098 Jiangsu, China
e-mail: zhenghu@hhu.edu.cn

H. Zheng
Department of Geotechnical Engineering, College of Civil Engineering, Tongji University, Shanghai 200092, China

Granular materials are commonly encountered in nature and in various industries. Different kinds of experiments and numerical studies have been conducted to investigate deformation mechanism of granular materials (Goodman and Cowin 1972; Deniela et al. 2007; Huang et al. 2011; Huang and Zheng2013). Coulomb (Goodman and Cowin 1972) presented the first yield criterion in mechanics, a yield criterion for soils, including granular materials, in 1776. Daniel et al. (2007) developed a constitutive model to represent large deformation of granular solids based on their experiments. Huang et al. (2011) presented the large deformation configurations of granular material for different chute angle cases. However, most of these traditional experimental studies could only describe macro scope of granular deformation. The inside properties of granular materials are very difficult to investigate because most of granular are non-perspective.

In this paper, the ways of granular relative displacement would be discussed to explain granular position changes in micro scope. Two latest novel experimental techniques would be introduced, which are widely used in granular studying nowadays.

G. Lollino et al. (eds.), *Engineering Geology for Society and Territory – Volume 4*,
DOI: 10.1007/978-3-319-08660-6_20, © Springer International Publishing Switzerland 2014

20.2 Deformation Mechanisms of Granular Materials

The essential deformation mechanism of granular materials is relative displacement between particles, while there are three different types of the relative displacement between particles, including granular slide, granular rotation and granular fracture. Actually, both the failure and shear band developing during the load applying on granular materials are the macro performance. In the following part of this section these three relative displacement types would be explained respectively.

20.2.1 Granular Slide

The schematic diagram of granular slide is shown in Fig. 20.1, initial state as showing in left part, and arrows are the marks of particle rotation. The displacement of particle would be generated, when the force applied. One particle would slide over others without rotation as the torque is zero, and then get to anther sate as showing in right side of Fig. 20.1, the dash circle indicate the initial position of this particle. The distance between dash circle center and moved particle center is the displacement of a single particle. The integral of all the particles of granular materials would be the deformation.

The classic numerical simulation method Discrete Element Method (DEM) is based on the slide model (Cundall and Strack 1979; Compbell and Brennan 1985). The advantage of this method is tracking the trajectory of each granular, therefore, the shear band developing could also be checked during granular materials failure (Kanatani 1981; Oda and Iwashita 1999, 2000). However, granular slide model is too simple to descript the complex mechanical properties of granular materials during loading applying (Jiang et al. 2005, 2006). The particle rotation has to be taking into account during the deformation prediction of granular materials (Utili and Nova 2008).

20.2.2 Granular Rotation

Oda and Iwashita (1999, 2000) showed the particle rotation could also effect the shear strength and shear band developing during loading applying on granular materials according to there granular rolling experimental results. Figure 20.2 shows the schematic diagram of granular rotation, the left part is initial state. The particle would move to another position when the torque applied as showing in right side of Fig. 20.2, and the dash circle marks the initial position of this particle. The same way as granular slide can be applied to calculate the deformation of granular materials.

The simulation results show the deformation prediction is much more veracious as the particle rotation affection be considered during calculation (Jiang et al. 2005, 2006). Jiang et al. (2005) proposed a novel two-dimensional DEM for granular materials with rolling resistance, which can predict granular deformation very well. The combination of granular slide and rotation model can well be used in granular deformation calculation. However, the defect of this method displays when particle broken phenomenon appears. The granular fracture during granular deformation calculation has to be taking into account for the particle broken case.

20.2.3 Granular Fracture

If the strength of particles of are very high and the confining pressure is relative low, particles would not be broken during granular deformation. On the contrary, for the low strength particles at a very high confining pressure environment, they would break very easily. The schematic diagram of granular fracture as showing in Fig. 20.3, the left side is initial state. The particle would divide into two pieces or more than two pieces. The displacement of this particle becomes very complicated for granular fracture situation (Hardin 1985; McDowell et al. 1996). Usually, the continuum mechanics is used to describe the deformation properties of granular materials.

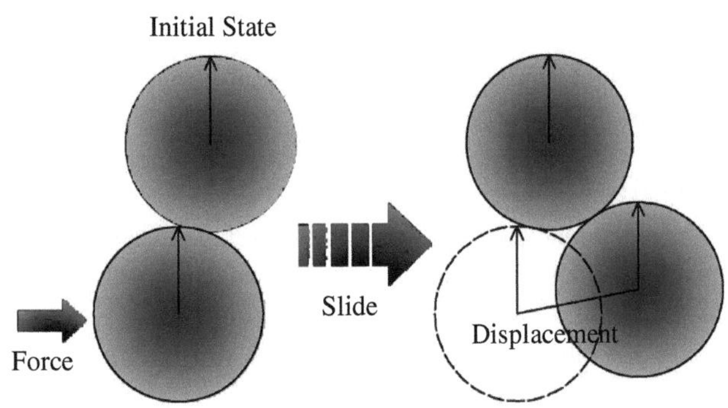

Fig. 20.1 The schematic diagram of granular slide

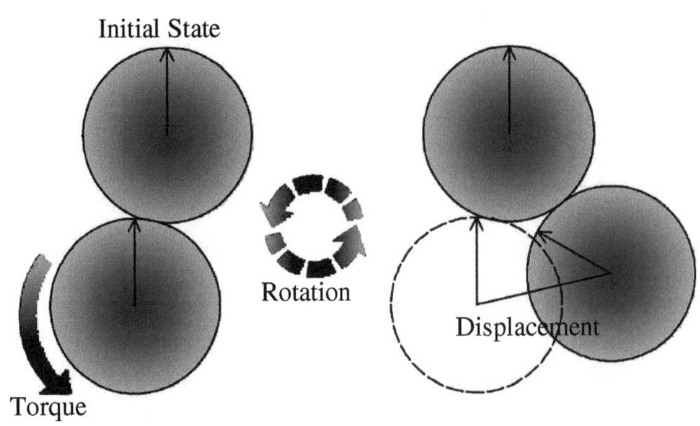

Fig. 20.2 The schematic diagram of granular rotation

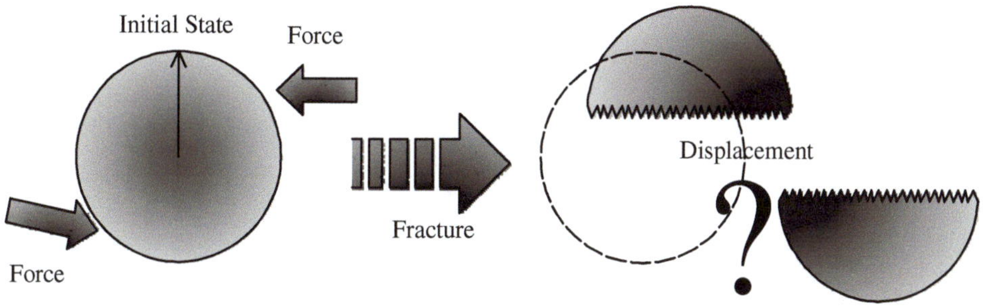

Fig. 20.3 The schematic diagram of granular fracture

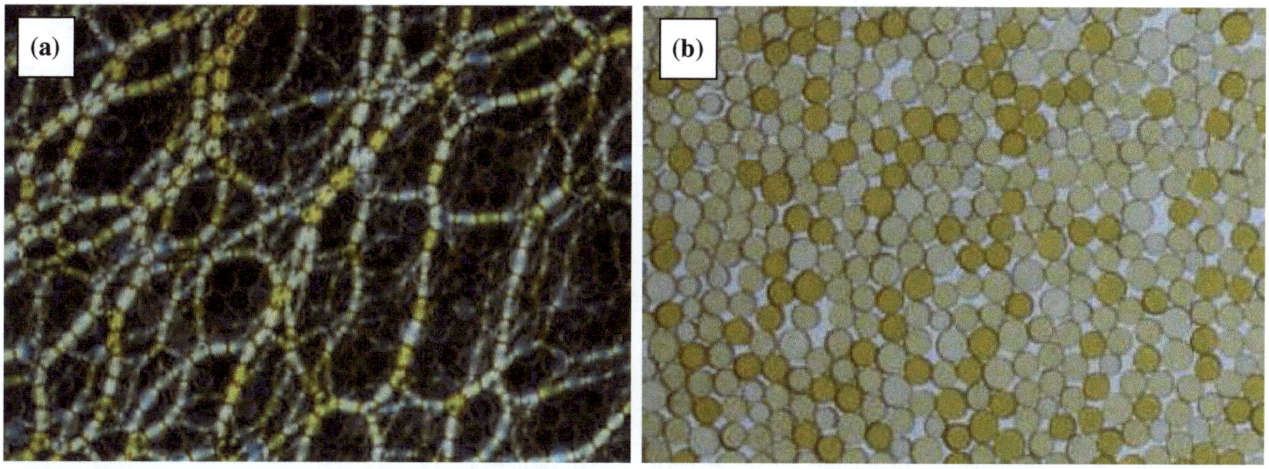

Fig. 20.4 Photograph: **a** photoelasticity phenomenon, **b** particle tracking

Vesic and Clough (1968) and Indraratna et al. (1993) found granular fracture have a significant influence on the strength of granular materials. Norihiko and Sukeo (1979) presented the increment of specific area during granular fracture. Due to the complexity of granular fracture, to propose a effective granular fracture constitutive model is still a big challenge.

20.3 Novel Experimental Techniques on Granular Studying

Many experimental methods have been applied into granular materials studying, including simple shear, bi-tri-axial test and ring shear et al. However, most traditional experimental methods could only get macro scale information during granular materials deformation (Zhang et al. 2010; Zheng et al. 2013; Dijksman et al. 2013). Numerical methods could well overcome this disadvantage. The micro scale information during granular materials deformation could be ascertained by this method, but there is scope for dispute the correction and veracity of simulation results (Jiang et al. 2005, 2006). Hence, some novel experimental methods are needed to investigate the micro scale information during granular materials deformation. The experimental results could also provide reference for simulation veracity checking.

Photoelasticity is an experimental method to determine the stress distribution in a material. The methods has been well applied into granular studying, it's a 'non-contact' technique that can be used to probe the internal stresses in a material by means of photo-elasticity, the motion of each single particle could also been tracked during granular deformation (Fig. 20.4) (Dijksman et al. 2013). However, photo-elasticity method could only be applied in 2 dimensional granular studying nowadays. Whereas, the Refractive Index Matched (RIM) method could explore single particle information inside of granular pile. The principle of RIM is using a laser to scan granular materials which put in the same index solution, so that the laser goes though the mixture without refraction (Sladen et al. 1976; Dijksman et al. 2012, 2013). All particles are dyed by fluorescent. There is a camera fixed perpendicularly on another side of granular materials. Only the cross section of particle pile could be seen during laser scanning. The 3 dimensional models could be reconstructed by using the photograph of

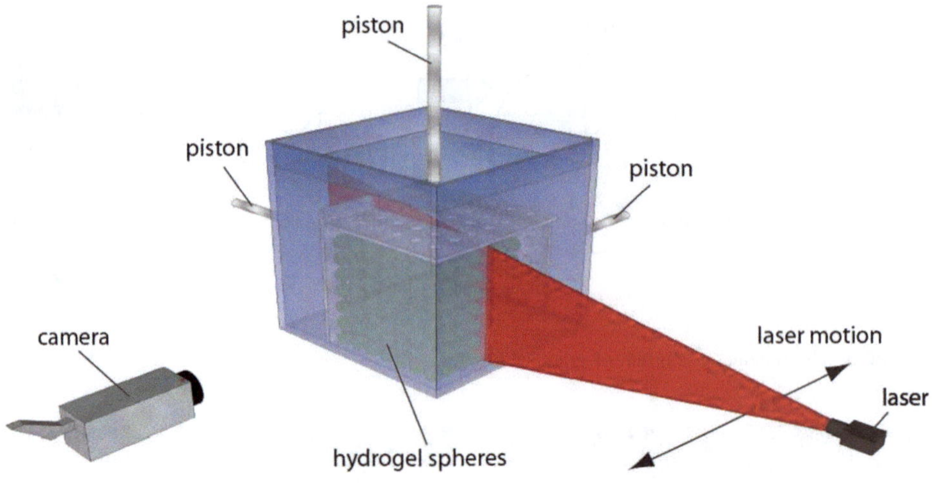

Fig. 20.5 The schematic diagram of tri-axial RIM experimental setup (Dijksman et al. 2013)

each cross section of granular pile (Inouye and Kawata 1994). The schematic diagram of RIM is shown in Fig. 20.5. The more introduction of RIM method can be found in Ref. Dijksman et al. (2012).

20.4 Conclusions

There different types of granular displacement were explained in detail. Meanwhile, two latest novel experimental techniques on granular materials studying were introduced, including photoelasticity and RIM methods, which could explore the micro properties of each particle during granular materials deformation and overcome the disadvantages of traditional experimental methods and numerical simulation.

Acknowledgments This research is supported by the National Natural Science Foundation of China (No. 41172204) and the Fundamental Research Funds for the Central Universities (No. 2014B03414). The first author would like to express his gratitude to China Scholarship Council (CSC), Hohai University and Tongi University.

References

Campbell CS, Brennan CE (1985) Computer simulation of granular shear flows. J Fluid Mech 151(167):88
Cundall PA, Strack OD (1979) A discrete numerical model for granular assemblies. Geotechnique 29(1):47–65
Daniela RC, Poloskia AP, Eduardo Sáezb A (2007) A continuum constitutive model for cohesionless granular flows. Chem Eng Sci 62(5):1343–1350
Dijksman JA, Rietz F, Lorincz KA, van Hecke M, Losert W (2012) Invited article: refractive index matched scanning of dense granular materials. Rev Sci Instrum 83(1):011301–011312
Dijksman JA, Zheng H, Behringer RP (2013) Imaging soft sphere packings in a novel triaxial shear setup. In: AIP conference proceedings, vol 1542, p 457
Goodman MA, Cowin SC (1972) A continuum theory for granular materials. Arch Ration Mech Anal 44(4):249–266
Hardin BO (1985) Crushing of soil particles. J Geotech Eng 111(10): 1177–1192
Huang Y, Zheng H (2013) Mechanical characteristics of a lunar regolith simulant at low confining pressure. Earth Environ Sci. doi:10.1007/s12665-013-2763-7
Huang Y, Zheng H, Mao W, Li G, Ye B (2011) Numerical simulation of air–soil two-phase flow based on turbulence modeling. Nat Hazards 58(1):311–323
Indraratna B, Wijewardena LSS, Balasubramaniam AS (1993) Large-scale triaxial testing of grey wacke rockfill. Geotechnique 43(1): 37–51
Inouye Y, Kawata S (1994) Near-field scanning optical microscope with a metallic probe tip. Opt Lett 19(3):159–161
Jaeger HM, Nagel SR, Behringer RP (1996) Granular solids, liquids, and gases. Rev Mod Phys 68(4):1259–1273
Jiang MJ, Yu HS, Harris D (2005) A novel discrete model for granular material incorporating rolling resistance. Comput Geotech 32(5): 340–357
Jiang MJ, Yu HS, Harris D (2006) Bond rolling resistance and its effect on yielding of bonded granulates by DEM analyses. Int J Numer Anal Meth Geomech 30(8):723–761
Jop P, Forterre Y, Pouliquen O (2006) A constitutive law for dense granular flows. Nature 441(7094):727–730
Kanatani KI (1981) A theory of contact force distribution in granular materials. Powder Technol 28(2):167–172
Liu AJ, Nagel SR (1998) Nonlinear dynamics: Jamming is not just cool any more. Nature 396(6706):21–22
McDowell GR, Bolton MD, Robertson D (1996) The fractal crushing of granular materials. J Mech Phys Solids 44(12):2079–2101
Norihiko M, Sukeo O (1979) Particle-crushing of a decomposed granite soil under shear stresses. Soils Found (JSSMFE) 19(3):1–14
Oda M, Iwashita K (1999) Mechanics of granular materials. A. A. Balkema, Rotterdam
Oda M, Iwashita K (2000) Study on couple stress and shear band development in granular media based on numerical simulation analyses. Int J Eng Sci 38(15):1713–1740
Rao KK, Nott PR, Sundaresan S (2008) An introduction to granular flow. Cambridge University Press, Cambridge

Sladen FME, Payne DN, Adams MJ (1976) Determination of optical fiber refractive index profiles by a near-field scanning technique. Appl Phys Lett 28(5):255–258
Utili S, Nova R (2008) DEM analysis of bonded granular geomaterials. Int J Numer Anal Meth Geomech 32(17):1997–2031
Vesic AS, Clough GW (1968) Behavior of granular materials under high stresses. J Soil Mech Found Div 94(SM3):661–688
Zhang J, Majmudar TS, Sperl M, Behringer RP (2010) Jamming for a 2D granular material. Soft Matter 6(13):2982–2991
Zheng H, Dijksman JA, Behringer RP (2013) Novel experimental apparatus for granular experiments on basal friction. In: AIP conference proceedings, vol 1542, p 465
Zhou Z, Wang J, Huang Y, Xu H (2013) Conceptual data model and method of settlement calculation for deformation and water release from saturated soft soil. Environ Earth Sci. doi:10.1007/s12665-013-2818-9

Dynamic Processes of the Benthic Boundary Layer in the Subaqueous Yellow River Delta, China

21

Xiaolei Liu, Yonggang Jia, Jiewen Zheng, Lei Guo, and Hongxian Shan

Abstract

The benthic, or bottom, boundary layer (BBL) is the region of the ocean adjacent to the ocean floor. The complex dynamic process of the BBL is closely related to seabed stability and geological hazards. In this study, the subaqueous Yellow River delta, where submarine landslide and erosion occur frequently, was chosen to conduct in situ observations. Hydrodynamic data, including wave and current parameters, and seawater turbidity, was observed using a tripod observation system. Besides, the excess pore water pressure was observed simultaneously below the seabed surface at depths of 30 and 60 cm. Data analysis showed that wave-induced shear stress can lead to the slight seabed erosion, while wave-induced pore water accumulation inside the seabed can greatly contribute to the severe seabed erosion.

Keywords

Benthic boundary layer • Sediment re-suspension • Seabed erosion • Sediment liquefaction • Pore water pressure

21.1 Introduction

As is known for its extremely heavy sediment load, the Yellow River discharges more than hundreds of millions of tons of sediment into sea each year, and approximately 80–90 % of these sediment loads is deposited within the range of 30 km around the current estuary (Saito et al. 2001). However, the modern Yellow River delta is subject to much more serious erosion after the estuary has been abandoned than the other deltas in the world (Meng et al. 2012). Dynamic processes of the Benthic Boundary Layer (BBL) in the Subaqueous Yellow River delta can play a significant role in the mechanism of sediment erosion. Therefore, we conducted the in situ BBL observations to study the variation in suspended sediment concentration, seabed interface, and pore water pressure under different hydrodynamic conditions.

21.2 Instrumentation and Experimental Sites

The shallow-water seabed we investigate in this study is situated in the northern subaqueous delta of the Yellow River forming during the period from 1964 to 1976 (Fig. 21.1). The tidal cycle for the test site is irregularly semidiurnal, with a mean tidal range varying from 0.7 to 1.7 m, and an extreme of 2.17 m. The maximum tidal flow rate is above 120 cm/s. The average wave height is less than 0.5 m, and the largest is 3.3 m. Under extreme conditions, the wave height can reach 5.8 m (Chu et al. 2006). The water

X. Liu · Y. Jia (✉) · L. Guo · H. Shan
Department of Environmental Science and Engineering, Ocean University of China, Qingdao 266100, China
e-mail: yonggang@ouc.edu.cn

X. Liu
e-mail: ffnlxl_001@sina.com

J. Zheng
First Institute of Oceanography, SOA, Qingdao 266061, China

G. Lollino et al. (eds.), *Engineering Geology for Society and Territory – Volume 4*,
DOI: 10.1007/978-3-319-08660-6_21,

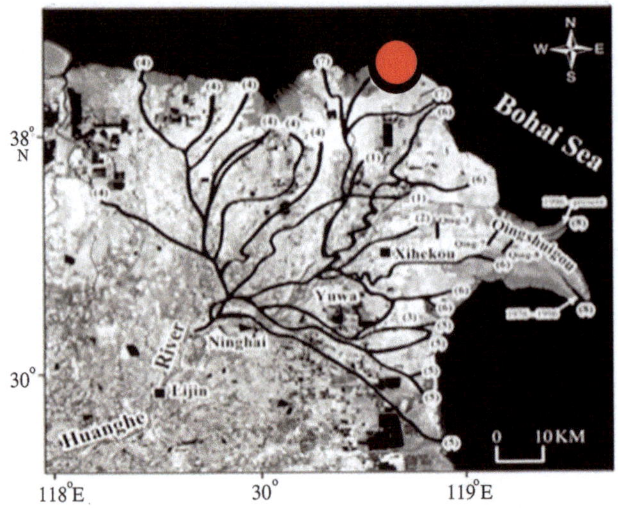

Fig. 21.1 Location of the study site

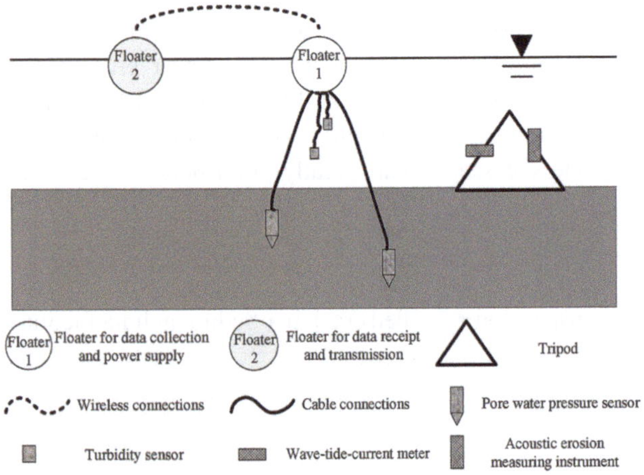

Fig. 21.2 The in situ observation system

depth at the experimental site ranges from 0.2 m to 2.6 m during the in situ observation period.

Over a 5 days period beginning on November 21, 2011 and ending on November 26, 2011 we deployed an array of hydrodynamic and sediment dynamic sensors on a tripod frame, including two turbidity sensors, two pore water pressure sensors, one sea gauge wave tide recorder, one kinemometer, and one acoustic erosion measuring instrument (Fig. 21.2).

21.3 Observational Results

21.3.1 Marine Hydrodynamics

In the observation region, the water depth ranged from 0.47 to 3.96 m, and the tidal range varied from 0.31 to 2.46 m, which was obviously influenced by north and northeast winds in winter (Fig. 21.3a). Rough sea conditions were recorded on November 24 and 25, 2011, during which the maximum wave height can nearly reach 2.0 m (Fig. 21.3b); the maximum significant wave height was 1.62 m (Fig. 21.3c); the significant wave period ranged from 4 to 8 s (Fig. 21.3d); and the current velocity varied prevalently from 20 to 80 cm/s with the maximum value of 120 cm/s (Fig. 21.3e).

On November 22, another strong northeasterly wind struck this sea area, which was weaker than that mentioned above (Fig. 21.3). During this event, the maximum water depth was 2.71 m; the maximum wave height was 1.32 m; the significant wave height ranged from 0.6 to 0.9 m; the significant wave period ranged from 3 to 7.5 s; and the maximum current velocity varied from 5 to 45 cm/s, with the maximum of 84.2 cm/s.

Under the normal sea conditions, the tidal range was small, varying from 0.4 to 1.4 m. The maximum water depth reached no more than 2 m, and the minimum value was 0.47 m. The maximum wave height was 0.69 cm. The significant wave height was no more than 0.6 m, and the significant wave period ranged from 3 to 6 s. The current velocity was mostly less than 20 cm/s.

21.3.2 Suspended Sediment Concentration

The seawater turbidity above the seabed was observed to indicate the variation of suspended sediment concentration under different hydrodynamic conditions. As is shown in Fig. 21.4, under the normal sea conditions, turbidity remained less than 200 FTU; while under the rough sea conditions on November 24 and 25, 2011, it reached 745 FTU; and under the other rough sea conditions on November 22, 2011, the maximum turbidity was 400 FTU.

According to the field seawater turbidity and sediment properties, it can be estimated that the suspended sediment concentration under rough sea conditions was ten times more than that under normal sea conditions, which is consistent with the field measurements conducted by Yang and Wang (1993).

21.3.3 Seabed Interface

Dynamic processes of seabed erosion and deposition both occurred during the in situ observation period (Fig. 21.5). In the normal sea conditions, sediment deposition predominated with the total deposition mass of 77.5 mm closely related to the sediment settling process. In the rough sea conditions, erosion predominated with the maximum erosion mass of 144.2 mm closely related to the sediment re-suspension.

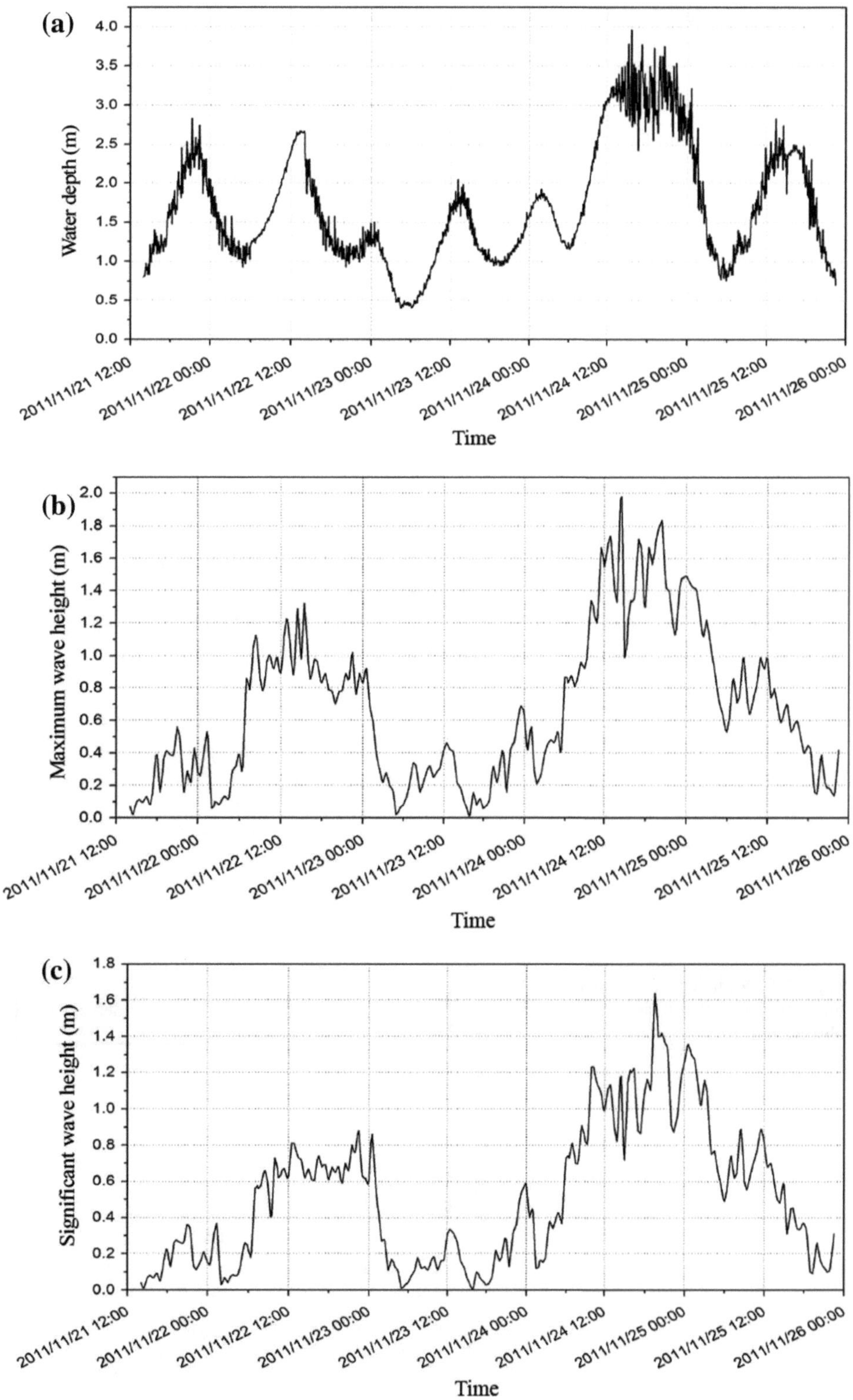

Fig. 21.3 Variations of marine hydrodynamic conditions. **a** average water depth; **b** maximum wave height; **c** significant wave height; **d** significant wave period; **e** bottom current velocity

21.3.4 Pore Water Pressure

The pore water pressure inside the seabed nearly remained stable except under the rough sea conditions with the maximum accumulated excess pore water pressure of 3.9 kPa when significant erosion occurred and seawater turbidity increased obviously (Fig. 21.6). The seabed sediment can be liquefied when the excess pore pressure builds up progressively and become equal to the effective weight of the overburden during cyclic loading (van Kessel and

Fig. 21.3 continued

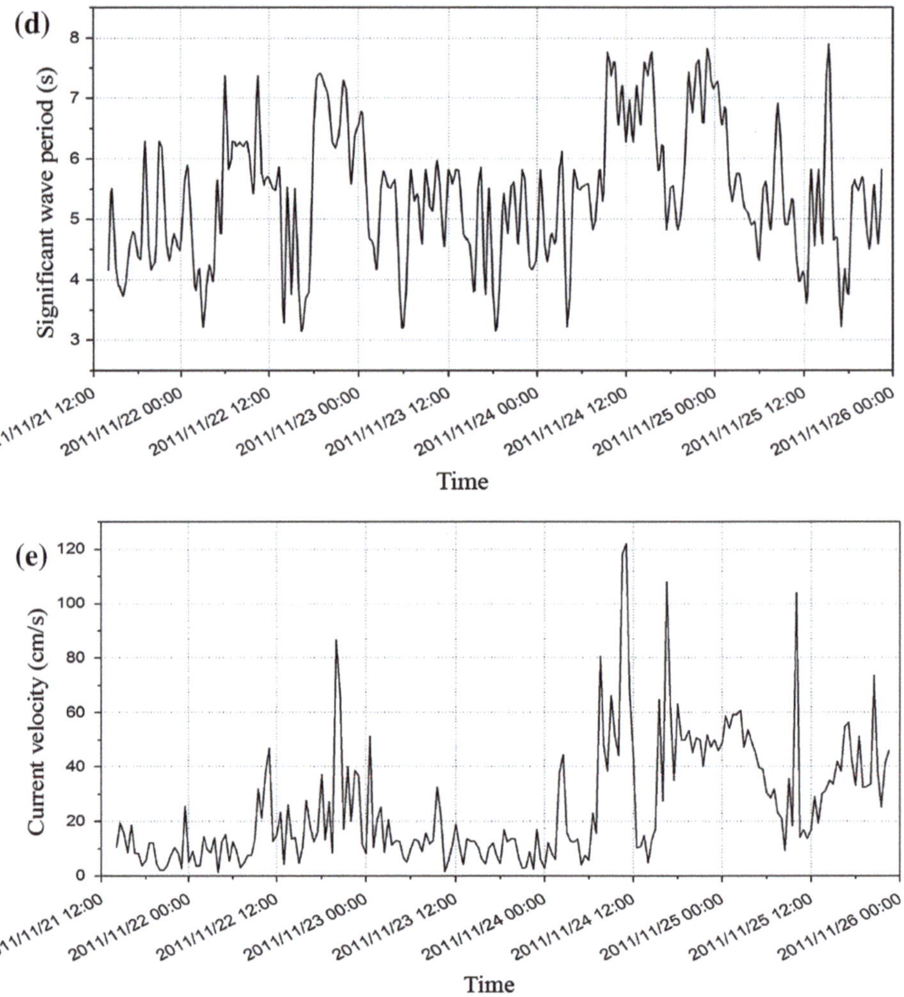

Fig. 21.4 Variation of seawater turbidity during the in situ observation

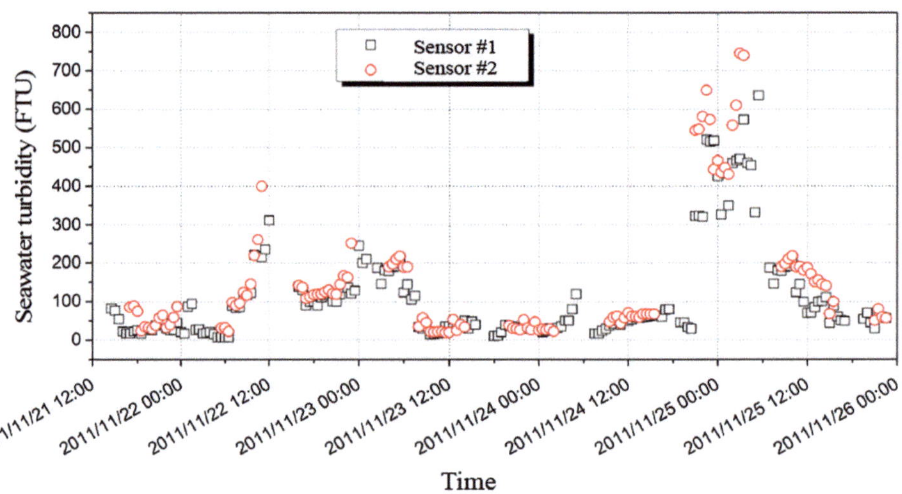

Kranenburg 1998; Jia et al. 2011). Therefore it can be assumed that wave-induced sediment liquefaction plays a significant role in sediment erosion and re-suspension under rough sea conditions. That is, sediment liquefaction in different degrees may lead to completely different erosion mass quantities.

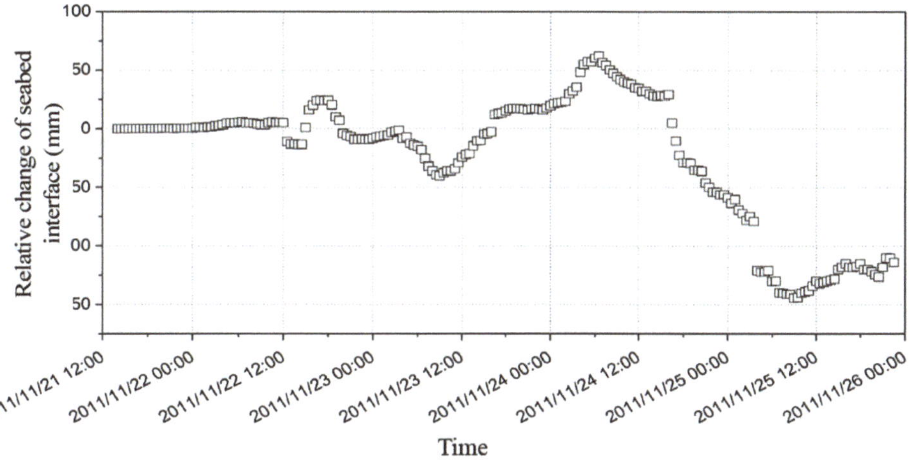

Fig. 21.5 Variation of seabed interface during in situ observation

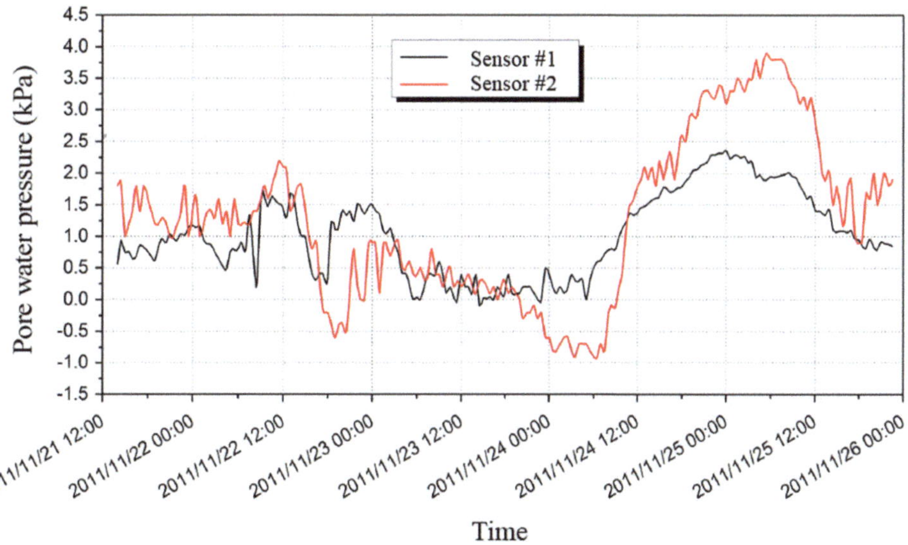

Fig. 21.6 Variation of pore water pressure during in situ observation

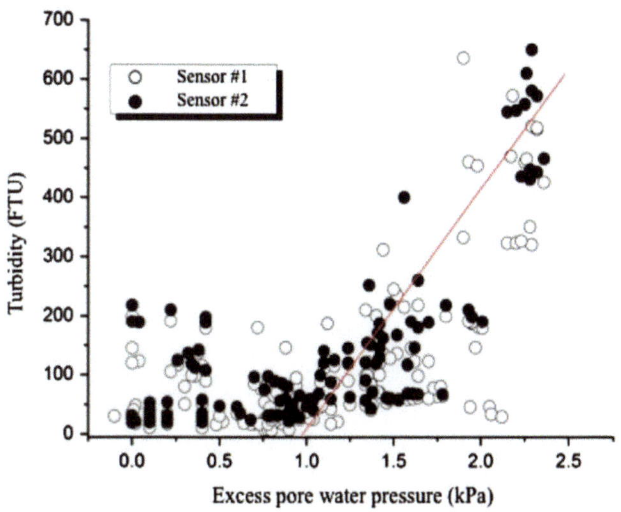

Fig. 21.7 Scatter plot of water turbidity versus excess pore water pressure

21.4 Discussion

A good positive correlation existed between turbidity and excess pore water pressure caused by wave action in stormy sea conditions when the excess pore water pressure exceeds 1.0 kPa (Fig. 21.7). This indicated that the erosion mass increased with the increasing of sediment liquefaction degree. To better understand the mechanism of seabed erosion, the correlation between turbidity and bottom shear stress induced by waves and currents was also constructed, but no obvious relationship can be achieved for the rough sea conditions. Therefore, we can conclude that wave-induced sediment liquefaction is the primary factor affecting seabed erosion behavior during the significant sediment re-suspension event.

21.5 Conclusion

Overall, this study reveals the facters affecting the BBL dynamic processes in the Subaqueous Yellow River delta. Bed shear stresses under combined waves and currents exert a strong control on the routine events of seabed erosion and sediment re-suspension, which is consistent with previous studies. While sediment liquefaction generated by wave-induced accumulation of excess pore water pressure plays a critical role in seabed erosion and sediment re-suspension, especially under extreme events like storm or tsunami.

Acknowledgements This work was jointly funded by the National Natural Science Foundation of China (Contract Nos. 41272316 and 41072215) and Comprehensive Investigation Plan of Marine Geology Security Engineering (Contract No. GZH201100203).

References

Chu ZX, Sun XG, Zhai SK et al (2006) Changing pattern of accretion/erosion of the modern Yellow River subaerial delta, China: based on remote sensing images. Mar Geol 227(1–2):13–30

Jia YG, Shan HX, Yang XJ et al (2011) Sediment dynamics and geologic hazards in the estuary of Yellow River, China. Science Press, Beijing

Meng XM, Jia YG, Shan HX et al (2012) An experimental study on erodibility of intertidal sediments in the Yellow River delta. Int J Sedim Res 27:240–249

Saito Y, Yang ZS, Hori K (2001) The Huanghe and Changjiang deltas: a review on their characteristics, evolution and sediment discharge during the Holocene. Geomorphology 41(2–3):219–231

van Kessel T, Kranenburg C (1998) Wave-induced liquefaction and flow of subaqueous mud layers. Coast Eng 34:109–127

Yang ZS, Wang T (1993) Marine environment of Chengdao. Ocean University of Qingdao Press, Qingdao

Airborne Unmanned Monitoring System for Coastal Erosion Assessment

22

P. Bellezza Quater, F. Grimaccia, and A. Masini

Abstract

Large scale and quantitative erosion assessment is one of the most important tools in the development of coastal management policies. In fact, any defense program or resource research planning to be carried out by the Public Administration needs updated knowledge of beach evolution and of the factors influencing this environment. Remote observation techniques have improved greatly during recent years and the availability of high-resolution images now allows systematic monitoring of beach and coastal environment evolution. Shoreline detection needs several digital processing and ancillary data, such as tide values, barometric pressure and wash zone slope. Nowadays data collection can be performed by unmanned aircrafts that can act as collector of different data from a ground layer. In this context hybrid Unmanned Aerial Vehicle (UAV) platforms can be used as an effective tool to perform accurate data acquisition. They are easy to deploy, cost effective and able to cover long distances performing different tasks at the same time. In this work, a real test case performed over a Marine Protected Area (MPA), located in the northern coast of Tuscany (Italy), is reported to prove effectiveness and reliability of such technology.

Keywords

Environmental monitoring • MPAs • Coastal erosion • UAV • Sensor technology • Data processing • Marine applications

22.1 Introduction

The monitoring of coastal areas plays a fundamental role in all territorial policies along the coast. The coastal zone is known to concentrate around 2/3 of the world's population and supports intense tourism activity in many countries. Beaches represent the element of highest economic value in the coastal system, but are also the most fragile one being subject to morphological variations. These dynamic phenomena can alter the protective function they provide to the hinterland as well as the potential use for tourism and recreation. The study of beach evolution is therefore fundamental for the planning of this densely populated zone, which is economically important, but also complex and not static.

Integrated Coastal Zone Management (ICZM) is a continuous decision-making process that aims at solving the conflicts arising from the diversity of uses of this complex part of the territory. Beach management, control of coastal erosion and management of natural hazards are among the many aspects that an Integrated Coastal Zone Management

P. Bellezza Quater (✉) · F. Grimaccia
Nimbus S.r.l., Via Del Boschetto 2/1, 10040, Turin, LO, Italy
e-mail: p.bellezza@nimbus.aero

F. Grimaccia
e-mail: f.grimaccia@nimbus.to.it

F. Grimaccia
Politecnico Di Milano, Via La Masa 34, Milan, Italy

A. Masini
FlyBy S.r.l., Via Puini 97, 57128, Livorno, Italy
e-mail: andrea.masini@flyby.it

G. Lollino et al. (eds.), *Engineering Geology for Society and Territory – Volume 4*,
DOI: 10.1007/978-3-319-08660-6_22, © Springer International Publishing Switzerland 2014

program must address, and specific data input is necessary in order to conduct such program in a scientifically-sound manner. Some typical management problems include the control of beach erosion, the design of artificial nourishment projects, the construction and monitoring of defense structures, as well as the definition of a set-back line for construction and of the risk zone, all requiring an accurate evaluation of coastal morph-dynamic evolution.

Seen as essential assets, many different functions are performed by shores, namely:

- Provide protection against erosion and coastal flooding;
- Dissipate wave energy, thereby reducing the risk of failure of coastal defenses;
- Provide recreational space, including access to the sea for recreation;
- Are the habitat, feeding ground or resting place of many species;
- Give access to the sea for small fishing vessels.

Beaches are also crossed by the entrances to ports, harbors, rivers or estuaries and have been encroached upon by ports and towns (particularly to form promenades). They provide space for further potential developments in urbanized or industrial areas. Therefore, the study of erosion phenomena is fundamental for any proper marine and coastal management plan.

22.2 Marine and Coastal Erosion Management

Research and management activities related to coastal zones need to perform geospatial analysis and acquire data to build and refine accurate models, which represents an essential background in environmental policy with respect to marine and coastal management. For example, much of Europe's coastline is eroding and erosion threatens some of the values and functions of the coast. Coastal erosion is the process of wearing away material from a coastal profile and the net loss of material from the section (Marchand 2010). A description of the processes involved in the erosion of sandy beaches and dunes has been provided by Van Rijn (2010a) together with the methods that are used to model these processes; in (Van Rijn 2010b) it can be found can be found a similar report for gravel/shingle beaches and barriers.

Satellite remote sensing is a recent technology for the monitoring of the earth and its resources. The technique of mapping and monitoring coastal sediments with the satellite requires high-resolution current measurements (Romeiser 2007). Satellite remote sensing of the coastal environment is important for environ-mental impact assessment because of its temporal and spatial characteristics.

According to (Chavez 1997) satellite images of coastal waters provides information about the quality of suspended sediment entering, residing and leaving a given system. The significance of satellite monitoring of coastal sediment is crucial for flood and erosion control management as well as to ensure conservation of natural resources that abound in the coastal zone.

Because of the challenges involved in monitoring river discharges in the coastal zone, it is fundamental to carry out observations using optical sensors on board satellites for easy identification and analysis of sediment movement pattern along the coastline. The characteristic nature of the inter-tidal zone is a major factor for consistent monitoring of river discharges in the coastal environment.

Caballero et al. (2012) emphasized that satellite images is useful in areas with high ecological value, for determining the dynamics of sediments in the coastal areas. Sustainable management of the coastal zone is essential because of the dynamic nature in the area.

Duffy et al. (2004) also stated that erosion conditions of sediment transport can be estimated by the spatial gradient of the calculated bed load transport rate. The advantage of using UAV technology in monitoring coastline erosion present significant advantages with respect to data accuracy and multiple scale resolution.

As reported also in (d'Oleire-Oltmanns et al. 2012), the UAV-based remote sensing enables user-controlled image acquisition and it promise to overcome the gap between ground observations and imagery acquired from conventional satellite sensors. The unmanned technology is also a cost effective method that allows adapting to the different size of the observed objects and to the speed of phenomena.

In this paper, the authors propose a new hybrid Unmanned Aerial System (UAS), which presents peculiar characteristics that make it particularly suitable for advanced marine and coastal monitoring purposes.

22.3 Hybrid Light UAV Platform for Marine Applications

In the last decade, new flying platforms have been developed in order to combine characteristics of dirigible and conventional aircrafts (Cafaro et al. 2008). Moving from military to civil sector, UAS can open up multiple research and commercial fields in the next future. Nimbus platforms have been already used in different research projects performing monitoring mission for multiple purposes (Neri et al. 2011; Grimaccia et al. 2011; Antonio et al. 2012). They combine heavy payload, easy piloting, and high safety with cost effective and environmentally responsible approach to aerial work.

Fig. 22.1 The UAV platform image and optical sensors

This hybrid platform called "metaplano" can be effectively used for info-mobility, monitoring and surveillance tasks in normal and emergency scenarios. It combines some peculiarities of dirigibles and traditional aircrafts with short take-off and landing capabilities. It is important to notice that the specific static lift of this platform allows reaching stall conditions ideal for monitoring missions, and this makes it competitive with respect to traditional manned aircrafts.

Moreover his unique concept permits rapid deployment to wide areas without any airport infrastructures and can be operative in very fast time. This platform can be easily customized with different payloads and the actual configuration has been achieved after the development of many prototypes.

In the last 5 years multiple permits to fly has been obtained from the Italian national flight authority and many research programs have been completed both at national and European level. Using this kind of technology allows us to reach and collect much more accurate images approximately at 150 m altitude in a very flexible and "on-demand" way. Moreover, any planned monitoring campaign can satisfy multiple requirements at the same time since a complex sensor bay can be mounted onboard. If compared with satellite images the UAV image capture results in a more cost effective solution and no image-based atmospheric correction is required in this case (Fig. 22.1).

The relation and work done beside the Italian flight regulation authority has allowed to obtained permissions to test new payloads and elastic suspension devices to minimize vibration and to optimize optical results. However, additional work is still required to test wing improvement and compact leaf spring undercarriage to move towards a bay installation. This improvement will require to change the link to the ground control station and to qualify the new flight control system compliance according to DO178B and DO254 specifications (Fig. 22.2).

The selected sensors for this case study were a video camera TASE with daylight and infrared capability, assumed as a key monitoring equipment and multiple high-resolution photo cameras suitable to take required pictures.

The combination of these sensors was selected according to the limitations in weight, cost and development level. The authors aim to achieve a better definition of the data collected during the missions in terms of recorded details of the picture as key item of the required investigation both as sequence (video camera) and as discrete details (photo cameras). Moreover the identification of the position through GPS information correlated with the air data telemetry output and ground points calibrated as position reference allowed to match the air photo with the coastline reference.

22.4 A Case Study

As explained in the previous sections, the constant increase in research on unmanned systems and relative technologies have allowed to satisfy monitoring task in different fields such as surveillance and environmental monitoring with high-resolution image processing and data services.

Some first flight tests have been already performed in the San Rossore marine protected area (Fig. 22.3). In this area, the wave activity is consuming the total area of the natural

Fig. 22.2 The optical sensors

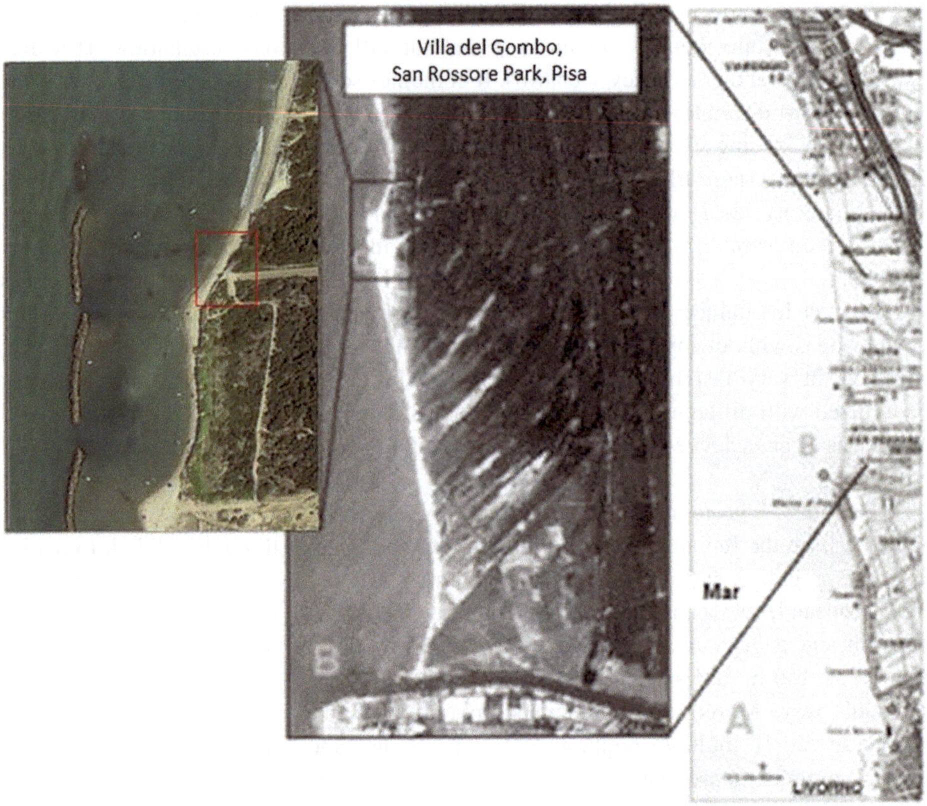

Fig. 22.3 The area of interest

park. Thus to monitor this coastal erosion effect we use the hybrid UAV to perform some monitoring flights over the described area.

The collected HD pictures have been georeferenced using both sensor and flight status data and different ground control point was selected to match the air photo with known ground target coordinates.

Figure 22.4 show the obtained georeferenced mosaic where the coastline is detected in red.

Using an advanced mosaicking technique in the UAV sensor data, the exact costal line was detected in the post-processing phase.

Fig. 22.4 The geo-referenced mosaic with the coastline highlight in *red*

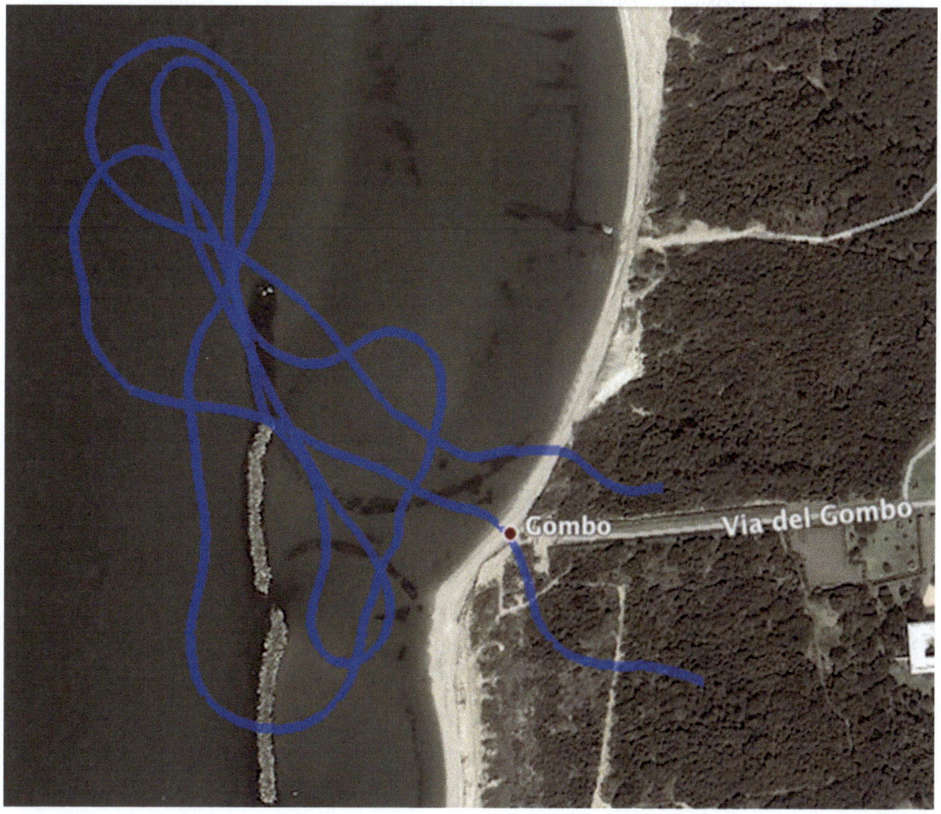

Fig. 22.5 The UAV flight path over the marine area

In these preliminary tests, we performed one first measure campaign but in the next future other missions will be performed with the aim to compare acquired data related to the ground status and thus evaluate coastal evolution and erosion phenomenon rate (Fig. 22.5).

22.5 Conclusion

In this work a novel hybrid unmanned monitoring system is proposed for efficient coastal and marine environmental monitoring. The specific effect of coastal erosion was observed by means of onboard E/O sensors in order to acquire data for advance ICZM purposes.

A real case study is presented using ad hoc image mosaicking with high definition geo-referenced pictures to detect the exact costal line from the UAV post-processed data. Further study will aim to repeat these experiments and evaluate the evolution of coastal erosion in the same area.

References

Marchand M (ed) (2010) Concepts and science for coastal erosion management. Concise report for policymakers. Deltares, Delft. http://www.conscience-eu.net/documents/concise-report-final.pdf. Accessed 30 Nov 2013

Van Rijn (2010a) Coastal erosion control based on the concept of sediment cells. CONSCIENCE deliverable D13a. http://www.conscience-eu.net/documents/deliverable13a-controlling.pdf. Accessed 30 Nov 2013

Van Rijn (2010b) Modelling erosion of gravel/shingle beaches and barriers. CONSCIENCE deliverable D13b. http://www.conscience-eu.net/documents/deliverable13b-modelling.pdf. Accessed 30 Nov 2013

Romeiser R (2007) High-resolution imaging of current fields from satellites. Sea Technol 48(9):44

Chavez PS Jr (1997) Mapping suspended sediments using remotely sensed satellite images. In: Proceedings of the U.S. geological survey (USGS) sediment workshop, 4/7 Feb 1997

Caballero I, Edward M, Gabriel N (2012) DEIMOS-1 satellite provides imagery for coastal management. Sea Technol 53(2):10–13

Duffy G, John H (2004) Detecting migration of sand dunes using spatial correlation. Sea Technol 45(10):53–58

d'Oleire-Oltmanns S, Irene M, Peter KD, Ries JB (2012) Unmanned Aerial Vehicle (UAV) for monitoring soil erosion in Morocco. Remote Sens 4(11):3390–3416

Cafaro E, Stantero L, Taurino R, Capuani A, Mercalli A, Fantino S (2008) Nimbus meta-plane, a new aerodynamic concept for unmanned hybrid aerial vehicle application. SAE Int. doi:10.4271/2008-01-2243

Neri M, Campi A, Suffritti R, Grimaccia F, Sinogas P, Guye O, Papin C, Michalareas T, Gazdag L, Rakkolainen I (2011) UAV-based capturing of HD/3D content with WSN augmentation for immersive media experiences. In: Proceedings of the IEEE international conference on multimedia and expo, pp 1–6

Grimaccia F, Bellezza Quarter P (2011) Hybrid UAS for HD-3D multimedia system for entertainment purposes. In: Proceedings of AUVSI unmanned systems North America, pp 90–94 (ISBN: 978-1-61839-348-7)

Antonio P, Grimaccia F, Mussetta M (2012) Architecture and methods for innovative heterogeneous wireless sensor network applications. Remote Sens 4:1146–1161 (ISSN: 2072–4292)

Test on Effect of Temperature on Area of Thermal Expansion of HDPE Geomembrane

23

Ping Yang, Xue-wen Zhu, Min-hui Wu, and Xue-qing Sun

Abstract

Thermal expansion of High Density Polyethylene geomembrane usually results in wrinkles in HDPE geomembrane in seepage control system of the landfill site. In order to analyze the effect of temperature on the thermal expansion of HDPE geomembrane, the area of thermal expansion of HDPE geomembrane under different environmental temperature and different heating time was measured. Test results show that the area of HDPE geomembrane changes with increasing of environmental temperature and heating time. Under same temperature, the area of HDPE geomembrane increases with increasing of heating time. The increase rate of area of thermal expansion decreases nonlinearly with the increasing of heating time. A fast increase rate of area of thermal expansion is observed during 30 min in incipient stages of heating time, the area of thermal expansion accounts for 34 % of the total area of thermal expansion. At the same heating time, the difference of increase ratio of area of thermal expansion of HDPE geomembrane, between 60 and 40 °C, increases nonlinearly with the increasing of heating time. The difference of increase ratio of area of thermal expansion increases rapidly during 30 min in incipient stages of heating time. Increase ratio of area of thermal expansion of HDPE geomembrane at 60 °C is from 2.5 to 3.9 times that of at 40 °C.

Keywords

HDPE geomembrane • Environmental temperature • Area of thermal expansion

23.1 Introduction

Geomembrane is a kind of ideal impervious materials for low permeability coefficient and is been widely used in seepage control system of landfill site. The main function of seepage control system is to collect leachate, prevent leachate to penetrate to the ground, which would pollute underground water. Due to the complexity of the environment, the geomembrane laying on the surface of landfill site slope will produce deformation, which may affect the seepage control function. Among of the factors causing the geomembrane deformation, temperature is one of important factors. Klein (2001) recorded the temperature of city solid waste landfill substrate during the period of 3 years. Rowe and Hoor (2009) and Rowe and Isiam (2009) pointed out that monitoring data of waste landfill shows, for a typical urban solid waste landfill, temperature of landfill liner can reach 30–40 °C. If water in landfill site is more, temperature of landfill may reach 60 °C. Giroud (1990) pointed out that if temperature of geomembrane reaches 50 °C, geomembrane will form a 0.1 m high wrinkles. Giroud et al. (1992) analyzed the relationship between geomembrane wrinkles

P. Yang (✉)
Key Laboratory of Geotechnical and Underground Engineering of Ministry of Education, Tongji University, 200092 Shanghai, China
e-mail: pingyang@tongji.edu.cn

P. Yang · X. Zhu · M. Wu · X. Sun
Department of Geotechnical Engineering, Tongji University, 200092 Shanghai, China

G. Lollino et al. (eds.), *Engineering Geology for Society and Territory – Volume 4*,
DOI: 10.1007/978-3-319-08660-6_23,

height, width, spacing and geomembrane type and temperature. Collins (1993) and Rowe (1998) pointed out that high temperature will increase the hydraulic conductivity of clay liner in the barrier system and the diffusion coefficient between geomembrane and clay liner. Akpinara (2005) studied the effect of temperature on geomembrane-geotextile interface shear strength. Rowe (2005) pointed out that the temperature of the landfill site, especially the temperature of landfill site basement and slope will affect the long-term performance of composite geomembrane lining system. Rowe and Hoor (2009) pointed out that the small and medium-sized lining can be affected by high temperature in the landfill site with double lining. Sabir and Brachman (2012) studied the effects of time and temperature on the stress of geomembrane. Rowe studied the change of the wrinkling of a geomembrane at different times of the day on a compacted clay liner on a slope, Xu et al. (2009) pointed out that geomembrane temperature stress increased linearly with the decrease of ambient temperature. Xu et al. (2010) put forward the evaluation method of geomembrane tension caused by temperature and interface friction characteristics of each material in the liner system of landfill slope. Although there are some researches about the effect of temperature on performance of geomembrane, the study on the area of thermal expansion of HDPE geomembrane under different environmental temperature and different heating time is few. Therefore, it is necessary to study the area of thermal expansion of HDPE geomembrane under different environmental temperature and different heating time.

23.2 Test Materials and Methods

23.2.1 Test Materials

The material used in this experiment is HDPE geomembrane with 2 mm thickness. The characteristics of HDPE geomembrane are shown in Table 23.1.

23.2.2 Test Equipment

In this experiment, oven is used to control the temperature of the geomembrane and simulated the environmental temperature of geomembrane in landfill.

23.2.3 Test Methods

This experiment was conducted under the temperature of 40 and 60 °C. The changes of geomembrane area at 0.5, 1, 2 and 4 h under each temperature was measured. Photo of geomembrane heated was taken by digital camera, then the photo was processed by MATLAB to obtain the area of geomembrane heated.

Table 23.1 Characteristics of HDPE geomembrane

Density/(g/ cm^3)	Tensile strength/MPa	Elongation rate/%	Modulus of elasticity/MPa
0.95	22	560	675.6

Linear expansion coefficient of HDPE geomembrane is 1.2×10^{-4} $\varepsilon/^{\circ}C$(Xu et al. 2009)

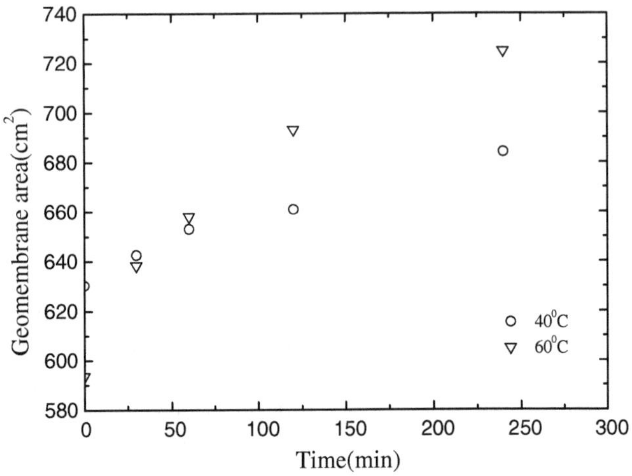

Fig. 23.1 Changes of geomembrane area over time

23.3 Analysis of Test Results

23.3.1 Changes of Geomembrane Area Over Time under Different Temperature

The change of HDPE geomembrane area over time was shown in Fig. 23.1. When environment temperature is 40 and 60° C, the HDPE geomembrane area increases with increasing of heating time. Under the same heating time, the increased area of HDPE geomembrane at 60 °C is greater than that of at 40 °C. The increase rate of area of thermal expansion of HDPE geomembrane becomes slow gradually with the increasing of heating time. A fast increase rate of area of thermal expansion is observed during 30 min in incipient stages of heating time, the area of thermal expansion accounts for 34 % of the total area of thermal expansion.

23.3.2 Effect of Temperature on the Change of Geomembrane Area

Ratio of geomembrane area are shown in Fig. 23.2. When the environment temperature is 40 and 60 °C, the ratio of geomembrane area increases nonlinearly with increasing of

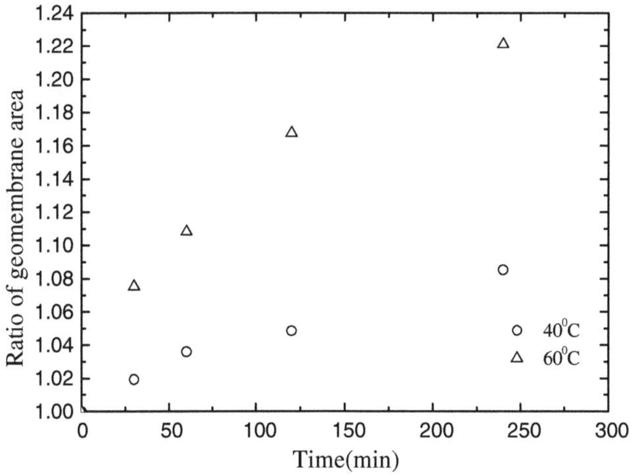

Fig. 23.2 The ratio of geomembrane area over time

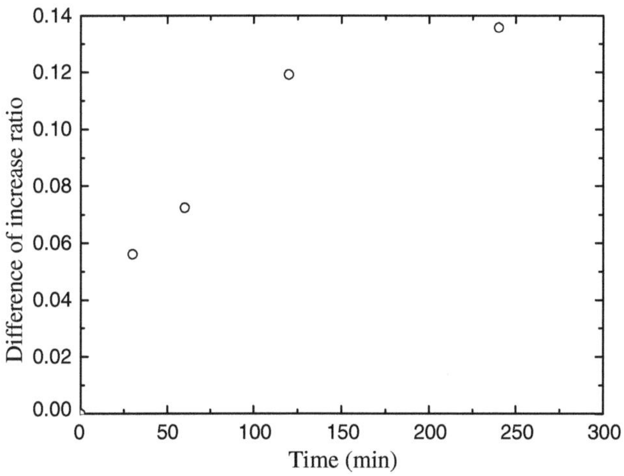

Fig. 23.3 Difference of increase ratio of area of thermal expansion at 60 and 40 °C

time. Under the same heating time, the ratio of geomembrane area at 60 °C is greater than that of at 40 °C. The ratio of geomembrane area is more than 1.2 when heating time reached 4 h at 60 °C. The slope of curve of ratio of geomembrane area decreases gradually with increasing of heating time, which suggests that the ratio of geomembrane area in unit time decreases gradually with increasing of heating time.

The difference of increase ratio of area of thermal expansion of HDPE geomembrane at 60 and 40 °C is shown in Fig. 23.3. The difference of increase ratio of area of thermal expansion of HDPE geomembrane at 60 and 40 °C increases nonlinearly with increasing of time. The slope of the curve decreases gradually with the extension of heating time. The difference of increase ratio of area of thermal expansion of HDPE geomembrane increases rapidly during 30 min in incipient stages of heating time. When heating time is over 120 min, the difference of increase ratio of area of thermal expansion of HDPE geomembrane increases slowly. Under the same heating time, the increase ratio of area of thermal expansion of HDPE geomembrane at 60 °C is greater than that of at 40 °C. Increase ratio of area of thermal expansion of HDPE geomembrane at 60 °C is from 2.5 to 3.9 times that of at 40 °C.

23.4 Conclusions

When the temperature is at 40 and 60 °C, the area of HDPE geomembrane increases with increasing of heating time. A fast increase rate of area of thermal expansion is observed during 30 min in incipient stages of heating time, the area of thermal expansion accounts for 34 % of the total area of thermal expansion of HDPE geomembrane.

Difference of increase ratio of area of thermal expansion of HDPE geomembrane between at 60 and 40 °C increases nonlinearly with increasing of time. The difference of increase ratio of area of thermal expansion of HDPE geomembrane increases rapidly during 30 min in incipient stages of heating time.

Under the same heating time, increase ratio of area of thermal expansion of HDPE geomembrane at 60 °C is greater than that of at 40 °C. The increase ratio of area of thermal expansion of HDPE geomembrane at 60 °C is from 2.5 to 3.9 times that of at 40 °C.

HDPE geomembrane can produce large area of thermal expansion after a short time of heating, which will result in wrinkles and increase the risk of leakage.

Acknowledgments This investigation was supported by the National Natural Science Foundation of China, Project No: 41002093, Opening fund of State Key Laboratory of Geohazard Prevention and Geoenvironment Protection (Chengdu University of Technology), Project No: SKLGP2014K013, the Fundamental Research Funds for the Central Universities, Shanghai Leading Academic Discipline Project, Project No: B308, Tongji University and the Program for Young Excellent Talents, Tongji University. The authors are deeply indebted to the five financial supporters.

References

Collins HJ (1993) Impact of the temperature inside the landfill on the behaviour of barrier systems. In: Proceedings of 4th international landfill symposium. CISA, Cagliari, pp 417–432

Giroud JP, Peggs ID (1990) Geomembrane construction quality assurance, waste containment systems: construction, regulation, and performance. American Society of civil engineering geotechnical special publication no. 26, pp 190–225 (Bonaparte R, ed.)

Giroud JP, Morel N (1992) Analysis of geomembrane wrinkles. Geotext Geomembr 11(3):255–276

Klein R, Baumann T, Kahapka E, Niessner R (2001) Temperature development in a modern municipal solid waste incineration (MSWI) bottom ash landfill with regard to sustainable waste management. J Hazard Mater 83(3):265–280

Akpinara MV, Benson CH (2005) Effect of temperature on shear strength of two geomembrane–geotextile interfaces. Geotext Geomem 23:443–453

Rowe RK (1998) Geosynthetics and minimization of contaminant migration through barrier systems beneath solid waste. In: Proceeding of 6th international conference in geosynthetics, pp 27–103

Rowe RK (2005) Long-term performance of contaminant barrier systems, 45th Rankine lecture. Geotechnique 55(9):631–678

Rowe RK, Hoor A (2009) Predicted temperatures and service-lives of secondary geomembrane landfill liners. Geosynth Int 16(2):71–82

Rowe RK, Islam MZ (2009) Impact on landfill liner time–temperature history on the service life of HDPE geomembranes. Waste Manage 29(10):2689–2699

Rowe RK, Yang P, Chappel MJ, Brachman RWI, Take WA (2012) Wrinkling of a geomembrane on a compacted clay liner on a slope. Geotech Eng J SEAGS AGSSEA 43(3):11–18

Sabir A, Brachman RWI (2012) Time and temperature effects on geomembrane strain from a gravel particle subjected to sustained vertical force. Can Geotech J 49(3):249–263

Xu S, Cai Q, Yang Y, Yu C, Wu W (2009) Experimental study on temperature stress of HDPE geomembrane. New Build Mater 04:75–78 (in Chinese)

Xu S, Wang G-C, Wang Z (2010) Evaluation of tensile forces of geomembrane placed on waste landfill slope due to temperature variation and filling height. Rock Soil Mech 31(10):3120–3124 (in Chinese)

24 Engineering-Geology Zoning and Risk Assessment of Hazards in Tianjin Central Fishing Port

Yiqun Tang and Jia Liu

Abstract

For the coastal cities of China, hydraulic reclamation has become an effective way to solve the problem of land resources shortage. Some cities select mucky soil dredged from channels instead of sea sand and river sand with high cost. However, the muck has a lot of adverse engineering characteristics, such as high natural water content, high void ratio, high compressibility and low bearing capacity. Therefore, engineering geological zoning and risk assessment are essential to the subsequent design and construction. Under the background of Tianjin central fishing port project which is a typical reclamation project, this paper conducts engineering-geology zoning with fully considering the requirement of the bearing capacity and settlement in this project. The following zoning principles are established: (1) the site is zoned based on the uniformity of soil layer; (2) the site is zoned based on the distribution of soft soil. The existence, category and thickness of soft soil are taken into account respectively. Then, based on the zoning map, this paper also conducts risk assessment of potential engineering-geology hazards in these zones. At the same time, risk assessment is verified by calculating the settlement of soft soil.

Keywords

Tianjin coastal new-region • Soft soil • Dredger fill • Engineering-geology zoning • Engineering-geology hazard

24.1 Introduction

Rapid development of urban has led to increasing demand for land resources, so Peng et al. (1995) concluded that reclamation land has become an effective way to release contradictions between construction land and agricultural land in China's coastal city. *Currently, for establishing new coastal land, muck dredged from adjacent channels is transported to reclamation areas by using medium-sized cutter suction dredger* (Li and Yang 2012). Considering the special engineering geological properties of dredger fill, such as high compressibility and poor consolidation, engineers conduct engineering-geology zoning before design and construction in order to avoid too large subsidence which can causes ground instability. However, due to

Y. Tang (✉) · J. Liu
Department of Geotechnical Engineering, Tongji University, 1239 Siping Road, 200092 Shanghai, China
e-mail: Tangyiqun2@tongji.edu.cn

J. Liu
e-mail: jidajgyblj@163.com

Y. Tang
Key Laboratory of Geotechnical and Underground Engineering of Ministry of Education, Tongji University, 1239 Siping Road, 200092 Shanghai, China

Y. Tang
Joint Research Center of Urban Environment and Sustainable Development, Tongji University, 1239 Siping Road, 200092 Shanghai, China

G. Lollino et al. (eds.), *Engineering Geology for Society and Territory – Volume 4*,
DOI: 10.1007/978-3-319-08660-6_24,

complex property of dredger fill and reclamation technology immaturity, there is no uniform method of engineering-geology zoning and risk assessment which is conductive to subsequent ground treatment and foundation design.

Under the background of Tianjin central fishing port project which is a typical reclamation project, this paper gives engineering-geology zoning principles based on sources of reclamation material, geomorphic conditions and soil conditions. According to requirement of bearing capacity and settlement in this project, the site is zoned based on both the uniformity of soil layer and the distribution of soft soil. The existence, category and thickness of soft soil are taken into account respectively. In the end, based on the zoning map, settlement of soft soil is calculated to verify risk assessment of potential engineering-geology hazards.

24.2 Brief Introduction of Project and Natural Geographical Conditions

Located in Hangu and the coastal city development belt of Tianjin coastal new-region, central fishing port is approximately 160 km from Beijing, 60 km from Tianjin, 55 km from Tianjin Binhai International Airport. The total central fishing port can be divided into two parts: one is a 10 km^2 fishing port used for industry and living, and the other is an 8 km^2 area mainly for working and leisure.

Center fishing port is in the east of Bohai Sea and possesses about 33 km coastline in Hangu district, so its coast belongs to a typical muddy coast. Center fishing port terrain is alluvial plain which is banded distribution along eastern coastal zone of Tanggu, Hangu and Dagang. This plain with 4–10 m width, 1–3 m elevation and thick marine layer is tendency to sea level slightly. Having salt flats, wetlands, being affected by the tide, the plain extends to the river.

24.3 Reclamation Material Resource of Central Fishing Port

With short depositing time, dredger fill consolidation under its own weight is less than the natural soft soil, so it has greater un-consolidation feature and compressibility. Currently, there are about three main sources (Zhou et al. 2012): (1) marine sedimentary muck; (2) marine sedimentary sand; (3) construction waste. According to the project needs (bearing capacity and deformation requirements), the property and cost of the filler, the difficulty degree of reclamation construction, reclamation material is determined.

Analyzing unique natural and geographical conditions of the central fishing port and project requirements, the two reasons for selecting marine sedimentary muck of adjacent sea as reclamation material is as follows:

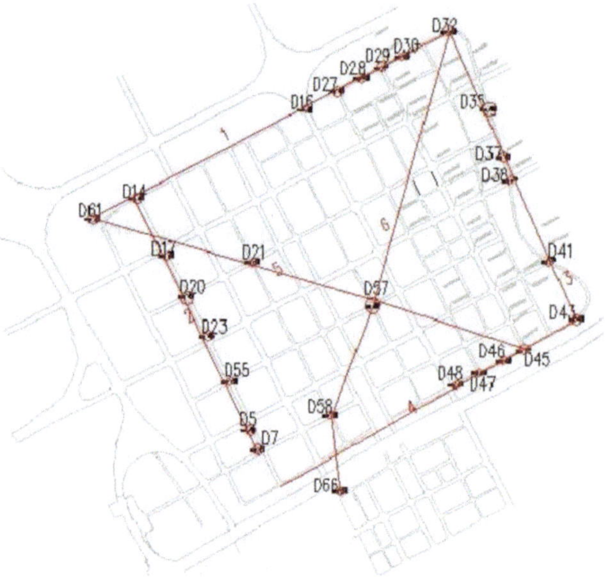

Fig. 24.1 Drilling distribution and engineering-geology hatches diagram (*red circles* stand for drillings and *red lines* stand for hatches)

(1) Meeting the project requirements. Material composition of the marine sedimentary muck meets the requests of forming slurry and consolidation feature. Relevant researches (Zhang et al. 2008) show that muck of adjacent sea is beneficial for dredger fill to drainage, fall and consolidate.
(2) Selecting marine sedimentary muck as reclamation material is low-cost and environmentally friendly. Center fishing port is in the east of Bohai Sea and possesses 33 km coastline, so using marine sedimentary muck as material of reclamation land can not only reduce transporting cost but also make full use of resource.

24.4 Soil Conditions in the Central Fishing Port

Because engineering-geology profile can intuitively reflect the engineering-geology conditions of the site, especially soil distribution, the engineering-geology profile is drawn to get type, color, density, plastic state of each soil layer statistically.

24.4.1 Determining Engineering-Geology Profile

According to total drilling distribution diagram, we select these drillings (red circles marked in Fig. 24.1) which can fully reflect engineering-geology properties of the site to drawing

Table 24.1 Category and thickness of soft soil in the site

Soil layer		Profile						
		One	Two	Three		Four	Five	Six
Mucky soil	Hole number	D14–D32 (unless D29)	D14–D55	D32–D37	D41–D43	D43–D48		D32
	Depth	9–14 m	5–14 m	9.5–15.4 m	11.3–15.3 m	11.5–17.3 m		12.3–15.3 m
	Thickness	2 m	1.5–2 m	4 m	2.9 m	3.4 m		4 m
Mucky silty soil	Hole number	D61	D5–D7				D61	
	Depth	10–15 m	9–15 m				10–15 m	
	Thickness	4 m	5.8 m				4 m	
Clay	Hole number			D32–D43		D43 and D48		
	Depth			17.6–25 m		20 m		
	Thickness			2.5 m				
Mucky soil intercalation	Hole number						D21	D58
	Depth						14 m	14 m
							1 m	1 m

engineering-geology profile (red line marked in Fig. 24.1). Six profiles are chosen to draw engineering-geology profiles by which engineering-geology condition is collected.

24.4.2 Soil Condition

We get soil engineering-geology conditions based on the engineering-geology profiles. Table 24.1 gives information about soil conditions of some sections. Soil in this site was formed in Quaternary, mainly containing Q^1m, Q_4^{3N}al, Q_4^2m, Q_4^1h, Q_4^1al, Q_3^eal, Q_3^dmc, Q_3^cal.

24.5 Engineering-Geology Zoning of Soil Containing Dredger Fills

Based on category and thickness of each soil layer, this section conducts engineering-geology zoning with fully considering the requirement of the bearing capacity and settlement in this project. According to uniformity of soil layer and distribution of soft soil, zoning is conducted.

24.5.1 Zoning Based on Uniformity of Dredger Fill Soil Layer

> Wu pointed out that in the reclamation process, we need to adjust line space in time, measure dredger fill elevation, extend muck mouth timely, adjust the direction of the muck mouth, so that muck surface flatness can be guaranteed to meet the requirements (p. 25).

However, due to various factors, the actual operation may encounter some uniform layer, so the uniformity of soil should be identified and zoned.

Having known soil distribution of every drilling, position of dredger fill layer in the three-dimensional space can be determined by using Matlab (shown in Fig. 24.2). Figure 24.2 shows the thickness of most soil is 2 m while regional area presents thinner layer (approximately 1 m) and thicker soil (approximately 4 m). Soil layer in the area is relatively uniform, so the region is judged as a geological area based on its uniformity.

24.5.2 Zoning Based on Soft Soil Containing Dredger Fill

24.5.2.1 Zoning Principle

Some other soft soil exist expect for dredger fill in the site, so based on the distribution of soft soil, the following zoning principle are established: (1) finding out category of soft soil; (2) judging distribution of soft soil; (3) mainly zoning based on thickness of soft soil.

24.5.2.2 Category and Thickness of Soft Soil (Shown in Table 24.1)

24.5.2.3 Zoning

The zoning result is as followings:

a. Zoning the area with soft soil based on thickness and whether clay layer exists.

① Zoning mucky soil area The mucky soil area is zoned through its thickness and whether there exists clay layer. In one area, there are 2 m thickness mucky soil and no clay layer. In the other area, there are 3.5 m thickness mucky soil and 1 m thickness clay layer at 20 m depth.

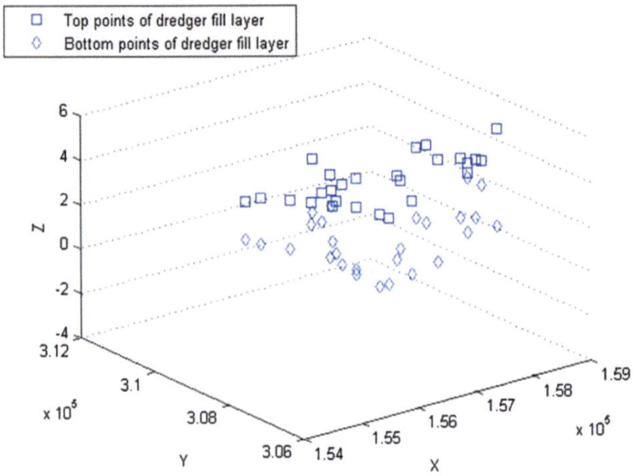

Fig. 24.2 Dredger fill position of each known drilling in the three-dimensional space (□ stands for top coordinates of the dredger fill layer and ◇ stands for bottom coordinates of the dredger fill layer)

② Zoning mucky silty clay area The mucky silty clay area is zoned as two sections based on its thickness. Its thickness is 4 m in one area and 6 m in the other area.

b. Zoning the area without soft soil.

According to whether there exists mucky soil intercalation, it is divided into two areas. 1 m thickness mucky soil intercalation exists in one area, and mucky soil intercalation doesn't exist in the other area. The zoning map is shown in Fig. 24.3.

24.6 Risk Assessments of Potential Engineering-Geology Hazards in the Zoning Areas

24.6.1 Risk Assessment of Engineering-Geology Hazard

Regarding 5 m thickness as standard of risk degree (Ye et al. 2000), risk assessment result of engineering-geology hazard in the site is shown in Table 24.2:

24.6.2 Settlement Verification of Risk Assessment

The hazard assessment is expected to be verified by calculating the settlement of soft soil at unit pressure in the zoning areas. Calculation formula of soil deformation is as follows:

$$S = \frac{p}{E_s} H \tag{24.1}$$

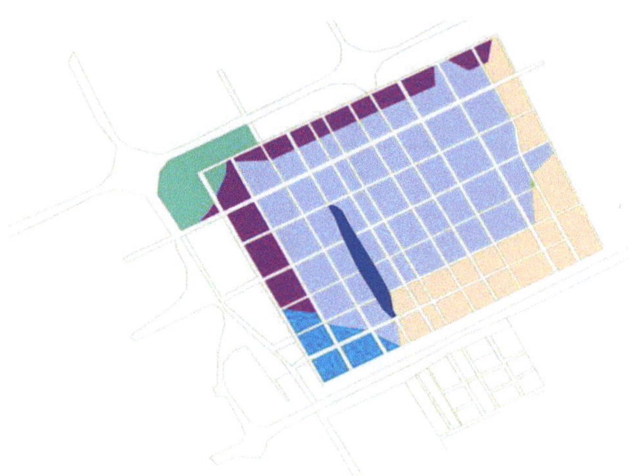

Fig. 24.3 Zoning map

In the formula:

P 1 Kpa;
Es Compression modulus (Kpa);
H Thickness of soil (m).

Soil settlement is determined by averaging compressive deformation of soft soil layers with considering settlement uniformity. The result is shown in Table 24.3.

According to the calculating result, we conclude that: hazard risk assessment based on thickness of soft soil is reliable.

24.7 Conclusion

By drawing engineering-geology profile, zoning is conducted with considering requirement of bearing capacity and settlement. Risk assessment is verified by calculating settlement. Zoning and risk assessment result are as follows:

① Soil layer (containing dredger fill layer) is uniform, and based on soil uniformity the site is zoned a area;

② Distribution and risk assessment of soft soil area is shown below:

Risk of central area with mucky soil intercalation of 1 m thickness is slight; risk of west and north area with 2 m mucky soil and no clay layer is medium; risk of northwest area with 4 m thickness mucky silty soil is a little serious; risk of east and south area with 3.5 m thickness mucky soil

Table 24.2 Risk assessment result of the site

Area	A	B	C	D	E	F
Risk degree	No	Slight	Medium	A little serious	Serious	Very serious

In the table
A–Central area without soft soil
B–Central area with mucky soil intercalation of 1 m thickness
C–West and north area with 2 m mucky soil and no clay layer
D–Northwest area with 4 m thickness mucky silty soil
E–East and south area with 3.5 m thickness mucky soil and clay layer of 1 m thickness at 20 m depth
F–Southwest area with 6 m thickness mucky silty soil

Table 24.3 Settlement result

Area	Existing mucky soil with 2 m thickness		Existing mucky soil with 3.5 m thickness	
Category	Dredger fill	Mucky soil with 2 m thickness	Dredger fill	Mucky soil with 3.5 m thickness
Area	Existing mucky silty soil with 4 m thickness		Existing mucky silty soil with 6 m thickness	
Category	Dredger fill	Mucky silty soil with 4 m thickness	Dredger fill	Mucky silty soil with 6 m thickness
Settlement (mm)	0.569	0.521	0.9776	0.9296
Settlement (mm)	0.449	1.2935	0.5538	1.378

and clay layer of 1 m thickness at 20 m depth is serious; risk of southwest area with 6 m thickness mucky silty soil is serious.

Acknowledgments This work is supported by National Key Technologies R&D Program of China (Grant No. 2012BAJ11B04) and Shanghai Leading Academic Discipline Project. Project Number: B308.

References

Li C, Yang J (2012) Reclamation amount analysis of reclamation land in Tianjin new coastal area. Haihe Water Resour 3:50–57
Peng T, Wei W, Huang S, Hou P (1995) Engineering-geology characteristics of dredger fill. Geotech Invest 5:1–5
Wu G (2010) Research on construction technology and quality control in land reclamation project. Port Technol Port Constr 7:17–25
Ye W, Tang Y-Q, Zhang X, Liu Y (2000) Engineering-geology hazard level zoning in Shanghai underground project. J Tongji Univ 28(6):726–730
Zhang D, Han F, Zhang L (2008) Land reclamation technology applications in Tianjin coastal new area. Shanxi Archit 34(30):115–116
Zhou D, Wu H, Chen L, Zhao W (2012) Engineering-geology research review of "Artificial land". Site Invest Sci Technol 2:1–11

Coastal Floods and Climate Change

25

J. Javier Diez, Efren M. Veiga, and Fernando Rodriguez

Abstract

The Earth has been undergoing a global warming trend since the Little Ice Age being mostly assumed to be progressively increasing along the last and a half centuries. But this secular warming and the previous Little Ice Age cold wave had precedents linked with sea level and other climatic actions. Populations have been driven by climatic conditions, and processes of founding and expanding cities in coastal areas, on the other hand, have undergone great changes over time, under increasing conditions of vulnerability and even hazard. Therefore, the risk of flooding can be attenuated, but never eliminated. And question goes further due to concomitant phenomena, as subsidence and isostasy, which are exacerbated by the transformation of terrain and by settlements. This paper pretends a syncretistic discussion to face the problem in its most realistic way. It reflects on the "cold drop" phenomenon as a whole, on the generation of its rains and on the different natures and consequences of its floods, and compare with other cyclone events. And it approaches the ways in which the maritime weather and the consequent sea level govern floods on the lowest hydrographical basin.

Keywords

Thermal machine • Settlement • Meteocean • Crustal movements • Tsunamis

25.1 Introduction: Coasts at Threat—Climate-Related and Tsunami Risks

Recent climate-related and tsunami catastrophic events in coastal areas have highlighted their increased society exposure and vulnerability; therefore, the necessity of strengthening the hydro-meteorological and tsunamis hazards extreme forecasting and mitigation capability is essential, capitalizing as possible on the knowledge gathered from recent dramatic events, to improve the assessment of risks. We can mainly focus on extreme hydro-meteorological events (e.g. extreme winds, storm surges, high precipitations, coastal floods) which are prone to increase impacts on urbanized coastal areas. They are relevant Climate expressions as the result of the Atmosphere-Ocean Thermal Machine (Diez 2000). But tsunami hazard is acquiring relevance with the expansion of vulnerable coastal developments on lowlands and deep bays, its frequency seems to be lower except along hazardous tectonic borders but it was described even in West Mediterranean Basin (Diez et al. 1982) due to an Alboran-African earthquake. Both kind of episodic events superimpose current eustasic sea level rising and its crustal concomitant phenomena, as general tectonics, isostasy, and subsidence, to provoke floods on coastal zones (Emery and Aubrey 1991).

J. J. Diez (✉) · E. M. Veiga · F. Rodriguez
Universidad Politecnica de Madrid, Madrid, Spain
e-mail: josejavier.diez@upm.es

E. M. Veiga
e-mail: efrenmartin@gmail.com

F. Rodriguez
e-mail: fernando.rodriguezl@upm.es

G. Lollino et al. (eds.), *Engineering Geology for Society and Territory – Volume 4*,
DOI: 10.1007/978-3-319-08660-6_25, © Springer International Publishing Switzerland 2014

Coastal flood is still usually considered as a simple type of flood due to inundation by sea waters. But most of coastal floods are the complex result of the concurrence of several former simple types. Actually, most of cyclone coastal flood events are the result of the concomitant effect of several simple types of floods, flash, fluvial, pluvial and groundwater floods and marine among them. Coastal erosion can besides be the origin of coastal floods through breaking littoral barriers (Diez 2000; Diez et al. 2009).

25.2 The Climate and Its Changes

Climate can be considered the result of the functioning in the hydrosphere and atmosphere of complex thermodynamic "machine" whose temperature gradient is caused by the uneven heating via solar radiation of condensed surfaces (the earth's crust and oceans) on all areas of the planet, depending upon location and natural conditions. Subsequent radiation from these surfaces results in the uneven heating of the atmosphere above them, along with the generation of convective movements and horizontal gradients producing the winds, the displacement of air masses and the atmospheric circulation.

Since the Little Ice Age the planet has been suffering a global warming change leading to sea level rise. In the Non Stationary (cycle-stochastic) evolution of that change, the tendency has likely suffered several changes in intensity being mostly assumed to be progressively increasing since the last quarter of the nineteenth Century. The idea of being from anthropogenic origin may be attributed to Arrhenius (1896), though it is of much later highlight (Brunn 1962, among others) since different scientific fields. Never before, in fact, the human factor had been able of such an influence on climate. Fossil fuels, which allowed men to escape the tyranny of the climate, may be currently destabilizing its evolution combined with the ensuing population qualitative increase. The role of solar activity, the greenhouse effect and biogeochemical activity in the oceans has yet to be satisfactorily discerned.

Current secular warming and the previous Little Ice Age cold wave had precedents, always linked with changes in sea level. That is why the observation and investigation of signs associated with ancient sea level always was a reference indicators source for estimating past climates. The obvious local signs of ancient levels permitted naturalists and geographers to estimate past climates since long time. In fact, the warming period following the Little Ice Age, whether natural or anthropogenic, produced the progressive sea level rise whose value represents the wildly different (even in sign) changes in sea level along the various coasts of the world. This variability is due to the effects of other concurrent factors. First question is to distinguish between average sea levels and sea level variation in different coasts, what requires the correct determination of the average sea level at any coastal point. A second issue is to identify the concomitant factors. As the history of the planet was finely tuned, other types of changes in sea levels became apparent, resulting from vertical movements of the crust, modifications of sea basins due to continents fracturing, drifting and coming together, or to different types of climate patterns (Fairbridge 1983).

25.3 Understanding Extreme Coastal Floods Events

According to their immediate cause, simple flood are usually sorted out into flash, pluvial, fluvial, groundwater and coastal types, considering the last as just caused by sea waters (marine flood would be better denomination). Coastal floods are most of times however the complex results of the concurrence of several former simple types. Actually, most of cyclone coastal flood events are the result of the concomitant effect of several of mentioned single floods within the coastal zone under the boundary condition imposed by the also concomitant stormy higher sea level. Littoral stormy phenomena are additionally the cause of erosive coastal barrier breaking that leads to single marine floods (Diez 2000; Diez et al. 2009). These climate littoral processes can be the origin of other kind of coastal threats when unleashed by marine transportation facilities or coastal settlement infrastructures; their importance here is due to converging on subsidence impact as on induced coastal erosion.

Continental precipitations start the hydrologic cycle, resulting in soil infiltration and surface runoff. The presence of urban development and landscape represent important, modifying factors: filtration is reduced due to the relative impermeability of the urban soil surfaces; the quantity and the speed of runoff consequently increases regardless of the slope because there is less friction; buildings become obstacles that creates damming, leading to localized flooding due to accumulation. Additional interference is caused by river and waste water drainage systems. Littoral water levels widely act, through its effect on basin levels as restrain factor for inland floods. They are not only caused by tides and other geoids' (Mörner 1976) and crustal effects but more frequently by climate: (a) winds and wind-waves set up temporary higher sea levels on coastal areas, and (b) littoral low pressures generate storm surges meaning also temporary higher levels. The sea level, controlling the mouth of the main channel of the basin's drainage system, is ultimately what governs flood levels on wide extensions of the lowest basin. A temporary rise in sea level acts as a dam at the mouth.

The coastal flood related to the rupture of the Tous dam, in the Spanish Mediterranean 20 October 1982, the *Sainte Irene* day (since its name), was a dramatic coastal (and inland) event indeed. It is well-known because of the unleashed trial (Audiencia Provincial de Valencia 1990) and was paradigmatically due to the very frequent "cold drop/bubble" (*gota fría*) phenomenon. It is a residual consequence of extra-tropical cyclone circulation in North Atlantic basin and the circumstances permit its presentation several times each year with variations in time, place and intensity, in a similar way to hurricanes, on Caribbean and western North-Atlantic areas, or to typhoons do (Diez 1992).

25.3.1 Meteocean Climate Versus Inland Rains and Coastal Floods

The referred Spanish Mediterranean "cold drop" can happen every year, and several times in a year, depending on the climate, and consists of an extremely cold and dry bubbled air mass dropped since polar to middle regions. It affects high the whole Troposphere and extends widely on the Peninsula and more, dynamically connecting to a North-Africa or Liguria cyclone warmer and humid air circulation. A situation like that constitutes in fact a very unstable system and leads to an authentic local "thermal machine". Cyclone drags dry and hot Sahara air to charge humidity over worm Mediterranean waters till Iberian littoral range, where it contacts cold drop formation. Both air masses fronts over the littoral mountain rift imposing a nearly vertical ascent to the lower warm, humid mass and leading to sudden and strong condensation, which keeps while gradient initial conditions persist (Diez 1992).

Rainfall produces frequent coastal and inland flash floods, but also generalized pluvial floods on great part of basin, which usually lead to fluvial floods. Major factor on the referred coastal zone flood was the temporary storm high sea level (80 cm at Valencia Port, 40 km northward of Jucar outlet and 150 cm at Denia Port, 30 km southward during several days). Rains and the sea level height are two faces of the same meteoceanic event. The greater strength and dimension of the cold drop the stronger and more persistent the thermal machine and its winds, and the greater and more durable the inland rains and the sea level height in front of the affected basin river outlet.

The stormy high sea level made impossible the river drainage into the sea during the time of permanence of the thermal machine generated by the meteocean phenomenon, even though direct rains cease on the river-mouth area. The flood along the river from the dam was, of course, very affected by the dam collapse of the 50 Hm^3 Reservoir, but the damages caused by the whole 900 Hm^3 flood were more notorious on the whole wide littoral plain.

25.3.2 Coastal Occupation and Settlements

Processes of founding and expanding cities in coastal areas have undergone great changes over time. Populations have been driven by climatic conditions and they have moved in both latitude and altitude with changes in climate. Coastal settlements looked for places above flood levels and away from swamps and other wetlands whenever possible. With time, as populations grew and expanded, cities developed and were extending since the original settlements trying to avoid low and wet lands. But no city has been able to limit its growth, and the systems for protecting against floods have not kept up with the pace of expansion of coastal cities. The problems in all of these areas are not just from Climate Change sea level rise and other associated climatic effects, but also due to subsidence and isostasy, which always affect these types of coasts, and which are exacerbated by the transformation of terrain and by settlements, and by the expansion of their cities into even lower areas than mean sea level, as the Netherlands may widely exemplify.

The increasing occupancy of coastal zones appears to any observer as a natural and logical process in settlement of populations, but it is a process very recent and of very heterogeneous occurrence over the world. It was totally constraint for safety and security reasons for centuries and millennia. Wars and maritime attacks or pillages restricted settlements to high structural coasts with suitable defensive place, and health necessities obliged to leave any sedimentary low coast for avoiding propagation of infection epidemics. Both problems changed their respective way of action along nineteenth and twentieth centuries as a general rule, and more accurately between 1850 and 1950. Cities and urban areas in coastal zones have so lately been spreading widely into areas below the probable average sea level, if not originally founded there.

As additional only apparent coincidence, that is the moment of the dawn of the intercontinental transport with its need of great ships and new ports. They led to the formation of very large urban areas and to coastal land transformation generating new extended processes of subsidence and, in parallel, to appearing of a great amount of large port structures affecting littoral processes. This line of evolution run in parallel with continental works affecting the sediment supplies to the littoral areas. Flood risks on these areas so are not just due to climate change and dynamic effects and associated sea level risings, but on subsidence and isostasy, which are exacerbated by land transformation and settlement evolutions. Cities have spread in every case onto areas below the threshold, if not originally founded there. The current climate change is generating besides an upward trend in average sea level.

25.4 Current Coastal Flooding. Decadal Changes

The flooding of coastal areas is today dramatically attributed to eustatic sea level rise caused by global climate change. This adscription can be somewhat inaccurate, however, even caused by sea waters. The current climate change is generating an upward trend in average sea level, but other regional and local factors result in this trend being accentuated in some places or attenuated, and even reversed, in others, as shown in precedent points.

Observation has widely shown for nearly all last century that the Spanish (Dynamic) Maritime Climate was following around 10–11 year cycles in its most significant figure, wind wave, despite it being better to register cycles of 20–22 years, admitting besides a stochastic presentation of the cycle wave. Those cycles were soon linked to sun activity and to the Solar System evolution. A new perspective may be opened looking at the climate and its changes as the expression of the complex "thermal machine" including fluid-dynamic and hydrological cycles.

A review of past flood events in Valencia (Turia and Júcar basins) has been specifically studied as case for the FP7 SMARTeST Project (2013). The historical record of floods affecting Valencia City or surroundings is highly large. The following years are considered as having registered great major floods: 1321, 1328, 1340, 1358, 1406, 1427, 1475, 1476, 1487, 1517, 1540, 1546, 1555, 1557, 1577, 1581, 1589, 1590, 1610, 1651, 1672, 1731, 1737, 1766, 1770, 1776, 1783, 1793, 1845, 1860, 1864, 1870, 1897, 1949, 1957, 1982 and 2000. Thinking on these floods as due to cold drop phenomena would be coherent though no concrete information is available yet.

25.5 Conclusions

- Coastal floods events show to be the result of the superposition within the coastal zone of flash, fluvial, pluvial and groundwater flood types under boundary condition imposed by the concomitant storm sea level rise.
- Climate change generates coastal flooding but not uniformly, depending on concomitant factors. The coastal settlements are additional relevant factor.
- The role of solar activity, greenhouse gases and biogeochemical activity in the oceans has not yet been well discerned.
- The current climate has particular meaning through its meteocean mechanisms and interferes with climate change itself. The Spanish cold bubble, as hurricanes or typhoons, explains the maritime weather governing floods on the whole lowest hydrographical basins.
- The tectonic factor (tsunamis) independent from climate and its changes is becoming increasingly important for the expansion of settlements.
- The inevitability of these risks leads to a more resilient resistant to these problems, from planning to occupancy protection measures.

References

Arrhenius S (1896) On the influence of carbonic acid in the air upon the temperature of the ground. Philos Mag J Sci 5(41):251

Audiencia Provincial de Valencia (1990). Sentencia en la causa penal 56/1982, por el desmoronamiento de la presa de Tous. 23 de octubre de 1990, Valencia

Bruun P (1962) Sea level rise as a cause of shore erosion. Proc Am Soc Civ Eng 88:117–130 (Waterways and Harbor Division)

Diez JJ, Arenillas M, Esteban V (1982). Shore management en Santa Pola Bay. Proceedings of the IV Congress. A.E.G.I.N., Vol. VII., pp 247–261. Nueva Delhi, 1982

Diez JJ (1992) El fenómeno meteorológico de Santa Irene. R. Obras Públicas 3315:77–81

Diez JJ (2000) A review of some concepts involved in the sea level rise problem. J Coastal Res 16(4):1179–1184 (Palm Beach)

Diez JJ, Esteban MD, Paz R (2009) Cancun-Nizuc coastal barrier. J Coast Res 25(1):57–68

Emery KO, Aubrey DG (1991) Sea levels, land levels, and tide gauges. Springer, New York

Fairbridge R (1983) Isostasy and Eustasy. In: Smith DE et al (ed) Shorelines and isostasy. Academic, London (Institute of British Geographer)

Mörner NA (1976) Eustasy and geoid changes. J Geol 84(2):123–151

SMARTesT FP7 Project (2013) http://www.floodresilience.eu/ (Retrieved 8 Julie 2013)

Part II

Coasts at Threat: Causes and Consequences of Coastal Settlements and Maritime Transportation

The increasing occupancy of coastal zones appears to any observer as a natural and logical process in settlement of populations. Wars and maritime attacks or pillages changed their respective way of action along nineteenth and twentieth centuries as a general rule, and more accurately between 1850 and 1950, at the same time that both the Industrial Era begins and the Oceanography was born, and, as additional only apparent coincidence, the dawn of the intercontinental transport with its need of great new ports. They led to the formation of very large urban areas and to the destruction of coastal zones. Welcome researches and reflections on the impacts of new and all settlements and ports on beach evolution.

17 Years of Changes (1989–2006) in Bathymetry and Sediments Texture in Segura River Mouth

26

Aragonés Luis, Pagán J. Ignacio, López Pilar, and J. C. Serra Peris

Abstract

Segura river (Guardamar, Spain) has always been a source of great floods that has cost considerable damage. In order to avoid this phenomenon, the channeling of the river basin was performed. In this study bathymetry and sedimentological data collected before and after the channeling have been analyzed in order to see the possible modifications that this construction might have caused in sediment distribution, bathymetry and plant species in the area near the river mouth. A data digitization process was carried out in order to create sediment distribution maps from both periods of time. Results from 1989 and 2006 have been analyzed separately and compared between them. It has been observed a high reduction of the sediments with diameter lower than 0.0039 mm, what has led to a depth increase of 1 m. Besides, a reflection area has appeared as a consequence of the generation of a regression zone next to the river mouth.

Keywords

Segura-river • Sediments • Grain-size • Bathymetry

26.1 Introduction

The knowledge of erosion, transport and sedimentation processes of sediments, as well as its spatial and temporal distribution, is essential to know the morphological changes that are taking place in our costs. That information will allow us to predict natural changes expected and so to act in order to avoid them when needed. In this research, the surroundings of the Spanish Segura river are the center of this investigation. This river, placed at the east cost of Spain, flows into the Mediterranean Sea. Throughout history, several floods of this river have occurred and they have caused material damage and even loss of human lives. As a consequence, a plan for the protection against floods in Segura basin was carried out. Construction works finished in 1994, five years after the last big flood in 1989. No considerable floods have been recorded ever since. In this work it has been studied the likely changes in bathymetry, grain size sediment distribution and plant species surface caused by the channeling along a Mediterranean area near the river mouth. Sediment samples, bathymetry data and plant species surface data have been collected during two periods of time, just after the last flood (1989) and over 10 years after the channeling of the river (2006).

A. Luis (✉) · P. J. Ignacio · L. Pilar
AlicanteUniversity, Carretera San Vicente del Raspeig s/n, 03690 San Vicente del Raspeig, Alciante, Spain
e-mail: mp.lopez@ua.es

J. C. Serra Peris
Valencia Polytechic University, Camino de Vera s/n, 46022 Valencia, Spain

26.1.1 Area of Study

This research is focused on an area around the Segura river mouth, which is placed in the province of Alicante (Spain). This area has a longitudinal length of approximately 35 km

G. Lollino et al. (eds.), *Engineering Geology for Society and Territory – Volume 4*,
DOI: 10.1007/978-3-319-08660-6_26, © Springer International Publishing Switzerland 2014

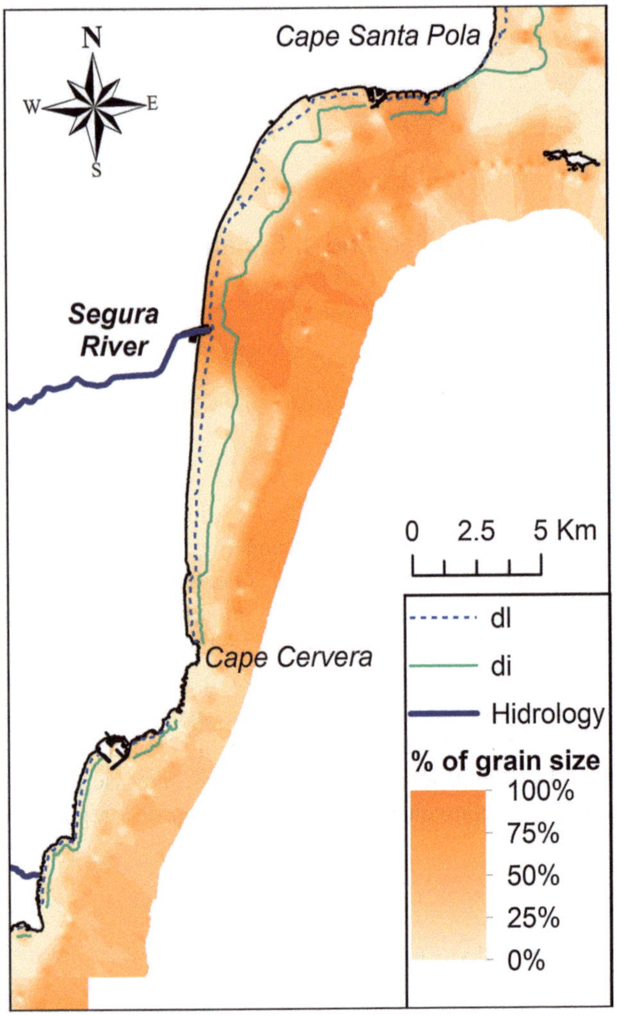

Fig. 26.1 Percentage distribution of 0–0.0039 mm sediments (1989)

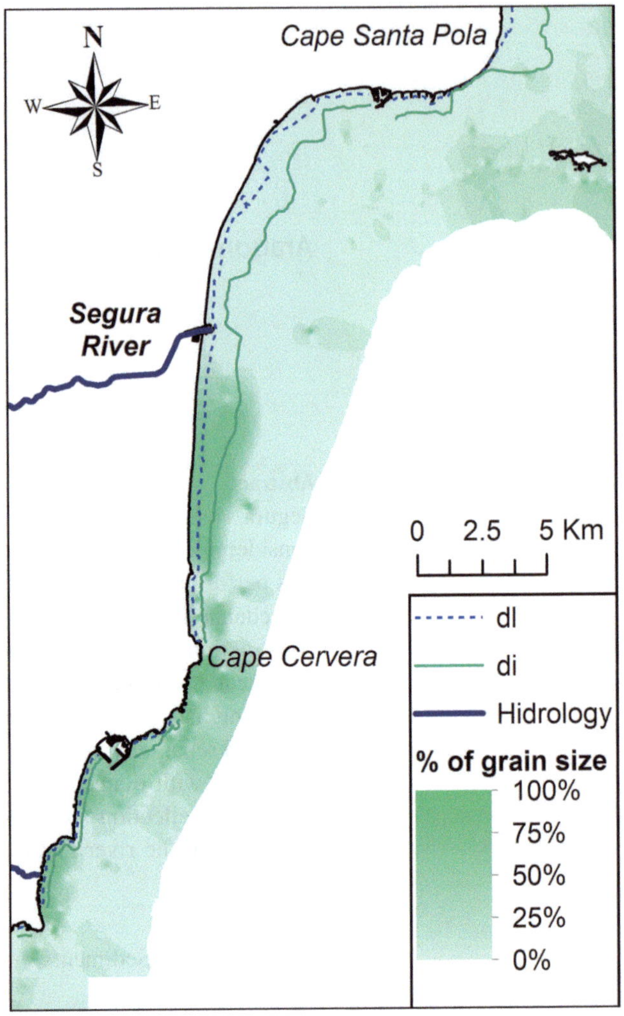

Fig. 26.2 Percentage distribution of 0.0039–0.063 mm sediments (1989)

and it goes from Cape Santa Pola (at north) to Cape Cervera (south). In this area, the East-Northeast swell is the most energetic one, providing a significant wave height of 3.3 m. Besides, the most frequent swell comes from the East and it has a significant wave height of 2.8 m.

26.2 Methodology

Data was obtained from two different studies carried out by the Spanish Directorate-General for Coasts. The first one took place in 1989, just after the last flood registered, with the purpose of finding sand for beach nourishments. The second one was developed in 2006. Its objective was to have a general knowledge of the Mediterranean coast state. As all the information was collected on paper format, a digitization process was performed. After that, all the information was uploaded in a computer by using a GIS software. Finally, the analysis of all the information was carried out, including the likely changes occurred between 1989 and 2006 (Molinaroli et al. 2009). Previous works (Diez González et al. 1982) (Chapapría 1987 Doctoral Thesis), in this area have studied backshore sediment characteristics as well as the coastal dynamics. However, this study is focused on the sediments changes inside and outside the depth of closure.

26.3 Results and Discussion

26.3.1 Results of 1989 Campaign

Based on the granulometric classification data available (ASTM sieve), a spatial distribution of grain size was done. Particularly, for each grain size, a map representation of that grain size percentage in every sample was done. As it can be

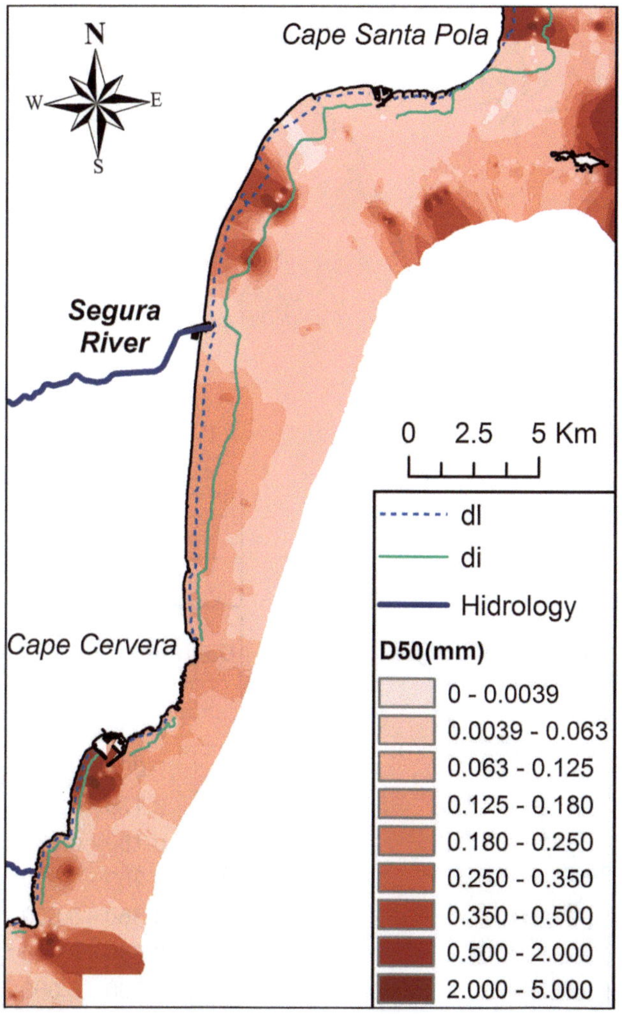

Fig. 26.3 Percentage distribution of 0.063–0.125 mm sediments (1989)

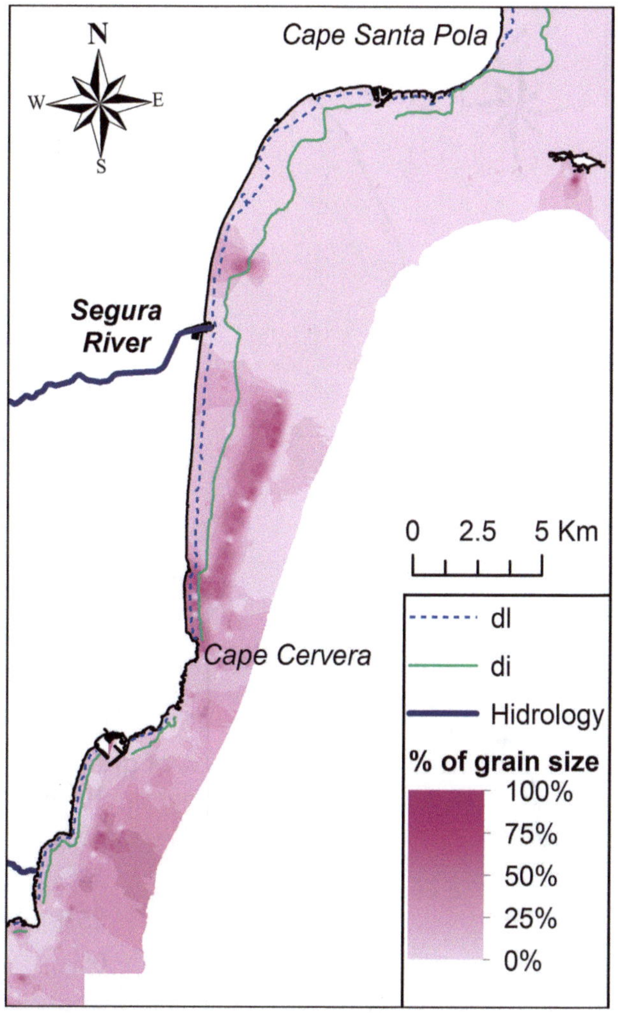

Fig. 26.4 Representation of D50 values in mm (1989)

observed in Fig. 26.1, the percentage of silts (0–0.0039 mm) in 1989 was considerable. Samples with percentage of silt above 60 % delimit the area affected by the river flow. Figures 26.2 and 26.3 show how percentage of sediments with 0.0039–0.063 mm and 0.063–0.125 mm diameter remain almost constant inside the depth of closure (di/dl). Thus, there is a transversal transport in which fines leave the di limit and move into offshore. Figure 26.4 represents the D_{50} parameter, that is the average value of the particle size distribution of each sample. This average diameter of the samples is directly influenced by the percentage of silts.

Approximately, 70 % of Alicante coastline beaches have posidonia (Jordá Guijarro et al. 2013). However, in the area of study, this percentage is reduced to 56 %. Besides, no beach with posidonia can be found from the river mouth to Cape Cervera. The transport of silts that come from the river might have reduced considerably the quality and transparency of water in this area, what would have negatively affected the growth of this species.

26.3.2 Results of 2006 Campaign

As for the sediments results, the percentage of 0–0.0039 mm sediments is considerably reduced to values under 10 % (Fig. 26.5). This is due to the lack of the great sediment supply that used to come after river floods. This fact is reflected in the increase of the D_{50} parameter value (Fig. 26.6). Grain size and D_{50} distribution analysis show an area, right under the river mouth (Fig. 26.6), where the average sediment size is smaller, as we come closer to the cost, and higher as we move away from it. Thus, in this area, higher sediments are being removed from the coast to offshore. This phenomenon wasn't observed in 1989. When analyzing and comparing the orthophotos of this area from 1989 and 2006, it can be noticed that there is a regression zone (Fig. 26.7) next to the river mouth and an accretion zone (Fig. 26.8) at the south. This shows the presence of a longitudinal transport from north to south. The loss is so high

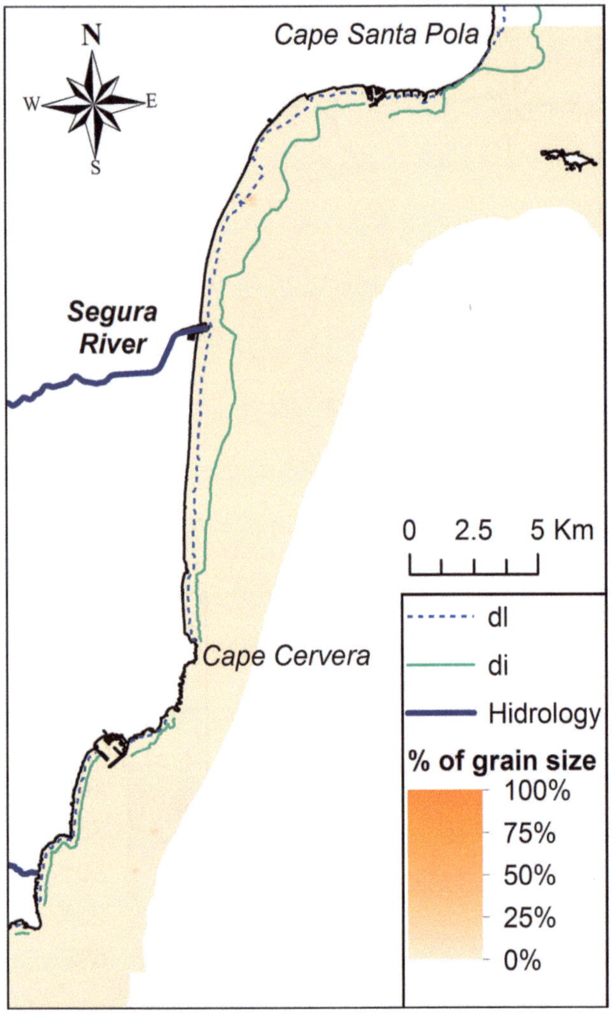

Fig. 26.5 Percentage distribution of 0–0.0039 mm sediments (2006)

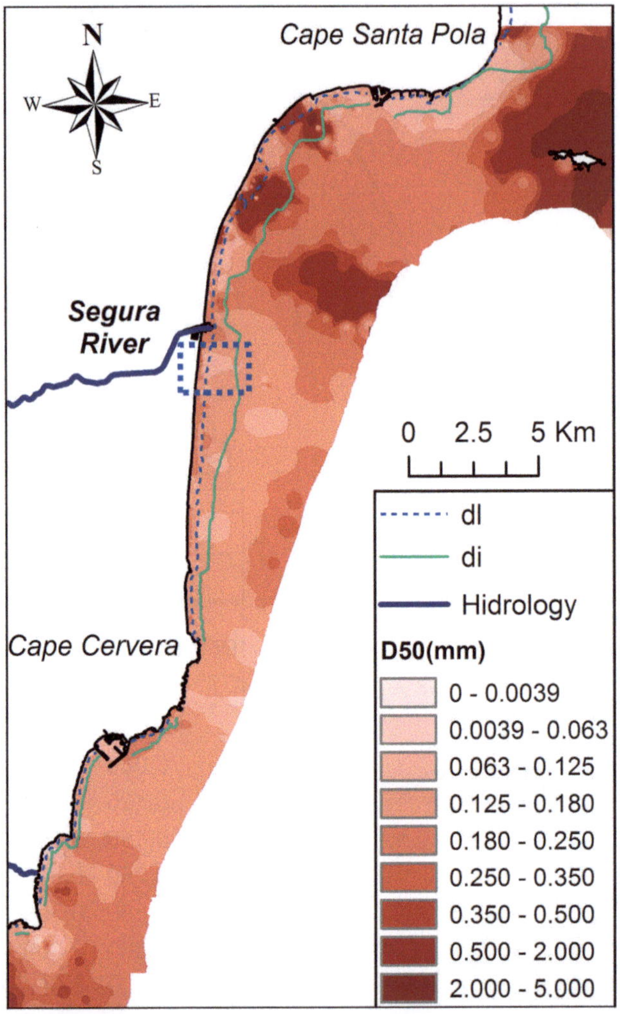

Fig. 26.6 Representation of D50 values in mm (2006)

that water even gets to the housing what leads in the generation of a reflection area. This phenomenon tends to extract the sediments from the coast.

26.3.3 Comparison Between 1989 and 2006 results

The bathymetry analysis shows that depth in the study area has increased 1 m from 1989 to 2006. The percentage of 0–0.0039 mm diameter sediments has decreased to values beyond 10 %. Thus, the increased of the depth could be due to the absence of that layer of sediments that used to cover the bottom. Finally, the lack of silt supply from the river after the channeling, it is also reflected in the growth of the Caulerpa Prolifera specie (Figs. 26.9 and 26.10) between 1989 and 2006. There is not a considerable growth of Posidonia because the growth rate of this species is very slow. The difference of Posidonia surface between

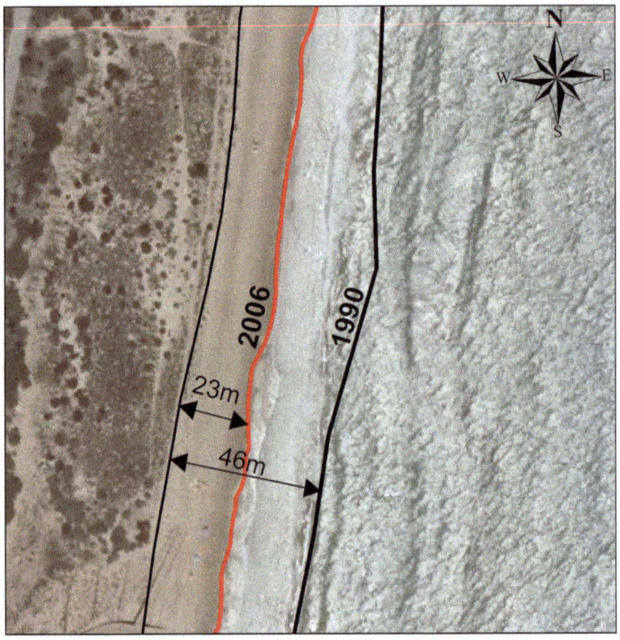

Fig. 26.7 Regression area

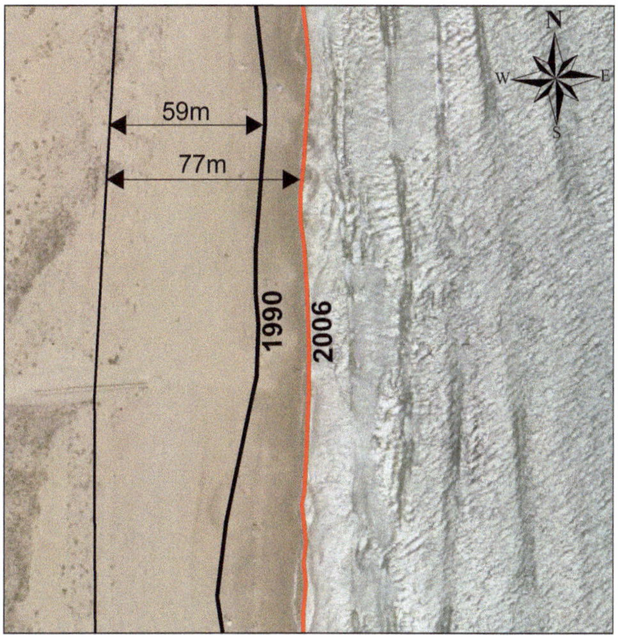

Fig. 26.8 Accretion area

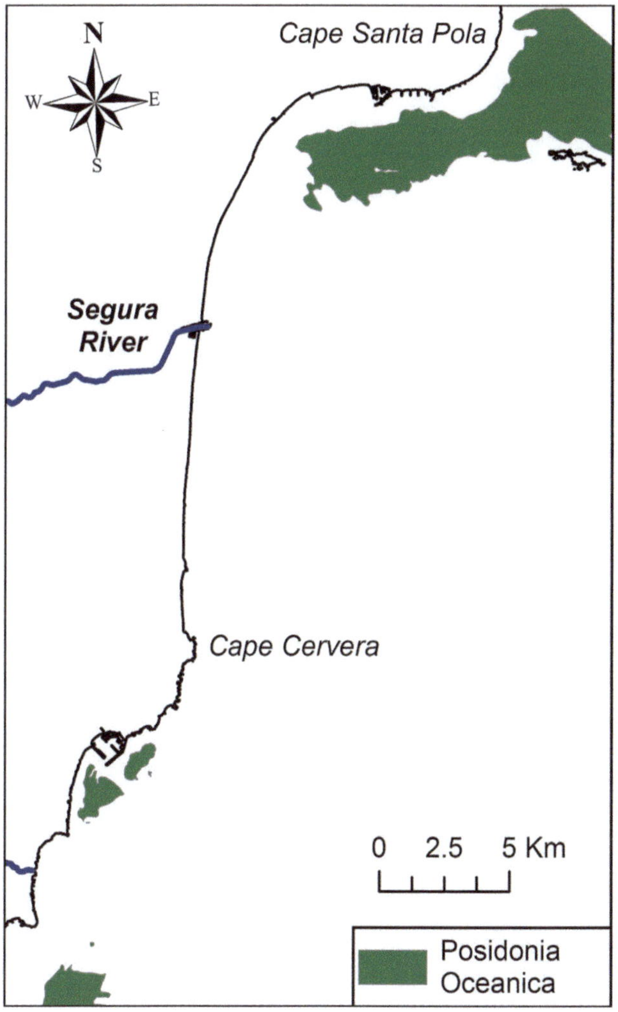

Fig. 26.9 Plant species map in 1989

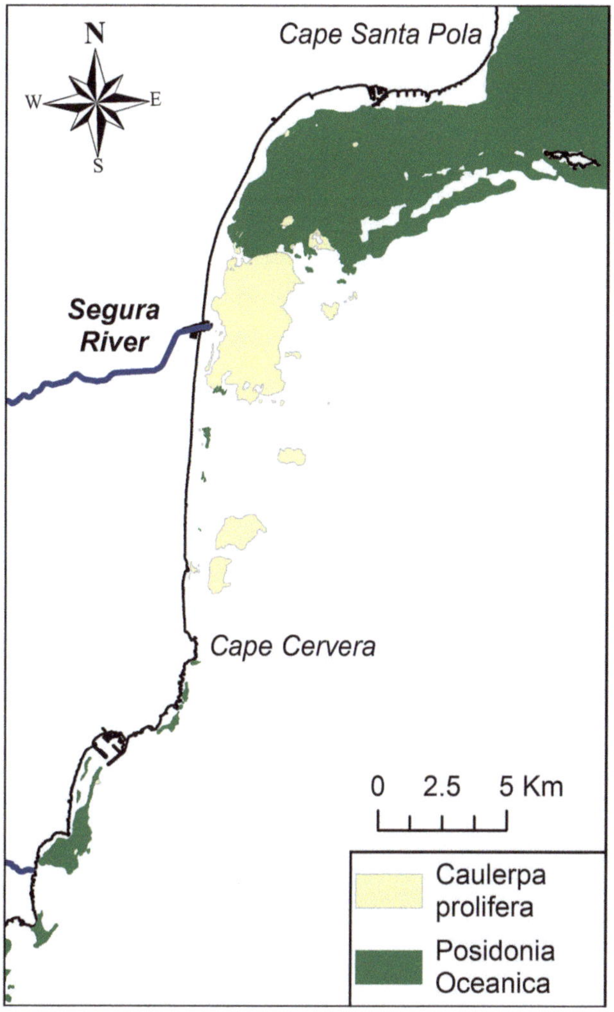

Fig. 26.10 Plant species map in 2006

1989 and 2006 is probably due to the fact that in 1989 part of that area was covered by a layer of silts and so it couldn't be detected.

26.4 Conclusions

The channeling of the Segura river has reduced considerably the amount of silts in the studied area what has led to the increase of the D_{50} sediment values in this zone.

It has been observed a longitudinal transport of sediments from north to south that has caused a regression and accretion areas. Besides, in the regression zone, the water has even reached the houses near the coast and so a reflection zone has appeared.

Finally, the Caulerpa Prolifera species has grown from 1989 to 2006 due to the improvement of the water quality due to the silts lack.

References

Diez González JJ, Arenillas Parra M, Serra Peris J (1982) Shore protection in Almazora coasts (Castellón). In: 4th International congress. Association of engineering geology, Nueva Delhi, India

Molinaroli E, Guerzoni S, Sarretta A, Masiol M, Postolato M (2009) Thirty-year changes (1970 to 2000) in bathymetry and sediment texture recorded in the lagoon of Venice sub-basins, Italy. Mar Geol 258:115–125

Jordá Guijarro MA, Aragonés Pomares L, Serra Peris JC, Así Berenguer A (2013) Clasificación de los estados morfodinámicos de las playas de la costa blanca. XII Jornadas españolas de ingeniería de costas y puertos. Ed:Autoridad Portuaria de Cartagena. Cartagena, Spain

Part III

Monitoring and Measurement of Seabed Dynamic Process

The seabed dynamic processes to wave and current affect sediments situation and stability on the ocean floor such as sediments consolidation, porewater pressure accumulation, liquefaction, mass movement, erosion, sediment relifted up, transportation, and depositation. The effective monitoring and measurement to seabed process play a central role to understand the marine engineering dynamic geological processes. In this section, contributions regarding to innovative instruments, methods of monitoring and measurement for seabed dynamic process will be collected, with particular pattention to the case studies related to in-situ seabed dynamic process observation under extreme climate conditions.

In-Situ Fluid Mud Motion Investigation in Channel After Storm

27

Pang Qixiu and Zhang Ruibo

Abstract

The motion of fluid mud is extremely complex, especially at the early stage after storm. To study the movement of storm-induced fluid mud, the vertical distribution of mud density was measured with the densitometer at several stations in Lianyun channel and Xuwei channel of Lianyungang Harbor, Jiangsu Province, China, after the Typhoon Damrey, and this measure was repeated every several days. With those data, some characteristics of fluid mud motion were analyzed, including the time variation of vertical density distribution of fluid mud, and the vertical average density value, and the thickness of fluid mud at each station. Further, the storm-induced fluid mud both in two waterways disappears after 9 days later since the storm landed. The main manner of motion was consolidation, and only about 20–30 % fluid mud was eroded, except two special segments where mainly due to bed slopes and hydrodynamic forces. The fluid mud consolidates fast in the early stages after forming, and then slows down gradually. The investigation of fluid mud motion was recommended to be arranged on the 1st, 2nd, 3rd, 5th, 7th, and 9th day. The time arrangement and methods might be references for subsequent field observation and study on fluid mud movement.

Keywords

Fluid mud • Motion • In situ investigation • Storm

27.1 Introduction

Fluid mud now appears to be a common, perhaps even ubiquitous, feature of water bodies laden with fine-grained sediment (William et al. 2007). In China, there are also many muddy harbors where exist fluid mud effecting the navigation safety, such as Tianjin Harbor, Lianyungang Harbor, Guangzhou Harbor, Shenzhen Harbor, Zhuhai harbor, etc. (Pang and Yang 2010). Fluid mud is a special existing status of sediment, typically with the density between 1,030 and 1,250 kg/m^3. The storm is usually the main hydrodynamic factor in formation of fluid mud, although there are other cases, for example, in the lower North Passage of the Yangtze Estuary, the fluid mud phenomena may fall into three categories: slack water, storm and salt wedge (Li et al. 2004). There are a lot of studies focusing on the transport of fluid mud (e.g., Traykovski et al. 2002; Puig et al. 2004; Hsu et al. 2007; Sequeiros et al. 2009; Cantelli 2009), which shows that the fluid mud can flow down bottom slopes as a density current or horizontally as streaming under current or wave forcing, and then it converges in the harbor basin or waterway. But after that, how long can the fluid mud exist in the waterway? And what way does the fluid mud move? Does it suspend or flow away under the tidal current and wave, or only consolidate

P. Qixiu (✉) · Z. Ruibo
Key Laboratory of Engineering Sediment of Ministry of Transport, Tianjin Research Institute for Water Transport Engineering, Tianjin 300456, China
e-mail: pangqixiu@163.com

G. Lollino et al. (eds.), *Engineering Geology for Society and Territory – Volume 4*,
DOI: 10.1007/978-3-319-08660-6_27, © Springer International Publishing Switzerland 2014

under the gravity? However, there are little reports on these questions, because of the shortage of field survey data, which needs to be surveyed repeatedly many times in field after fluid mud forming. In this paper, the time variation of thickness and density of fluid mud, and the hydrodynamic conditions are analyzed to determine the motion mode, based on the field survey performed in Lianyun channel and Xuwei channel of Lian yungang harbor, Jiangsu Province, China, after Typhoon Damrey.

27.2 General Information About the Survey

27.2.1 Location

Port of Lian Yungang is in the Northeast coast of Jiangsu Province, locating at latitude 34°44′ and longitude 119°27′. It contains three districts, i. e. Lianyun District, Xuwei District, and Ganyu District. Lianyun District is the main harbor area, which channel level has been promoted gradually in the past decades, and now it is 2,50,000 dwt and the designed water depth is 20 m. The Xuwei District is on the south side of Lianyun District, as illustrated in Fig. 27.1. Now Xuwei waterway has been dredged to be approximately 10 m, and the work of dredging is still ongoing.

Lian Yungang is a typical muddy coast, which average particle size of the sediment (D_{50}) is approximately 0.006 mm.

27.2.2 Typhoon Damrey

Tropical storm Damrey was strengthened to be a typhoon near Japan on August 1, 2012. After that, it moved quickly to China East Sea, and landed at Xiangshui County about 20 km from the area where the investigation was carried out at 22:00 on August 2. According to the data recorded by the Xuwei Marine Station marked in Fig. 27.1, the largest wind speed was 40.5 m/s, and the gales with a speed of larger than 17.2 m/s lasted about 10 h. At the Xuwei Marine Station, there were two instruments named as No. 1 and No. 2 to survey the wave height and period. Based on the wave data, the time variation of efficient wave height $H_{1/3}$ was drawn in Fig. 27.2. It shows that the largest $H_{1/3}$ was 3.86 m, and the waves with $H_{1/3}$ larger than 2.0 m lasted 9 h.

27.2.3 Methods and Apparatus

27.2.3.1 Time Arrangement

Since the thickness and density of fluid mud change with time, it is necessary to survey repeatedly many times. But after the storm, especially in the first few days, the wave is still too strong to carry out the work. And hence, the in situ survey of fluid mud density was conducted on August 5, i.e. 3 days after the typhoon landing, and then the work was repeated on August 7 and 11.

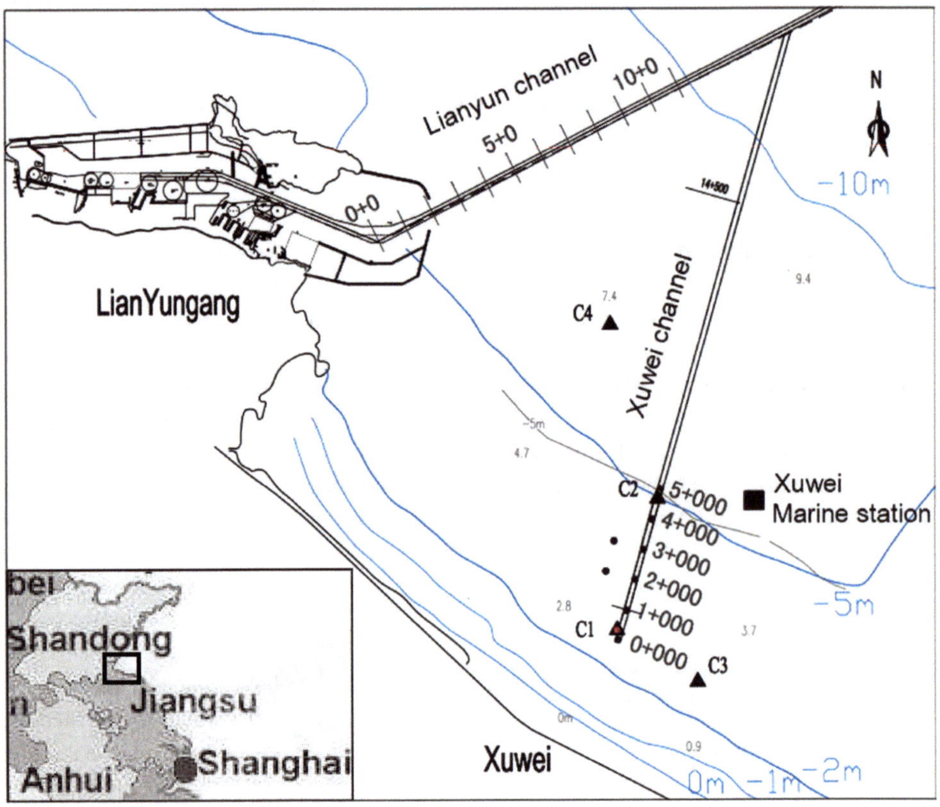

Fig. 27.1 Bathymetric map of Lian Yungang sea showing the location of channel and points deployed for the fluid mud survey

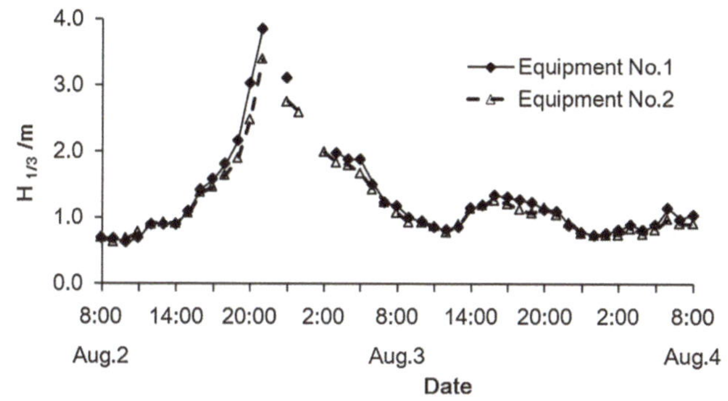

Fig. 27.2 Time variation of efficient wave height $H_{1/3}$ surveyed by the equipment Nos. 1 and 2 at the Xuwei marine station

27.2.3.2 Survey Positions

The survey points in the Lianyun channel were mainly deployed at the integral channel mileage along the central axis, which were marked in the Fig. 27.1 as 0 + 0, 1 + 0, 2 + 0 etc. However, in Xuwei channel, we focused mainly on the inshore segment, namely from the 0 + 000 to the 5 + 000. And the points were also set in the central axis of the channel, and the distance between points was still 1 km.

27.2.3.3 Apparatus

The vertical densities of mud were measured with the density meter named as Rheotune produced by Stem Company in Netherland, as shown in Fig. 27.3. The density range of the instrument is 1,000–1,800 kg/m^3, and the measurement accuracy of density is 1 % of the measured value. The measure tilt angle can be determined by two tilt sensors. In this survey work, the density meter was calibrated with 11 samples compound with the sediment collected in the channel, which density values were between 1,032 and 1,461 kg/m^3. In addition, the temperature module and depth module of the density meter were calibrated respectively using one-point method and two-point method.

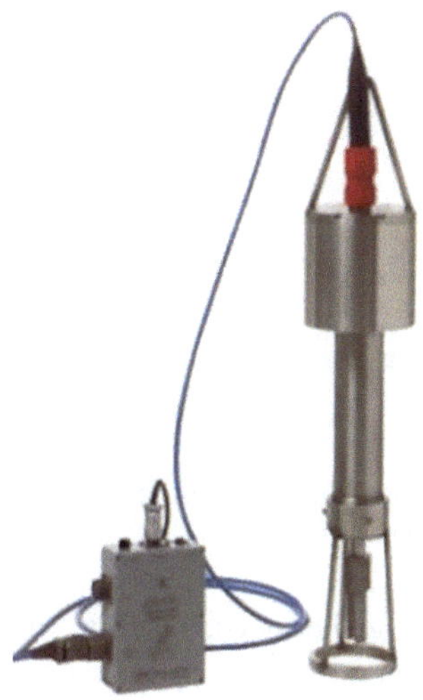

Fig. 27.3 Rheotune density meter

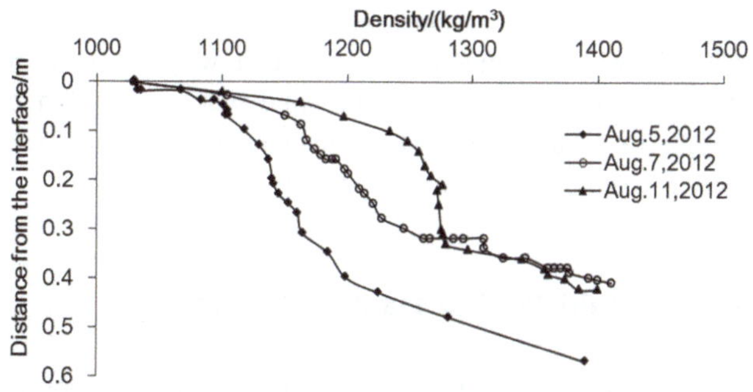

Fig. 27.4 Vertical distributions of mud density surveyed at the point 4 + 000 of Xuwei channel on different dates

27.3 Variation of Thickness and Density of Fluid Mud

27.3.1 Fluid Mud in Xuwei Channel

When the Rhoetune density meter submerged in the sea water, the values of density were about 1,026 kg/m^3, but the density increased quickly so long as the tuning forks touched mud. We describe the interface of mud and water as the layer which density is 1,030 kg/m^3. The densities of mud at different layers can be obtained from the surveyed data. The relationship curves of the relative distance from the interface and the density surveyed at the point 4 + 000 of Xuwei channel on different days were drawn in Fig. 27.4. It shows that the vertical density distribution was not uniform, and the densities increased with the depth increasing. Meantime, the thickness of mud layer with density between 1,030 and 1,400 kg/m^3 decreased with time, especially the one in the upper layer with lower density.

By referring to other research results, we define the mud which density values are between 1,025 and 1,030 kg/m^3 as fluid mud. Then the thickness of fluid mud at each site on different dates can be calculated, as illustrated in Fig. 27.5. It presents that a lot of fluid mud existed in the channel after storm, especially in the segment between 2 + 000 and 4 + 000, which thicknesses were larger than 0.3 m, and the largest thickness was approximately 0.5 m. However, there was not fluid mud at the point 5 + 000. The bathymetry map surveyed with the high frequency echo sounder before the storm indicates that a steep slope about 1:100 exists in the segment 4 + 800 ~ 5 + 100, and the outer channel depth is 13.0 m, meaning that the outer is about 3 m deeper than the inner. Therefore, the fluid mud pours into the

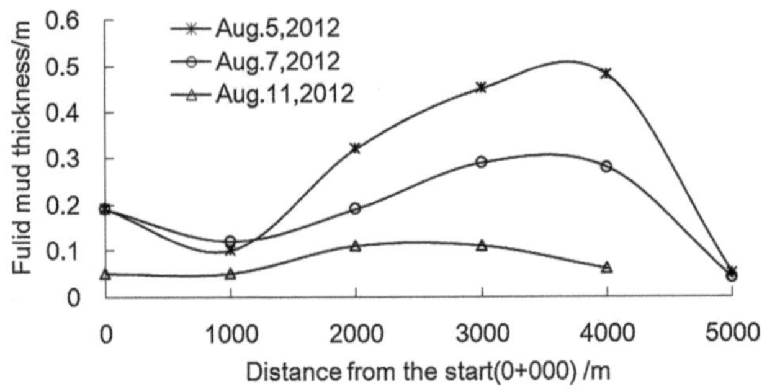

Fig. 27.5 Fluid mud thickness distribution along Xuwei channel surveyed on different dates

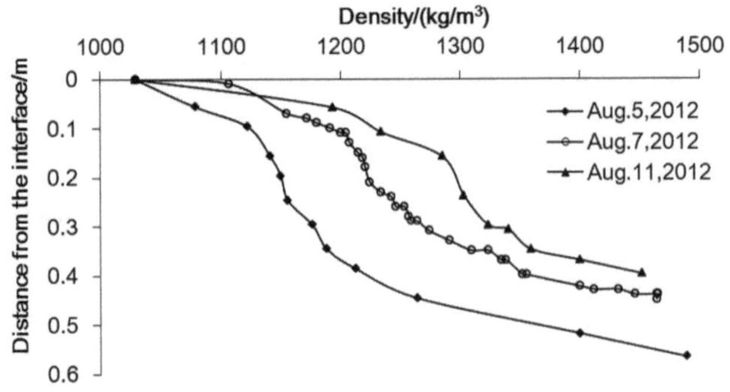

Fig. 27.6 Vertical distributions of mud density surveyed at the point 5 + 0 of Lianyun channel on different dates

deeper area along the slope, resulting that no fluid mud exists at the point 5 + 000.

The thickness of fluid mud decreases with time. On August 7, i.e. 5 days after the storm, the thickness of fluid mud was smaller obviously. From Fig. 27.5, it indicates that the fluid mud in the channel segment between 2 + 000 and 4 + 000 was still much thicker, and it was about 0.3 m. However, the thickness of fluid mud was only 0.1 m on August 11, meaning that the storm-induced fluid mud in the Xuwei channel vanished 9 days later.

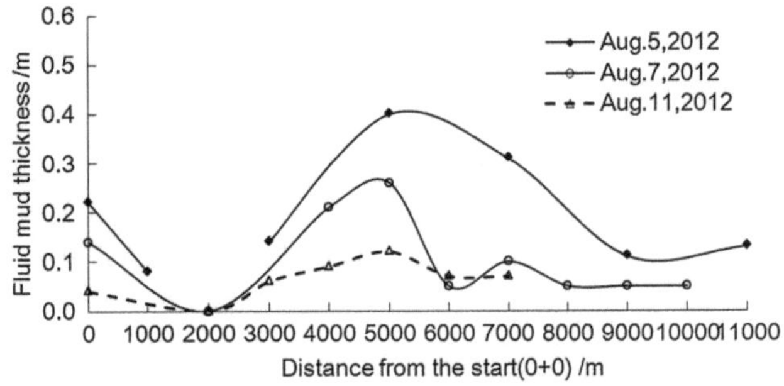

Fig. 27.7 Fluid mud thickness distribution along Lianyun channel surveyed at different dates

27.3.2 Fluid Mud in Lianyun Channel

Based on the density data surveyed in Lianyun channel, the relationship curves of the relative distance from the interface and the density were drawn. For example, Fig. 27.6 is the one at the point 5 + 0 of Lianyun channel. Similarly, the vertical distribution of mud density was not uniform, and the densities increased with the depth increasing.

The thickness of fluid mud at each site on different dates was calculated, as illustrated in Fig. 27.7. The fluid mud was thicker in the segment between 4 + 0 and 8 + 0, which thicknesses were larger than 0.3 m, and the largest thickness was approximately 0.4 m. But there was not fluid mud in the 2 + 0 point at all, where is exactly the port entrance, as shown in Fig. 27.1. The width of the entrance is only 1,000 m, and it constrains the current flow to be very strong, which maximum vertical average velocity is about 1.0 m/s. And hence the fluid mud could not stay here. In addition, the thickness of fluid mud in the Lianyun channel decreased with time, too.

27.4 Motion of Fluid Mud

Fluid mud moves mainly due to bed slopes and hydrostatic pressure gradients caused by the slope of the mud–water interface or water surface (Whitehouse et al. 2000). In the section 4 + 800 ~ 5 + 100 of Xuwei channel, the fluid mud moves to the deeper bottom of the slope, and there was not fluid mud in this segment. But in the port entrance of Lianyun channel, the fluid mud moves mainly under the action of hydrodynamic forces. The sediment is difficult to deposit in the conditions of strong current flow, and even in the slack periods, the fluid mud forms, and it will be driven away by the strong current flow in the next flooding or ebb periods. And hence, the two main types of motion (slope, flow shear) happened in this observation.

Except these two special sections, the fluid mud both in Xuwei channel and Lianyun channel moves mainly under gravity, namely consolidation. After the storm, Shanghai Waterway Engineering Design and Consulting Co Ltd surveyed the current velocity at four fixed stations illustrated in Fig. 27.1 on August 5 and 6. The C_2 station was deployed in the central axis of Xuwei channel. And the current velocity in the channel was about 0.3 m/s in the flooding or ebb period. And the wave height recorded by the Xuwei marine station during the measure was smaller than 0.6 m. The total maximal shear-stress was less than 0.16 Pa in most time, which was calculated with the formulas (Whitehouse et al. 2000) considering wave-current coupling effect. And hence, the hydrodynamic condition was very weak, and consolidation under the gravity would be the main motion manner. And it's the same in the Lianyun channel.

The average density values of the siltation layer which densities are between 1,030 and 1,400 kg/m^3 were calculated using the weighted average method, according to the curve of vertical density distribution, Figs. 27.4 and 27.6 as an example. The results show that vertical average density of each station was between 1,162 and 1,185 kg/m^3 on third day after Typhoon, and it was 1,214 $\sim$ 1,228 kg/m^3 on 5th day, and 1,245 $\sim$ 1,273 kg/m^3 on the 9th day. Further, the vertical total amounts of sediment per unit area were calculated for August 5 and 7, and then, the decreasing amounts in the 2 days were calculated to be about 20 $\sim$ 30 %. If the fluid mud only consolidates under gravity, the total mass of sediment will keep constant. Therefore, the sediment decreased was mainly caused by erosion under flow and wave.

27.5 Conclusions

The investigation on the ninth day since the typhoon Damrey landed shown that the thickness of fluid mud was only approximately 0.1 m, meaning that the fluid mud forming during the storm could exist less than 9 days, both in Lianyun channel and Xuwei channel. Most of the decreasing amounts were caused by the consolidation of fluid mud under the gravity, and only 20 $\sim$ 30 % of sediment was suspended or flowed away by the tidal current and wave. In addition, there also were two special segments where fluid mud disappeared due to bed slopes and hydrodynamic forces.

The fluid mud consolidates fast in the early stages after forming, and then slows down gradually. Therefore, the measures should be carried out immediately as possible after the storm, and the measure frequency can be reduced a few days later, until the thickness of sludge keeps nearly steady. We recommend that the investigation is arranged on the 1st, 2nd, 3rd, 5th, 7th, and 9th day.

To analyze better the fluid mud motion, it is necessary to survey the density distribution at fixed stations using the density meter, as well as the water depth with the high and low frequency sounder along the fixed sections in the waterway.

Acknowledgments This paper is funded by *the National High Technology Research and Development Program("863"Program) of China (2012AA112509).*

References

Cantelli A (2009) Uniform flow of modified Bingham fluids in narrow cross sections. J Hydraul Eng 135(8):640–650

Hsu T-J, Traykovski PA, Kineke GC (2007) On modeling boundary layer and gravity-driven fluid mud transport. J Geophys Res 112:C04011. doi:10.1029/2006JC003719

Li Jiufa WAN, Xinning He Qing, Ming Ying, Lianqiang Shi, Hutchinson SM (2004) In situ observation of fluid mud in the north passage of Yangtze Estuary China. China Ocean Eng 18(1):149–156

Puig P, Ogston AS, Mullenbach BL, Nittrouer CA, Parsons JD, Sternberg RW (2004) Storm-inducedsediment gravity flows at the head of the Eel submarine canyon, northern California margin. J Geophys Res 109:C03019. doi:10.1029/2003JC001918

Qixiu P, Hua Y (2010) Research and application of technique of nautical depth in China. PIANC MMX Congress, Liverpool UK

Sequeiros OE, Naruse H, Endo N et al (2009) Experimental study on self-accelerating turbidity currents. J Geophys Res 114:C05025. doi:10.1029/2008JC005149

Traykovski P, Geyer WR, Irish JD, Lynch JF (2002) The role of wave-induced density-driven fluid mud flows for cross-shelf transport on the Eel River continental shelf. Cont Shelf Res 20:2113–2140

Whitehouse R, Soulsby R, Roberts W, Mitchener H (2000) Dynamics of estuarine muds a manual for practical applications. Thomas Telford Publishing, London, pp 63–74

William HM, Asce F, Friebrichs C, Hamiltou G et al (2007) Management of fluid mud in estuaries, bays, and lakes.I: present state of understanding on character and behavior. J Hydraul Eng 133(1):1–9

Nuclear Techniques for Monitoring Sediment Dynamics in the Coastal Zone

28

Jefferson Vianna Bandeira, Lécio Hannas Salim, Virgilio Lopardi Bomtempo, Rubens Martins Moreira, Patrick Brisset, Catherine E. Hughes, Harish Jagat Pant, Jovan Thereska, and Anders Wörman

Abstract

The knowledge of sediment transport in the coastal region is of vital importance to the management of this critical interface between land and sea, where most of the world population lives. For the assessment of sediment behaviour, hydraulic and sediment measurements, bathymetric survey, mathematical and physical models, are applied. Sediment tracers provide a unique capability for understanding sediment transport assisting in sediment management which cannot be obtained any other way, whether conventional monitoring or physical and numerical models because tracers integrate all the hydrodynamic actions in time and space. Computational fluid dynamics (CFD) is now an essential tool for the management of the natural systems and is increasingly used to study the fate and behaviour of particulates and contaminants. Tracer techniques are often employed to validate hydrodynamic models to enhance confidence in the predictive value of the models. In-situ detection afforded by the use of a radioactive tracer allows accurate quantitative determination of the rate of sediment transport, compared with other tracer methods. Also, it is cost effective and safe: its radiological impact to the environment is minimal. Several case studies, performed in different regions of the world, are presented.

Keywords

Coastal region • Sediment dynamics • Radiotracers • CFD modelling • Dredging

J. V. Bandeira (✉) · L. H. Salim · V. L. Bomtempo · R. M. Moreira
Environmental Department, Brazilian Nuclear Energy Commission (CNEN/CDTN), Belo Horizonte, Brazil
e-mail: jvb@cdtn.br

L. H. Salim
e-mail: salimlh@cdtn.br

V. L. Bomtempo
e-mail: vlb@cdtn.br

R. M. Moreira
e-mail: rubens@cdtn.br

P. Brisset
Department of Nuclear Sciences and Applications, International Atomic Energy Agency (IAEA), Vienna, Austria
e-mail: P.Brisset@iaea.org

C. E. Hughes
Australian Nuclear Science and Technology Organisation, Sydney, NSW, Australia
e-mail: Cath.Hughes@ansto.gov.au

H. J. Pant
Isotope Applications Division, Bhabha Atomic Research Centre (BARC), Trombay, Mumbai, India
e-mail: hjpant02@gmail.com

J. Thereska
Private Consultant, Vienna, Austria
e-mail: thereska@gmail.com

A. Wörman
Royal Institute of Technology, KTH, Stockholm, Sweden
e-mail: worman@kth.se

G. Lollino et al. (eds.), *Engineering Geology for Society and Territory – Volume 4*,
DOI: 10.1007/978-3-319-08660-6_28,

28.1 Introduction

In the coastal region the construction, maintenance and expansion of harbours, including dredging and dumping of the dredged material in the sea, are activities where knowledge of sediment transport is mandatory. The selection of a suitable dumping site for dredged material is very important to strike a balance between the cost of transport of the dredged material and its eventual return to the dredged areas. On the other hand, municipal and industrial effluents discharged into the coastal region are frequently a mixture of aqueous and particulate components. Information concerning the dispersion and settling behaviour of the particles has important implication for understanding the impacts of effluent discharges on the receiving water. Finally, many coastal areas are submitted to undesirable erosion or accretion processes and an understanding of the underpinning processes is necessary for mitigation.

For the assessment of sediment behaviour, hydraulic and sediment measurements, bathymetric surveys, and mathematical and physical models are applied.

The use of artificial tracers for sediment dynamic studies started in the 1950s and 1960s. From this time on, tracers with radioactive or fluorescent properties have been used to simulate sediment movement. The tracer must have similar hydrodynamic properties to that of the native sediment. The radiotracer is submerged underneath several metres of water. The tracer particles, as demonstrated by experiments, disperse rapidly among the natural sediment and a decrease in concentration quickly results. The tracer particles are then subject to the same transport processes as the native sediment.

Experimental tracing and CFD are complementary methods of studying complex systems. Tracer data are based on direct observation, but are limited to the labelled component of the system and to a restricted domain of space and time. CFD models are not restricted in space and time and can, in theory, accommodate all the important parameters, but are limited by their underlying assumptions. Individually both approaches have limitations, but together they offer a very powerful method of investigating complex systems. Over the past few years it has become clear that the synergistic modelling and tracer approach can make a significant contribution to addressing complex problems in natural systems.

Tracers, with radioactive properties (gamma emitters) can be used in two ways: as a radioactive or an activatable tracer. In the first case the sediment is labelled with radioactive material and measured in-situ by scintillation detectors towed by GPS positioned boats (open eyes detection), a great help in planning the detection strategy. In the second way the sediment is labelled with an element, not present in the natural sediment, which will be measured in laboratory after sampling and irradiation in a nuclear reactor. This technique is used to study the long term behaviour of fine sediment dredged and dumped in aquatic environments. Fluorescent sediment tracers also can only be measured in the laboratory, after sampling (IAEA 2008). This represents blind eyes detection, because no feedback is received in real time to guide the detection work.

Radioisotopes as sealed sources: gamma scattering and transmission gauges, are also used for sediment monitoring, being applied for measuring high concentration of sediment in suspension or for the measurement of bulk density of fine sediment deposited in reservoirs and in coastal region. This application is not addressed here and can be seen in Aun and Bandeira (1995) or (IAEA 2008).

28.2 Case Studies

Case studies showing the application of radiotracers for various purposes are mentioned. Only the first case study is reported in detail due to space limitation.

(a) *Radiotracer study of bedload transport at the Port of Songkhla, Thailand*

The Port of Songkhla lies at the entrance to Songkhla Lake, a brackish tidal lake covering 1040 km^2. Sedimentation of the dredged shipping channel poses a significant economic problem affecting Thai shipping. Large tidal flows pass through the 380 m wide channel to the Gulf of Thailand but long-shore transport supplies sand which is deposited in the channel.

To assist in addressing this problem a numerical model of the port entrance was developed and validated using radiotracer studies of actual sand transport in the area (Nielsen et al. 2001; Airey et al. 2003) under an IAEA/RCA funded project involving Australia and Thailand. The principal aim of the investigation was to validate the model prediction of the bedload transport in the vicinity of the Port of Songkhla using tracer techniques so the model could be used to evaluate engineering options for port redesign.

A finite element numerical hydrodynamic model describing the tidal discharge characteristics of the Songkhla estuary, with particular attention to the inlet, was developed using RMA2 (Norton et al. 1973) and based on detailed hydrographic survey data. It was also used to select radiotracer injection sites (Fig. 28.1). The model predicted that, on the ebb tide, current separation occurred on both sides of the inlet (Fig. 28.1). This induced an inlet-directed current close to shore that had a similar direction to that on the flood tide. Therefore, there were locations where the gross and net sediment transport was likely to be much

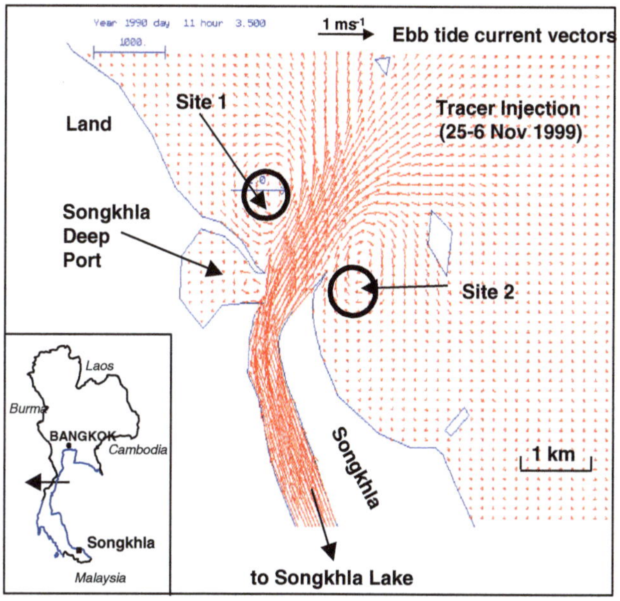

Fig. 28.1 Superimposed on the ebb tide current vectors in the vicinity of the Port of Songkhla are the locations of the two tracer injections

the same. For this reason, Sites 1 and 2 (Fig. 28.1) were chosen as injection sites for the tracer study.
The sediment transport model indicated clearly the enhancement of transport occurring under wave action. A 12-month period of Waverider Buoy records indicated that the best time to undertake the field tracer study was from early November to the end of March. This was during the period of the northeast monsoon when significant wave heights commonly reached 1.5–2.0 m with periods of around 6 s. At other times, the wave climate was benign, with wave heights rarely exceeding 1 m and with wave periods generally less than 4 s. The tracer study was performed between 25 November 1999 and 7 January 2000, a period of 6 weeks.
The mean grain size of the sediment off Samila Beach (near Site 1) was about 170 μ whereas off the northern breakwater (Site 2) the mean grain size was coarser at about 300 μ. The grading of the tracer material used at each site was designed to match the native sediments based on their fall velocities. The radioisotope tracer used was iridium-192 labelled synthetic sand made from a manufactured glass containing 0.03 % iridium (Vance et al. 1997).
Tracer injections at Sites 1 and 2 were conducted in approximately 3 m of water. Tracer surveys were conducted on four occasions including the injection dates. Over the 6 weeks following the tracer injections, the activity surveys (Fig. 28.2) detected little movement of the centre of activity at each site. At Site 1, the centroid of the tracer translated 76 m along the bearing 333°. At Site 2, the translation was 5 m along the bearing 168°. The translation of the centroid of the tracer (X) can be converted to the mass of sediment transport (Q) based on the depth of movement (D) and the in-situ dry density of the material (γ_d) as follows: $Q = X \cdot D \cdot \gamma_d$.
The stratigraphy of cores taken on the seabed indicated clearly that, at each site, the depth of movement was 0.19 m. The net transport results are given in Table 28.1.
The validated sediment transport model was used by the Thai Harbour Department to evaluate the effect of constructing a parallel breakwater on the northern side of the channel to reduce sedimentation and predicted that sand transport rates on the ebb tide would be an order of magnitude greater with the additional breakwater. This breakwater has now been constructed.

(b) *Suspended and bottom sediment transport studies with radiotracers in Santos Bay, Brazil* (Aun and Bandeira 1995)
Suspended sediment transport studies in Santos Bay and offshore were performed aiming at evaluating various dumping sites, to reduce the distance from the dredging region to the disposal areas. Fine sediment was labelled with ^{198}Au.
Hydraulic measurements and tracer studies using ground glass labelled with ^{192}Ir were intended to define the behaviour of the bottom fine sandy sediment that is found in some regions of the bay, on behalf of model studies.

(c) *Bottom sediment transport in the gulf of Durres, Albania* (IAEA 2013)
This study was performed aiming at the optimization of the maintenance works of the access channel to the Durres harbour. Ground glass in the range of 40–100 µm simulating the bottom sand, and doped with ^{192}Ir, was injected in four sites around the access channel. The results of sediment movement in three sites showed the contribution of the Durres gulf sediment to the accretion of the dredged channel. The results in the remaining site, located at the other side of the channel, indicated that it could be used to dump the dredged material since it is not transported in the direction of the access channel.

(d) *Bottom sediment transport in New Mangalore Port, Mangalore, India* (Sharma et al. 2010)
A radiotracer investigation was carried out during October 2007–January 2008 (post south-west monsoon period) to investigate movement of sediment. Scandium-46 as scandium glass powder (60–100 µm) was used as radiotracer. The results of the investigation showed that the sediment predominantly moves towards north-west direction and the proposed site was found suitable for dumping the dredged material as the

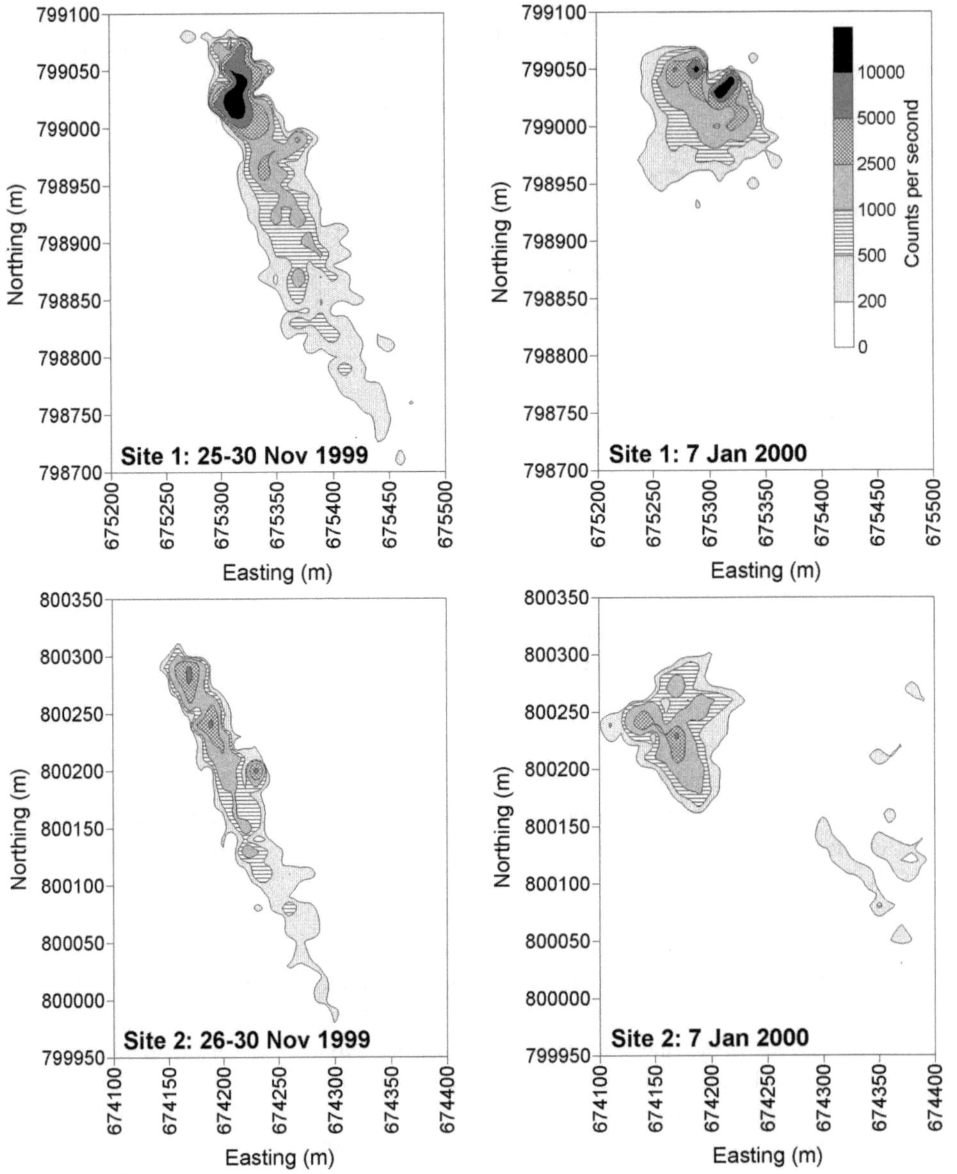

Fig. 28.2 Decay and background corrected tracer distribution immediately following injection and six weeks later (from Nielsen et al. 2001)

Table 28.1 ^{192}Ir sediment tracer characteristics, tracer study and modelling results

Site	Total mass of tracer (g)	Total activity (GBq)	Injection date	Transport (kg/m)		Bearing (°)	
				Model	Tracer	Model	Tracer
1	150	10.75	25/11/99	16,560	22,400	321	333
2	180	12.68	26/11/99	4,337	1,470	162	168

movement of sediment was away from the shipping channel.

(e) *Recycling of dredging product—the case of Zeebrugge Harbour, Belgium* (Brisset et al. 1996)
The global efficiency of dredging works depends on the efficiency of the dumping site, meaning the fraction of particles which stay on site a long time compared to the fraction recycled to the dredged areas. In-situ experiments using radiotracers have allowed us to quantify this return. The selected tracers (^{181}Hf and ^{160}Tb) were measured by gamma spectrometry low background on natural sediment samples. Four large scale experiments have been performed. More than 200 samples have been taken on 66 stations along the Belgian coast and analyzed during each experiment. It was concluded that the coastal zone appears as a trap for fine sediments. The dredging works products discharged inside this zone are recycled to the coast. A discharge outside this zone, more than 25 km offshore, is necessary to eliminate the recycling risk.

28.3 Conclusion

The use of radioactive labelled sediment tracers with in-situ detection is an advanced tool for monitoring sediment dynamics in the coastal zone. This is of great importance to define and evaluate the fate and behaviour of particulates and contaminants, as demonstrated in the case studies presented. Over the past few years it has become clear that synergism between modelling and tracer approaches can make a significant contribution in addressing complex problems in natural systems. This is the main message of this paper.

Acknowledgment To FAPEMIG, MG, Brazil for their support for the presentation of this work.

References

Airey P, Hughes C, Kluss T, Duran E, Miller B, Chiuenta S, Nielsen AF, Hollins S (2003) Evolving role of radiotracers in coastal zone studies. Appl Radiat Isot 58:401–406

Aun PE, Bandeira JV (1995) The role of nuclear techniques in sedimentological studies and some applications in Latin America. In: IAEA-TECDOC-828: use of nuclear techniques in studying soil erosion and siltation, Vienna, p 29–97

Brisset P, Hoslin R, Meyer G, Caillot A (1996) Study and optimization of dredging works using radioactivity—the case of Zeebrugge harbour. IAEA-SM-223/8

IAEA (2008) Radiotracer and sealed source techniques for sediment management. Report of the CM, held in Vienna, Austria, 21–25 April 2008, p 64

IAEA (2013) Radiotracer and sealed source applications in sediment transport study. Technical document. Brisset P (ed) Training course series, Vienna, August 2013, p 146

Nielsen AF, Kluss T, Hughes CE, Sojisuporn P, Chueinta S, Adamantidis CA (2001) Field verification of formulations for sand transport under wave and current action. In: Proceedings of the Ports and Coasts 2001, Surfers Paradise, Australia. IEAust, Canberra, pp 481–486

Norton WR, King IP, Orlob GT (1973) A finite element model for lower granite reservoir. Water Resources Engineers, Inc. (prepared for Walla District, U.S. Army)

Sharma VK, Pant HJ, Jagadesh HB, Girish BK, Gursharan S (2010) Radiotracer investigation for bed-load movement at New Mangalore Port, Mangalore, India. Indian J Mar Sci 39(1):49–56

Vance ER, Carter ML, Blackford MG (1997) Iridium distributions in silicate glass for sedimentology studies. J Mater Sci Lett 16:991–992

Numerical Modeling of the Submarine Debris Flows Run-Out Using SPH

29

Hualin Cheng, Yu Huang, and Qiang Xu

Abstract

Differing form subaerial debris flows, submarine debris flows transport sediment across the continental shelf into deep ocean overcoming the resistance of seawater. Recently, submarine debris flow has been concerned by increasing researchers, owing to the unique characteristics of large mass movement and long run-out even at gentle slopes. It is therefore a potential threat to the offshore and coastal facilities. To identify the dynamic behaviour and the areas potentially at risk of flows, an accurate numerical method is needed to simulate such submarine debris flows. In this paper, we proposed a mesh-free particle method called smoothed particle hydrodynamics (SPH) method to simulate the dynamic behaviour of submarine debris flows. The Navier-Stokes equations were adopted as the governing equations and a Bingham model was adopted to describe the rheology of the viscoplastic fluids. Besides, a viscosity coefficient was introduced to consider the resistance of sea water. In this paper, we applied the SPH method to a geometric model of submarine debris flow to verify the accuracy of the SPH method. Considering the validity of parameters, we have tested different set of parameters to calculate this submarine debris flow, and compared the simulation results with the existing data. The simulation results provide a better understanding of the flow mechanism involved in submarine debris flows.

Keywords

Submarine debris flows • Smoothed particle hydrodynamics • Bingham model • Large deformation

29.1 Introduction

As one of the common geological phenomenon, debris flows occur in both subaerial and subaqueous environments. The subaerial debris flows are usually induced by heavy rainfall (Xu et al. 2012). However, the submarine debris flows are often caused by the seismic and wave loading (Imran et al. 2001). Besides, the submarine debris flows could also be derived from the earthquake induced submarine landslides (Elverhøi et al. 2000; Jakob and Hungr 2005). Although it is not easy to be observed like subaerial debris flow, submarine debris flows occurred across the world frequently (Masson et al. 1993). While subaerial

H. Cheng · Y. Huang
Department of Geotechnical Engineering, College of Civil Engineering, Tongji University, Shanghai 200092, China

Y. Huang (✉)
Key Laboratory of Geotechnical and Underground Engineering of the Ministry of Education, Tongji University, Shanghai 200092, China
e-mail: yhuang@tongji.edu.cn

Q. Xu
State Key Laboratory of Geo-hazard Prevention and Geo-environment Protection, Chengdu 610059, China

G. Lollino et al. (eds.), *Engineering Geology for Society and Territory – Volume 4*,
DOI: 10.1007/978-3-319-08660-6_29, © Springer International Publishing Switzerland 2014

debris flow can cause serious disasters to the houses, bridges, and traffic facilities, submarine debris flows can also cause massive damage to property. Debris flows in the lakes may cause numerous sediment depositions, reducing the capacity of hydropower station and destroying the facilities of hydropower station and affecting the waterway transportation (Yu et al. 2006). In addition, debris flows in the sea may also cause destruction of submarine cables and the offshore facilities, and even trigger tsunami by generating surface waves (Tappin 1999; Jiang and LeBlond 1992). As a result, the submarine debris flows, characterized by long run-out and large-scale, are under increasing researchers' attention.

Although there are many difficulties in researching the dynamic behaviour of such submarine debris flows, over the past few decades, numerous studies have been done by researchers, including the theoretical, experimental and numerical simulation of submarine debris flows (Marr et al. 2001; Imran et al. 2001). Considering the accuracy and efficiency of the methods, numerical simulations have the significant advantage of performing the run-out analysis for the submarine debris flow over irregular terrains. The existing studies usually consider submarine debris flows as viscous model, including three types of fluid: Bingham, Herschel-Bulkley, and Bilinear (Elverhøi et al. 2005; Imran et al. 2001; Jiang and LeBlond 1992). Elverhøi et al. (2000) pointed out that submarine debris flows have mostly been described as Bingham fluids, and the stress-strain relation of submarine debris flows is laminar flow. It implies that no deformation takes place until a specified yield stress is applied to the material.

Considering the significant feature of submarine debris flows, large-scale and long run-out, we proposed a SPH modeling technique, a mesh-free particle method, to simulate the submarine debris flows. The SPH modeling technique includes the Navier-Stokes equations as the governing equations and a Bingham model to analyse the relationship between material stress rates and particle motion velocity. Compared with the traditional numerical methods, the SPH modeling technique is a true mesh-free method in pure Lagrangian description. It can instantaneously track the motion of each particle, accurately predict the velocity and simulate the large deformation for the long run-out of the submarine debris flows. In addition, the SPH method is based on continuum mechanics. It is therefore an efficient method to simulate large-scale submarine debris flows. In this paper, we applied the SPH method to a geometric model of submarine debris flow to verify the accuracy of SPH model. This paper reports the outline and results of the topographic analysis, which could provide a better understanding of the dynamic behaviour of submarine debris flows.

29.2 Simulation Approach

29.2.1 Basic Descriptions

Smoothed particle hydrodynamics (SPH) is a mesh-free method based on a pure Lagrangian description, and is originally applied to the continuum scale. It was first invented to solve astrophysical problems in three dimensional open space (Lucy 1977; Gingold and Monaghan 1977). The state of the mesh-free systems is represented by a set of arbitrarily distributed particles. Furthermore, as a pure Lagrangian method, the moving particles possess material properties (e.g., mass, density, and stress tensor) (Lucy 1977). Based on the material properties carried by the moving particles, the field variables (e.g., density, velocity, and energy) can be obtained in SPH form through two kinds of approximation: kernel and particle approximation. In the previous study, the kernel approximation and particle approximation have been detailed introduction (Huang et al. 2013).

29.2.2 Governing Equations

The governing equation we adopted is the Navier-Stokes equations, of which the equations for dynamic fluid flows can be written as a set of partial differential equations. It is therefore the kernel and particle approximation techniques mentioned above can be rewritten in forms of the SPH formula. In the previous study, a typical continuity equation and motion equation have been detailed introduced (Huang et al. 2013).

29.2.3 Constitutive Models

For a Bingham fluid, the relationship between the shear strain rate and shear stress is given by

$$\tau = \eta\dot{\gamma} + \tau_y \qquad (29.1)$$

where τ is the shear stress tensor, τ_y and η are the Bingham yield stress and viscosity. However, considering the resistance of seawater to the submarine debris flows, an additional viscosity coefficient is needed in the Bingham fluid, which was used in submarine debris flows to express the seawater resistance and in rock fall simulation to express the arborous resistance (Ma et al. 2009; Shinji et al. 1997). In actual, the η is expressed by

$$\eta = \eta_a - \eta_b \qquad (29.2)$$

where η_a is the viscosity of submarine debris flows, η_a is the viscosity coefficient of seawater resistance.

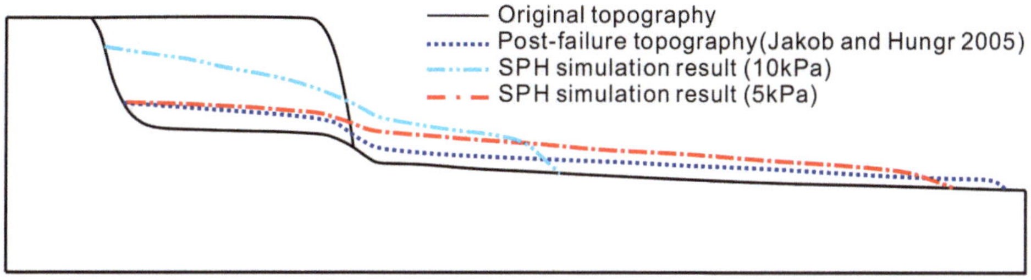

Fig. 29.1 Pre- and post-failure profile and of SPH simulation for the submarine debris flow

29.3 Simulation Application

In this paper, we applied the Bingham flow model to simulate submarine debris flow to verify the validity of the SPH modeling technique. As mentioned above, submarine debris flows are characterized by large mass movements and long travel distances. Almost all submarine debris flows could occur at gentle slopes. In this paper, we adopted a conceptual geometric model of submarine debris flows referred by Jakob and Hungr (2005). For the simulation of the submarine debris flow events, it is not possible to measure the rheological properties and the shear strength parameter of the flow directly. As a result, applying the Bingham flow model and its parameters correctly become the major challenge. Referring to the existing research of submarine debris flows, τ_y is a sensitivity parameter of submarine debris flows. Based on the previous research results (Elverhøi et al. 2000, 2005; Jakob and Hungr 2005), η adopted in this paper is 0.1 Pa s, and τ_y adopted in this paper is 5 kPa and 10 kPa, respectively. The numerical model included 6,748 particles in total, composed of 4,048 particles for the moving debris flow and 2,430 particles for the fixed boundary. Just as in the real situation, the debris flow particles can be deformed both horizontally and vertically with gravitation applied in the vertical direction only. Figure 29.1 show a comparison of the SPH simulated geometry and the simulate result by Jakob and Hungr (2005).

It is easy to obtain from the figures that: (a) the simulation results showed a good agreement with the observation data by Jakob and Hungr (2005), when τ_y is adopted in 5 kPa. (b) Compared with the 10 kPa of τ_y, it could flows longer when τ_y is adopted in 5 kPa. It is evident that the simulated run-out could be longer in a smaller τ_y.

29.4 Conclusions

Submarine debris flows had catastrophic impacts on coastal communities and infrastructures. Due to the huge mass movement and long run-out distance, the dynamic analysis of such submarine debris flows is important for the hazard assessment.

In this paper, a novel mesh-free method called SPH, which can readily accommodate the extremely large deformation of a great amount of soil mass, is used for the simulation of submarine debris flows. The Bingham rheological model was applied to describe the relationship between material stress rates and particle motion velocity. The simulation results had a high degree of similarity with previous configurations.

Acknowledgments This work was supported by the National Basic Research Program of China (973 Program, Grant No. 2012CB719803), the National Natural Science Foundation of China (Grant No. 41225011) and the Chang Jiang Scholars Program of China.

References

Elverhøi A, Harbitz CB, Dimarkis P, Mohrig D, Gary Parker JM (2000) On the dynamics of subaqueous debris flows. Oceanography 13(3):109–117

Elverhøi A, Issler D, de Blasio FV, Ilstad T, Harbitz CB, Gauer P (2005) Emerging insights into the dynamics of submarine debris flows. Nat Hazards Earth Syst Sci 5:633–648

Gingold RA, Monaghan JJ (1977) Smoothed particle hydrodynamics-theory and application to nonspherical stars. Mon Not R Astron Soc 181:375–389

Huang Y, Dai ZL, Zhang WJ, Huang MS (2013) SPH-based numerical simulations of flow slides in municipal solid waste landfills. Waste Manag Res 31(3):256–264

Imran J, Harff P, Parker G (2001) A numerical model of submarine debris flow with graphical user interface. Comput Geosci 27(6):717–729

Jakob M, Hungr O (2005) Debris-flow hazards and related phenomena. Springer, Chichester, UK

Jiang L, LeBlond PH (1992) The coupling of a submarine slide and the surface water waves which it generates. J Geophys Res: Oceans 97(C8):12731–12744

Lucy LB (1977) A numerical approach to the testing of the fission hypothesis. Astron J 82:1013–1024

Ma GC, Kankeo F, Hori S (2009) Application of DDA to evaluate the dynamic behavior of submarine landslides which generated tsunamis in the Marmara sea. In: Proceedings of the International Conference on Analysis of Discontinuous Deformation, Singapore, pp 185–192

Marr JG, Harff PA, Shanmugam G, Parker G (2001) Experiments on subaqueous sandy gravity flows: the role of clay and water content

in flow dynamics and depositional structures. Geol Soc Am Bull 113(11):1377–1386
Masson DG, Hugget QJ, Brunsden D (1993) The surface texture of the Saharan debris flow and some speculations on subaqueous debris-flow processes. Sedimentology 40(3):583–598
Shinji M, Ohno H, Otsuka Y, Ma GC (1997) Viscosity coefficient of the rock-fall simulation. In: Proceedings of ICADD-2, Kyoto, pp 201–210
Tappin DR (1999) Sediment slump likely cause 1998 Papua New Guinea tsunami. Eos, Trans Am Geophys Union 80(30):329–340
Xu Q, Zhang S, Li WL, van Asch TWJ (2012) The 13 August 2010 catastrophic debris flows after the 2008 Wenchuan earthquake, China. Nat Hazards Earth Syst Sci 12:201–216
Yu B, Wang SG, Zhang SC, Meng GC (2006) Preliminary study on the effect of debris flows from Goose Foot river on lake Qionghai, Sichuan, China. J Lake Sci 18(1):57–621 (in Chinese)

Dynamic Response of an Embankment Foundation to a Simulated Tsunami Wave

30

Yu Huang, Chongqiang Zhu, and Min Xiong

Abstract

The interaction between tsunami waves and an embankment is a complicated coastline phenomena and one of the least studied aspects of a tsunami disaster. When the tsunami wave breaks and hits the surface of a dike, the impact pressure, characterized by high intensity, repeated application and short duration, may cause the destabilization of the embankment and the liquefaction of its foundation. The hydrodynamic loads on a dike are therefore of considerable interest for the safety and design of dike foundations. The present paper estimates the collision forces on an embankment due to tsunami waves with the application of the moving particle semi-implicit (MPS) method, which has the advantage of tracking water free surfaces with the fragmentation and coalescence of water due to the absence of numerical diffusion. MPS is a mesh-free, purely Lagrangian particle method, where the discretization of the Navier-Stokes equation is based on the interaction of neighboring particles. The results are in agreement with the smoothed particle hydrodynamics simulation, which indicates that the MPS method is good at describing the interaction of tsunami waves and the dike. The impact of the tsunami waves on the dike is visualized and presented, providing qualitative understanding of the interaction between the tsunami and the embankment. In addition, we obtain hydrodynamic load data and analyze the stress distribution at the bottom of the foundation, which is under the highest load. The simulation can provide some valuable insight for the conceptual design of coastal defense facilities.

Keywords

Tsunami • MPS • Foundation • Embankment • Dynamic response

30.1 Introduction

Tsunamis are a series of destructive waves caused by a sudden large-scale movement of the seabed triggered by earthquakes, landslides, volcanic eruptions or explosions. They usually last for 200–2000 s and have a velocity of 540–900 km/h with wavelengths up to 5.0×10^5 m in the open ocean. The wave retains its high energy as the tsunami approaches the shore, which causes danger to inhabitants and severe damage to coastal structures. For example, the total number of people dead or missing reached 20,416 across Japan in the March 11, 2011 Great East Japan Tsunami, which was triggered by a Mw 9.0 earthquake

Y. Huang (✉)
Key Laboratory of Geotechnical and Underground Engineering of the Ministry of Education, Tongji University, Shanghai 200092, China
e-mail: yhuang@tongji.edu.cn

Y. Huang · C. Zhu · M. Xiong
Department of Geotechnical Engineering, Tongji University, Shanghai 200092, China

G. Lollino et al. (eds.), *Engineering Geology for Society and Territory – Volume 4*,
DOI: 10.1007/978-3-319-08660-6_30,

(Arikawa et al. 2012). An embankment, the common coastal structure protecting the coastal area from tsunamis, suffers serious damage during a tsunami. It is therefore important to study the dynamic response of an embankment foundation to minimize the loss of life and damage to coastal structures caused by a tsunami.

With the development of computer technology, numerical simulations play a significant role in analyzing tsunami behavior because of its advantages over disaster area field investigations and hydraulic model experiments. Tsunami waves can be characterized as large-scale and transient, with complex free-surface behavior upon contacting a seawall. The rapidly developing mesh-free methods have great advantages over the traditional mesh-based methods for such simulations. The mesh-free simulations allow the accurate tracking of the free surface including the tsunami wave propagation, fragmentation and breaking. The moving particle semi-implicit (MPS) method, which was originally developed by Koshizuka and Oka (1996) for fragmentation of an incompressible fluid and has proven valuable in several engineering fields, is particularly well suited for the simulation of the impact of tsunamis on seawalls (Gotoh et al. 2001). Few previous tsunami simulations have been systematic, provided qualitative results or studied the response of the seawall foundation. In this paper, we applied the MPS method to study the dynamic load caused by a tsunami and calculated the response of the seawall foundation under the maximum wave force using the ADINA program.

30.2 The Concept of MPS

30.2.1 Governing Equation

The continuity equation:

$$D\rho/Dt = 0. \tag{30.1}$$

The Navier-Stokes equation:

$$D\mathbf{u}/Dt + (\mathbf{u} \bullet \nabla)\mathbf{u} = -\nabla p/\rho + v\nabla^2\mathbf{u} + \mathbf{g} + \mathbf{F}, \tag{30.2}$$

where ρ is the density, t is the time, **u** is the velocity vector, ∇ is the gradient, p is the pressure, v is the kinematic viscosity, ∇^2 is the Laplacian, **g** is the gravitational acceleration vector, and **F** is external force. The right hand of Eq. (30.2) includes the pressure gradient, viscosity, gravity and external force terms. In this simulation the viscosity and external force terms are ignored.

30.2.2 Gradient and Laplacian Model

The gradient vector at particle i is calculated by the weight average of the gradient vectors:

$$\langle \nabla\phi \rangle_i = d/n^0 \sum_{i \neq j} \left[\phi_j - \hat{\phi}_i r_i - r_i w(|r_i - r_i|) \big/ |r_i -_i|^2 \right], \tag{30.3}$$

where d is the number of space dimensions, $\hat{\phi}_i$ is the minimum value of effective radius, and w(r) is the kernel function. The diffusion of ϕ at particle i is modeled as follows:

$$\langle \nabla^2\phi \rangle_i = 2d/\lambda n^0 \sum_{j \neq i} \left[\phi_j - \hat{\phi}_i) w(|r_j - r_i| \right], \tag{30.4}$$

where λ is the coefficient by which the variance increase is equal to the analytic solution and is expressed as:

$$\lambda = \sum_{j \neq i} w(|r_j - r_i|)|r_j - r_i|^2 \Big/ \sum_{j \neq i} w(|r_j - r_i|). \tag{30.5}$$

The weight function is defined as:

$$w(r) = \begin{cases} r_e/r - 1 & (0 \le r < r_e) \\ 0 & (r > r_e) \end{cases}, \tag{30.6}$$

where r is the distance between two particles and r_e represents the radius of the interaction area.

There are two sections in the time integration process. The first section is an explicit process, in which temporary velocity (**u***), temporary position (**r***) and the temporary number density (n*) are calculated, considering the given external force and gravity terms. Because the position of the particles is updated, mass conservation might not be satisfied. The pressure term, not included in the first section, is calculated in the second section and the Poisson equation of pressure is also solved implicitly. The Poisson equation is given as follows:

$$\frac{1}{\rho} \nabla^2 P_{k+1} = \frac{1}{\Delta t^2} \frac{n^* - n^0}{n^0}. \tag{30.7}$$

30.3 Numerical Simulations

Tatsuya et al. (2010) simulated the destabilization of a breakwater when struck by a tsunami using the smoothed particle hydrodynamics (SPH) method and presented the time history of pressure at different points on a seawall. However, they did not analyze the response of the foundation. We calculate and discuss the time history of the

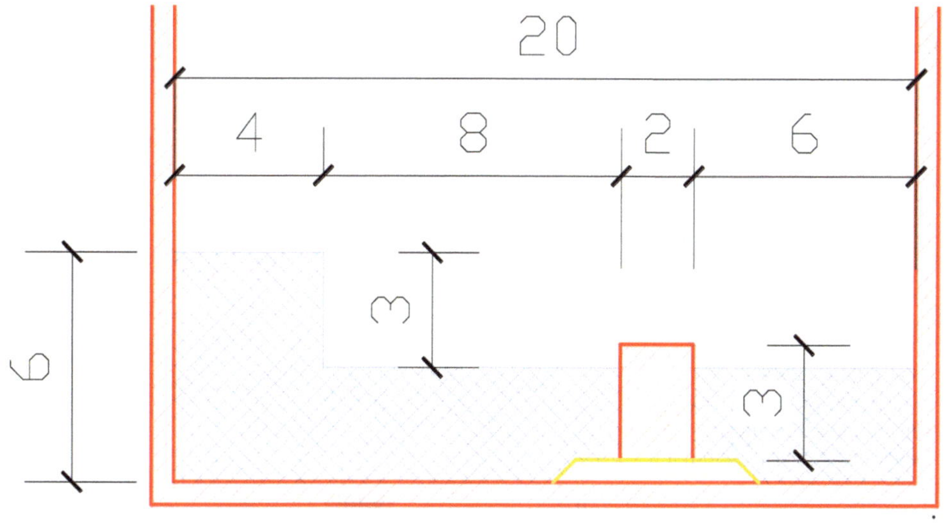

Fig. 30.1 Model layout (m)

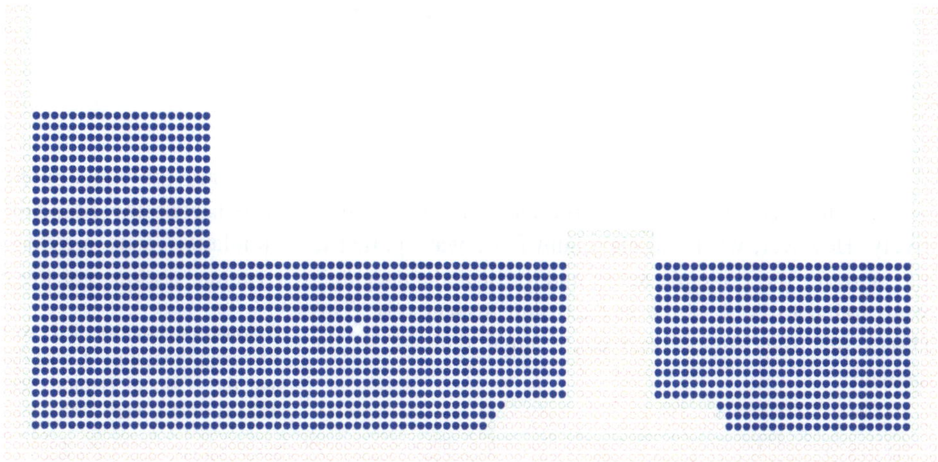

Fig. 30.2 Discrete model applied in the MPS simulation

resultant force using the MPS method and then apply the ADINA method to calculate the response of the foundation under the maximum force. The layout of the model is shown in Fig. 30.1. In the MPS simulation the initial distance between particles is 0.2 m. The simulation includes a total of 2,449 particles of which 1,816 particles represent water and 633 particles are for the boundary. The discrete model applied in the MPS simulation is depicted in Fig. 30.2.

The time history of the resultant force exerted on the embankment is shown in Fig. 30.3. The force on the embankment was calculated by integrating the pressure distributed across the front of the embankment. The maximum force in the MPS simulation was 68 kN/m width of wall at 1.2 s. In the SPH simulation, the maximum force was approximately 60 kN/m width of wall. The simulated force using the MPS method shows good agreement with the SPH simulated result reported in the literature (Tatsuya et al. 2010).

The calculated force is applied to the seawall to determine the stress and displacement of the foundation. The mesh size is set to 0.2 m and the seawall and foundation which is Fujian standard sand are regarded as elastic media. The parameters used in the ADINA method are listed in Table 30.1. The cloud pictures of principal stress and vertical displacement before and after the tsunami raids are shown in Fig. 30.4 panels (a), (b), (c) and (d) respectively.

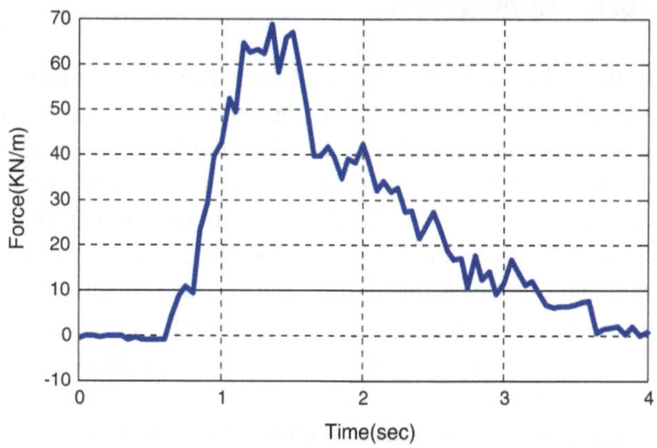

Fig. 30.3 Time history of force exerted on the embankment

Table 30.1 Parameters used in the ADINA simulation

	Embankment	Foundation soil
Density (kg/m^3)	2,450	1,600
Elasticity modulus (MPa)	2,000	3.5
Poisson ratio	0.15	0.30

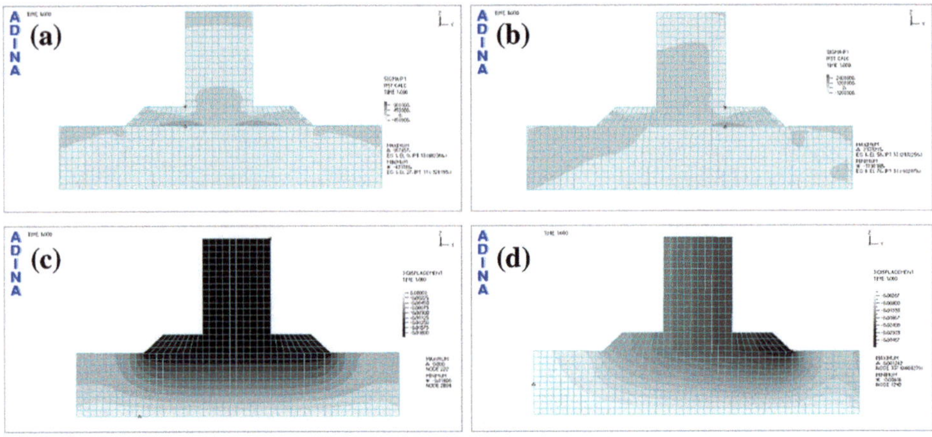

Fig. 30.4 Cloud pictures of principal stress and vertical displacement: **a** principal stress before tsunami raids, **b** principal stress after tsunami raids, **c** vertical displacement before tsunami raids, **d** vertical displacement after tsunami raids

Before the tsunami arrived, the principle stress and vertical displacement were 0.2 MPa and 0.18 mm respectively. However, when the maximum force was applied to the seawall, the principle stress and vertical displacement were found to be 2.3 MPa and 2.0 mm respectively. This shows that the tsunami can cause great damage to the seawall.

30.4 Conclusions

In this paper, the time history of force exerted on a coastal embankment is simulated using the MPS method. The result shows good agreement with the SPH simulation reported in previous literature, which indicates that the MPS method can be employed to model the tsunami-embankment interaction. A maximum force was applied to the seawall to calculate the principle stress and vertical displacement using ADINA software. The comparison of this data shows that the tsunami can cause considerable principle stress and vertical displacement, which finally leads to significant damage of the embankment. Although the conceptual model can only provide limited information about the dynamic response of the foundation induced by the tsunami, this simulation does provide a qualitative description of the dynamic response of the embankment foundation.

Acknowledgements This work was supported by the National Basic Research Program of China (973 Program, Grant No. 2012CB719803) and the National Natural Science Foundation of China (Grant Nos. 41211140042, 41372355).

References

Arikawa T, Ishikawa N, Beppu M, Tatesawa H (2012) Collapse mechanisms of seawall due to the March 2011 Japan tsunami using the MPS method. Int J Protect Struct 3(4):457–476

Gotoh H, Shibahara T, Sakai T (2001) Sub-particle-scale turbulence model for the MPS method Lagrangian flow model for hydraulic engineering. Adv Methods Comput Fluid Dyn 9:339–347

Koshizuka S, Oka Y (1996) Moving-particle semi-implicit method for fragmentation of incompressible fluid. Nuclear Sci Eng 123(3):421–434

Tatsuya I, Kenichi M, Miyake T, Tsurugasaki K, Sawada Y, Hiroko K (2010) Unstabilization of breakwater that originates in transformation of marine soil when tsunami raids. 22th Central Geotechnical Engineering Symposium, pp 101–108

Stages of Clayey Sediments Formation in the Presence of the Organic Matter

31

Zdobin Dmitry

Abstract

Results of long-term monitoring researches of physical and chemical properties of Holocene silts of Kandalaksha Bay of the White Sea are considered. On the basis of the comprehensive analysis of a ratio of the sizes W_e (water content), W_l (liquid limit) and C_{org} (content of organic matter) on depth of a continuous sequence of sea organo-mineral soil the conclusion about existence of nine stages in natural system a deposit—silt—fluid clay is made. The transition boundary sedimentgenesis—diagenesis (silt is offered—fluid clay) is to be determined by $W_e \geq W_l$ and $C_{org} \geq 3$ %.

Keywords

Clay silt • Organic matter • Physical-chemical properties • Stages of clayey sediments formation

31.1 Introduction

At present, while the object in soil systems studying, an attention pays to several classifications, such as: on the typological level—mineral soil; on the species level- clayey soils and also on the variety level-an individual properties. An interrelation of levels is a very suitable point for discussion of any questions in the soil system science.

The majority of works are devoted to study of various clay soils properties, because they are the most common entities of the upper horizons of the Earth's crust. Significantly less attention to organic and organic-mineral soil is paid, firstly, because of their well-known limited distribution on the surface, secondly, of their study complexity. However, the organic-mineral soils are the most important type of dispersed coherent sedimentary soil, is the geological basis, which is evolving in time and space, transformed, eventually, in all other types of clayey soils. According to author's opinion, it is extremely important to trace (and what is significant—to characterize) all stages of the process: from getting a separate mineral particles in the sedimentary basin, to a stable groundwater system.

The study of stages of formation of physical-chemical properties of clayey soils relate to the fundamental problems of modern soil system science. A sedimentological stage in a line of sedimentogenesis-diagenesis (early and late) lithogenesis-catagenesis, is most important, because the original mineral composition, presence of botanical and physical and chemical conditions determine the structure in the subsequent nature of any transformation in soil.

A large number of works is dedicated to a problem of clayey layer formations stage: beginning with fundamental works, considering the whole process in general, ending with private developers of certain provisions. The qualitative breakthrough in the research process of diagenesis and sedimentogenesis occurred since physical and chemical mechanics of disperse medium, which managed to build a two-phase (solid and liquid components) model formation of clay sediment have been developed and applied. The whole process (up to fluid clay) is divided into four stages:

Z. Dmitry (✉)
Saint–Petersburg State University, 7/9 Universitetskaya nab, 199034 Saint-Petersburg, Russian Federation
e-mail: zdobin_soil@mail.ru

G. Lollino et al. (eds.), *Engineering Geology for Society and Territory – Volume 4*,
DOI: 10.1007/978-3-319-08660-6_31, © Springer International Publishing Switzerland 2014

coagulation, aggregation, structure formation and physical and chemical consolidation (Osipov et al. 2004). However, under a natural conditions, this model does not fully correspond to the natural reality, as it does not take into account such an essential component of soil as an organic matter. At present, the total sequence of events, doesn't evoke a particular objection. And yet, there is still some points in the very issue requires separate study. These are the effects of organic matter: one way or another, in the early stages of the sedimentary processes, which are involved in the transportation, deposition, in the formation of the primary structure (coagulation) links in the system "particle-suspension-sediment (proto-soil)-silt".

31.2 Data and Results

During 1997–2011 field seasons the near-shore marine bay clayey silts of White Sea (North European part of Russia) were studied. Number of bottom sampling stations and sufficient time frame for repetitive observations for the same points were enough for obtaining confident values of physical-chemical properties of the silts.

We have attempted to divide this process into stages, taking into account the influence of an organic matter (OM) (Table 31.1). Instead of previously defined four stages (Osipov et al. 2004) here we firstly identified nine following stages based on the values of water content (W_e), plastic limit (W_p), liquid limit (W_l), plasticity index (I_p), liquidity index (I_c) and content of OM (C_{org}). The following names of the stages and newly proposed by the author and thus are not of common use yet.

I—Biomineral snow. The volume of H_2O significantly exceeds the volume of nonaqueous particles. This stage is identified from the water surface to level +7.0 mm. ("0" (zero-level) here corresponds to the hydrosphere-lithosphere boundary).

II—Sedimentary cloud (conditional water). Increase of nonaqueous particles concentration. The interval: +7.0 to +1.0 mm. $H_2O \approx$ to quantity of nonaqueous particles.

III—Sedimentary fog (suspension). The interval: +1.0 to 0.0 mm. Start of aggregation; coagulation effect of long-distance coagulative contacts. Start of orientation of clayey particles (cubic symmetry sediment). Thus there are no physical boundaries between the first three stages we believe that the definition an event horizon can be applied instead of term "boundary" here. The third stage is characterized by $W_e > 5W_l$, $C_{org} > 10$ %. The atoms and molecules of OM are likely represent some kind of a «bridges» between the clayey particles.

IV—Mineral liquid. The interval: 0.0–5.0 mm. The silica-alumina gel (sol-gel phase).

Table 31.1 The White Sea. Physical-chemical properties of silts

Horizon (sm)	Stage	W_e	W_l	W_t	I_p	I_L	n	e	ρ	ρ_s	C_{org}
−0.05 to 0.0	III	539	110	68	42	8.83	91	9.907	1.27	2.57	11.06
0.0–0.05	IV	235	109	69	40	3.71	85	5.726	1.29	2.59	8.89
0.05–1.0	V	205	102	67	35	3.94	84	5.147	1.30	2.62	6.85
1.0–2.0		181	90	68	22	5.14	82	4.620	1.31	2.62	6.47
2.0–3.0	VI	170	90	67	23	4.48	81	4.380	1.32	2.63	5.63
3.0–4.0		166	91	68	23	4.26	81	4.340	1.31	2.63	5.94
4.0–5.0		144	85	67	18	3.34	80	3.899	1.31	2.63	5.81
5.0–8.0		147	86	68	17	4.38	80	3.921	1.32	2.63	5.73
9.0–11.0		141	85	68	18	4.29	79	3.766	1.33	2.63	5.53
14.0–16.0	VII	145	89	70	19	3.94	79	3.827	1.34	2.64	5.61
16.0–19.0		137	92	68	24	2.88	79	3.652	1.35	2.65	5.68
19.0–21.0		139	87	69	18	3.88	79	3.657	1.36	2.65	5.32
21.0–24.0		124	91	70	21	2.57	77	3.333	1.37	2.65	5.30
24.0–26.0		123	94	69	25	2.16	77	3.345	1.36	2.65	5.22
26.0–29.0		118	93	69	24	2.04	76	3.217	1.37	2.65	5.18
29.0–30.0		116	92	70	22	2.09	76	3.148	1.38	2.65	5.06
34.0–35.0	VIII	112	95	71	24	1.71	75	3.071	1.38	2.65	4.86
35.0–36.0		98	98	71	23	1.08	73	2.762	1.40	2.66	4.79
36.0–39.0		96	95	70	25	1.00	73	2.697	1.41	2.66	4.83
40.0–42.0	IX	90	93	68	25	1.00	72	2.573	1.42	2.67	2.98

Long-distance coagulative contacts. Physical-chemical properties of soil: $2W_l < W_e < 5W_l$, 10 % > OM > 7 %. From the stage IV the soil itself appears so one can define its physical and chemical parameters using laboratory methods.

V—Proto-soil. The interval: 5.0–17.0 mm. The formation of near coagulative contacts beginning. The beginning of syneresis. Physical-chemical properties of soil: $2W_l < W_e$, 7 % > C_{org} > 5 %.

VI—The organic-mineral silt, black-colored. The interval: 17.0–105 mm. The beginning of the of chemical OM decomposition. Close coagulation contacts. Adhesion of greenish-gray silts. Physical-chemical properties of soil: $2W_l > W_e > W_l$, 7 % > C_{org} > 5 %.

VII—The mineral-organic, greyish-black silt. The interval: 105–310 mm. Physical-chemical properties of soil: $W_e > W_l$, 5 % > C_{org} > 3 %.

VIII—The mineral greenish-gray fluid silt. The interval: 310–460 mm. Physical-chemical properties of soil: $W_e \geq W_l$, $C_{org} \geq 3$ %. The boundary of silts and fluid clays.

IX—The greyish-green liquid clay. The interval: 460 mm and lower. Physical-chemical properties of soil: $W_e \geq W_l$, $C_{org} \leq 3$ %

We haven't studied stages, following the liquid clays (lower than IX).

31.3 Discussion and Conclusions

In the first three stages the linear quantities (such as thickness of layer) are important. In the subsequent stages the physical-chemical parameters of soils are of the point, because the properties of silts directly depend on the specific situation and specific sedimentary facies conditions.

If one adhere to this scheme of formation of clayey sediment, it can be assumed that the soil become a four-phase (mineral, liquid, organic and gaseous) at the bottom of IV stage (and probably in the early beginning of the stage V), where the H_2S concentration begins to influence soil properties significantly.

Our conclusion is in a good agreement with results of independent microbiological studies of upper horizons of marine sediments. Based on this research the three main steps of evolution of microorganisms in the geological environment.

The first step is an aerobic decomposition. If the concentration of O_2 drops below 1 mg/l, microaerophilic eubacterias participate in the aerobic decomposition. Though, the inflow of oxygen to sediment is limited, the aerobic stage is usually quickly replaced by anaerobic one (stages III–V).

The second step is hydrolytic (primary anaerobe). Among eubacterias of this cycle the main role is played by eubacterium of such a kind, which decompose polymeric substratum, especially biopolymers of carbohydrates. Hydrolysis of polymers leads to the creation of monomeric areas, which serves as substratum for hydrolytic organisms itself (stages VI–VIII).

The third step is sulphate reducer (eubacterium of secondary anaerobe). If the amount of sulphate is unlimited, it can lead to full oxidation of organic matter. Typically, in marine sediments the number of sulfates is exhausted as a result of its sulfur in metal sulfides transition, sulfur compounds or in the form of hydrogen sulfide (as in our case), which gradually evaporates (stage IX and lower).

Table 31.2 Stages of clayey soils formation

No	Stage	Horizon (sm)	Color	Type of contacts	Type of microstructure	Organic matter	Physical and chemical factors
I	Biomineral snow	$-\infty$ to -7.0 (-3.0)			Cellular	OM > 10 %.	$W_e > 5W_l$
II	Sedimental cloud	-3.0 to $-1.0(0.05)$					
III	Sedimental fog	-0.05 to 0.0	Fulvous	Start of DCC			
IV	Mineral liquid	0.0–7.0	Black	DCC		10 % > OM > 7 %	$2W_l < W_e < 5W_l$
V	Protosoil	5.0–17.0				7 % > OM > 5 %	$2W_l < W_e$
VI	Organic-mineral silt	17.0–112		DCC + NCC		5 % > OM > 3 %	$2W_l > W_e > W_l$
VII	Mineral organic silt	112–310	Grayish-black			5 % > OM > 3 %	$W_e > W_l$
VIII	Mineral silt	310–480	Greenish-grey	NCC + DCC		OM ≥ 3 %	$W_e \geq W_l$
IX	Fluid clay	480	Grayish-green	NCC		OM < 3 %	$W_e \geq W_l$
X	Fluid and plastic clay	–				OM < 3 %	$W_e \leq W_l$

DCC long-distance coagulative contacts, *NCC* near coagulative contacts

Thus, the following scheme of stages clayey rocks formation should be presented: sedimentogenesis—iagenesis—lithogenesis—catagenesis, without dividing the term "diagenesis" to the early and late periods. The sedimentogenesis—diagenesis (sediment, soil) border is to be determined after: $We \geq Wl$, and $C_{org} \geq 3$ %.

As the result of the study, the new scheme of organic-mineral (and clayey sediments also) soils formation is presented (Table 31.2). The most important point of the suggested scheme is its mathematical content.

Perhaps, if other climatic and sedimentary setting are provided the chosen stages can vary in depth, or even more to lack, but the numerical values of physical-chemical characteristics of soils as a whole will be kept.

Reference

Osipov VI, Sokolov VN, Eremeev VV (2004) Clay seals of oil and gas deposits. A.A.Balkema Publishers, Lisse, p 288

Part IV

Relative Land Subsidence in Transitional Coastal Environment: Causes, Effects, Quantification, Monitoring

Land subsidence combined with eustasy, i.e., relative land subsidence, represents one of the main hydrogeological threats for the survival of low-lying coastlands distributed worldwide. The process, both natural and man induced, is responsible for flooding, wetland loss, flow inversion in the drainage watercourses, coastline retreat, beach erosion, and saltwater intrusion. If the eustatic sea rise amounts up to 1–2 mm/yr, the lowering of the ground surface can peak few cm/yr. Therefore, an accurate quantification of land subsidence as well as the assessment of the components and the prediction of the effects on the natural transitional environments and coastal urban areas are of paramount importance. We invite contributions exploring regional or local case studies, the application of state-of-the-art monitoring and modeling techniques, the investigation of the natural/anthropogenic causes contributing to the measured relative land subsidence.

Relative Land Subsidence of the Venice Coastland, Italy

32

Luigi Tosi, Pietro Teatini, Tazio Strozzi, and Cristina Da Lio

Abstract

The Venice coastland, Italy, is an emblematic case of a coastal area prone to progressive submersion by the rising of the sea. Although the city of Venice represents the most relevant case in the northern Adriatic coast, the loss in ground elevation with respect to the sea level, i.e. the relative land subsidence (RLS), is particularly important not only in sectors directly affected by the marine ingression but also in the in-land coastal plain. In this work we quantify the RLS occurred over the period 1992–2010 in the Venice coastland by the superimposition of the Adriatic sea level rise to the ground displacements assessed by the Interferometric Point Target Analysis performed on ERS-1/2 and ENVISAT scenes. Results show that the lowering of certain portions of the coastland relative to the mean sea level amounts to several centimeters, which are values particularly severe for the maintenance and management of the coastal area.

Keywords

Land subsidence • Relative sea level rise • SAR-based interferometry • Venice coastland

32.1 Introduction

The Venice coastland extends along the about 15-km wide and 100-km long strip between the Adige and Tagliamento Rivers. It is formed by a series of estuaries, lagoons and wetlands developed during the prograding phase of the coastline driven by the interaction between eustatic rise and land subsidence during the Holocene sea-level highstand.

Today, a large portion of the coastland is composed of reclamation farmlands, which is made of reclaimed low-lying areas, up to 5 m below the mean sea level, estuaries, and the lagoon of Venice. Also minor sea level changes are vital for the survival of this transitional coastal environment.

The relative land subsidence (RLS), i.e. the loss in ground elevation with respect to the sea level, is particularly important not only in the city of Venice and in sectors directly affected by the marine ingression but also in the in-land coastal plain. Because of the complex hydraulic setting of the coastlands, RLS severely impacts and increases the

L. Tosi (✉) · P. Teatini · C. Da Lio
Institute of Marine Sciences, National Research Council, Venice, Italy
e-mail: luigi.tosi@ismar.cnr.it

P. Teatini
e-mail: pietro.teatini@unipd.it

C. Da Lio
e-mail: cristina.dalio@ve.ismar.cnr.it

P. Teatini
Department of Civil, Architectural and Environmental Engineering, University of Padova, Padova, Italy

T. Strozzi
GAMMA Remote Sensing AG, Worbstrasse 225, CH-3073 Gümligen, Switzerland
e-mail: strozzi@gamma-rs.ch

G. Lollino et al. (eds.), *Engineering Geology for Society and Territory – Volume 4*,
DOI: 10.1007/978-3-319-08660-6_32, © Springer International Publishing Switzerland 2014

risk of river flooding, damages to buildings and other man-made structures, and enhances the saltwater contamination of soils and aquifers causing land desertification. Hence, from a "regional scale" point of view, the evaluation of multi-year RLS patterns is fundamental and should be required for implementing appropriate water-management and land-use strategies.

Nowadays SAR-based interferometry methods are broadly used to measure ground displacements. Their high spatial resolution has allowed increasing the number of information acquired up to 1,000 times more than those obtained by the in-situ measures (leveling and GPS). Furthermore, for scales of investigation comparable to that of the present study, the millimetric accuracy recently reached by SAR interferometry makes the in situ measurements to be used only for calibration and validation of the former.

This study is aimed at quantifying the RLS occurred over the period 1992–2010 in the Venice coastland.

32.2 Ground Movements and Sea Level Rise

We assessed the ground movements from 1992 to 2002 and between 2003 and 2010 by the Interferometric Point Target Analysis (IPTA) on ERS1/2 and ENVISAT stripmap images, respectively. IPTA provides land displacements along the line-of-sight (LOS) between the satellite and the targets. Hence the knowledge of a movement of the reference, e.g., from leveling and GPS, is required to calibrate the SAR results and obtain "absolute" displacements (Teatini et al. 2012a). However, the use of SAR-based interferometry in large areas is also affected by the so-called flattening problem, i.e. the slight phase tilt resulting by the inaccuracy in estimation of the orbital baseline due to the not perfect knowledge of the satellite positions. In order to overcome this problem, IPTA results have been corrected by levelling and permanent GPS stations (CGPS) data (Teatini et al. 2012b). Finally, we assumed negligible both the difference between the LOS and the vertical component of the movements, which for ERS and ENVISAT is about 8 %, and the contribution of E-W displacements (e.g., due to tectonics). This latter because the network of CGPS stations installed around the lagoon has shown that the Venice coastland is moving in the horizontal direction almost uniformly (Tosi et al. 2013).

Figure 32.1 provides the annual average rate of ERS and ENVISAT LOS surface displacements. In general, both 1992–2002 and 2003–2010 velocity maps show similar behavior at the regional scale. The displacement rates exhibit significant spatial variability, ranging from a slight 1–2 mm/yr uplift, to a serious subsidence of more than 10 mm/yr.

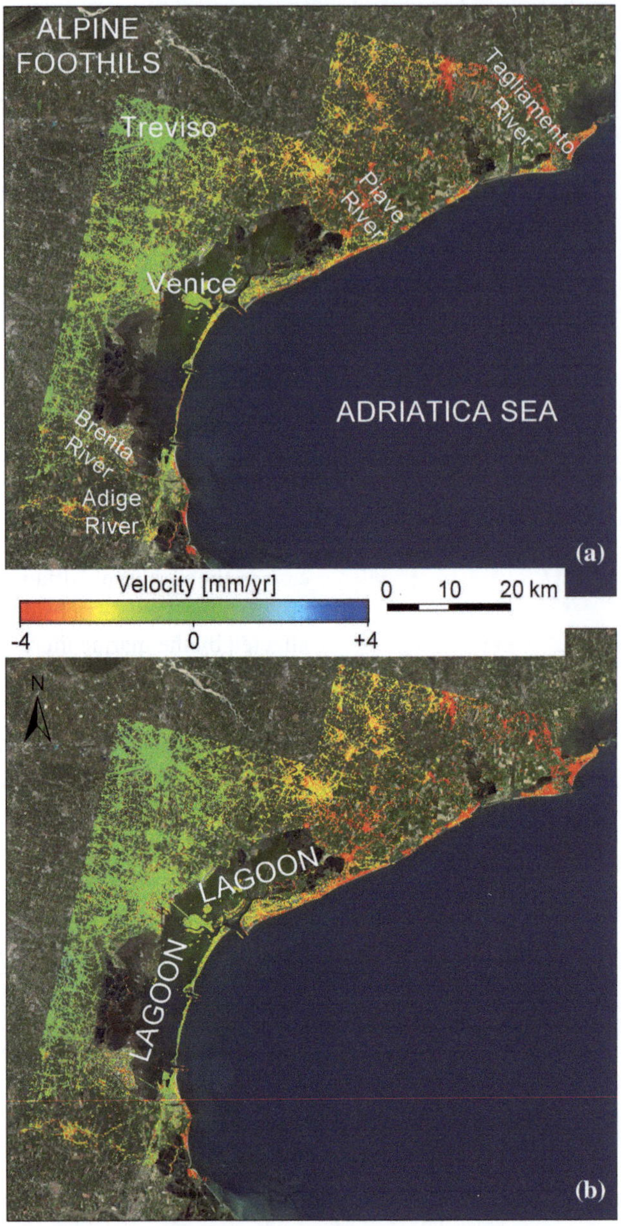

Fig. 32.1 Average LOS surface displacement (mm/yr) for the Venice coastland obtained by IPTA on: **a** ERS-1/2 images acquired between 1992 and 2002 and **b** ENVISAT images acquired between 2003 and 2010. Negative velocity values mean subsidence

The movement sources have been studied on the base of their area (Teatini et al. 2005) and depth (Tosi et al. 2009) of occurrence. Deep causes act and are recognizable in the movement of the pre-Quaternary basement. Medium causes include geological features, such as a major presence of compressible clay layers in the southern and northern portions of the study area and groundwater withdrawals, mainly in the north-eastern coastland and western mainland (Da Lio et al. 2013). Shallow causes are related to the architecture and geomechanical properties of the Pleistocene and Holocene

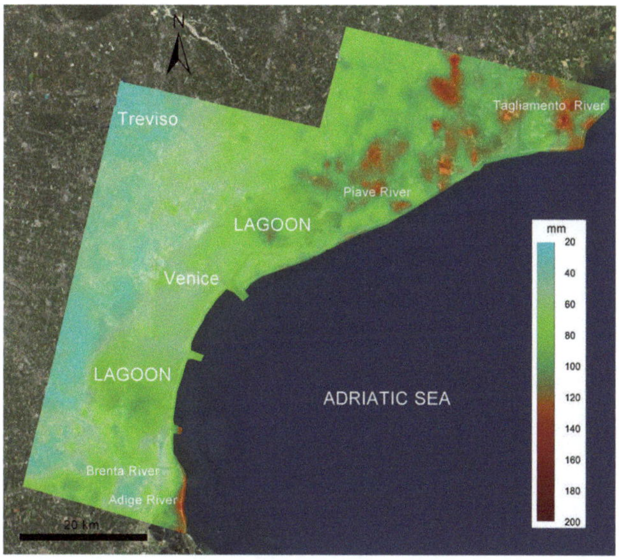

Fig. 32.2 Map of the RLS (mm) occurred from 1992 to 2010 in the Venice coastland

deposits, geochemical compaction and oxidation of the outcropping organic soils drained by land reclamation and the building loads in newly developed areas.

Concerning the sea level rise (SLR), it is well known that the long-term time series recorded at the Trieste tide gauge station likely corresponds to the eustacy component of the Northern Adriatic Sea because of the negligible contribution of land movement. Therefore we assumed of 1.2 mm/yr as the SLR at Venice, which is the average based on the 1890–2007 period (Carbognin et al. 2010).

32.3 Relative Land Subsidence

The RLS occurred from 1992 to 2010 has been quantify by interpolating using the kriging technique the IPTA 1992–2002 and 2003–2010 displacements ($d_{1992-2002}$ and $d_{2003-2010}$) on the same 1000 × 1000 m grid. For each grid cell, RLS $_{1992-2010}$ has been computed as:

$$RLS_{1992-2010} = SLR_{1992-2010} - (11 \times d_{1992-2002} + 8 \times d_{2003-2010})$$

where: $SLR_{1992\text{-}2010} = 19\times$ SLR.

Figure 32.2 shows the RLS computed at regional scale. A clear trend can be detected with values that increase from the central sector of the Venice lagoon northward and southward. The values range from less than 20 mm to more than 140 mm. In addition, local patches of higher RLS displacements superpose to the regional trend.

32.4 Conclusions

In this work we quantify the RLS occurred in the Venice coastland over the period 1992–2010. First, ground displacements are assessed by IPTA and then, the RLS are calculated by adding the component of the sea level rise to the ground displacements.

The results point out a high variability of the RLS with values that at regional scale range from less than 20–140 mm, and locally rising up to 200 mm.

In the hydrogeologic context of the Venice coastland, the largest effect of the loss in ground elevation with respect to the sea level concerns with the increasing exposure to flood during severe winter storms and saltwater contamination of agricultural soils and aquifers. The combined effect of both processes is producing an alarming social and environmental impact on the Venice coastland, also in relation to the expected global climate change. Both regional and local RLS have to be taken in account for appropriate planning of water-management, work interventions, and land-use strategies.

Acknowledgments This work has been developed in the framework of the Action 2 (SP3-WP1) funded by the Flagship Project RITMARE—The Italian Research for the Sea—coordinated by the Italian National Research Council and funded by the Italian Ministry of Education, University and Research within the National Research Program 2011–2013. Data courtesy: ERS and ENVISAT, INLET Project, funded by the Venice Water Authority through its concessionary Consorzio Venezia Nuova.

References

Carbognin L, Teatini P, Tomasin A, Tosi L (2010) Global change and relative sea level rise at Venice: what impact in term of flooding. Climate Dyn 35(6):1055–1063

Da Lio C, Tosi L, Zambon G, Vianello A, Baldin G, Lorenzetti G, Manfè G, Teatini P (2013) Long-term groundwater dynamics in the coastal confined aquifers of Venice (Italy). Estuarine, Coastal Shelf Sci 135:248–259

Teatini P, Tosi L, Strozzi T, Carbognin L, Wegmüller U, Rizzetto F (2005) Mapping regional land displacements in the Venice coastland by an integrated monitoring system. Remote Sens Environ 98(4):403–413

Teatini P, Tosi L, Strozzi T, Carbognin L, Cecconi G, Rosselli R, Libardo S (2012a) Resolving land subsidence within the Venice Lagoon by persistent scatterer SAR interferometry. Phys Chem Earth 40–41:72–79

Teatini P, Tosi L, Strozzi T (2012b) Comment on recent subsidence of the Venice Lagoon from continuous GPS and interferometric synthetic aperture radar by Y. Bock, S. Wdowinski, A. Ferretti, F. Novali, and A. Fumagalli. Geochem Geophys Geosyst 13(1)

Tosi L, Teatini P, Carbognin L, Brancolini G (2009) Using high resolution data to reveal depth-dependent mechanisms that drive land subsidence: the Venice coast, Italy. Tectonophysics 474(1–2):271–284

Tosi L, Teatini P, Strozzi T (2013) Natural versus anthropogenic subsidence of Venice. Sci Reports 3(2710), doi:10.1038/srep02710

Capability of X-Band Persistent Scatterer Interferometry to Monitor Land Subsidence in the Venice Lagoon

33

Pietro Teatini, Luigi Tosi, and Tazio Strozzi

Abstract

Land subsidence is a widespread phenomenon, particularly relevant to natural areas such as wetlands, deltas, and lagoons characterized by low elevation with respect to the mean sea level. The possibility of mapping vertical displacements of large coastal lowlying areas at very high resolution began in the 2000s with the development of the Persistent Scatterer Interferometry (PSI). This synthetic aperture radar (SAR)-based methodology detects the displacements of man-made structures within the landscape, such as buildings, utility poles, and roadways. Experience with SAR data of the European Remote Sensing satellites ERS-1, ERS-2, and ENVISAT, characterized by a spatial resolution of about 20 m, a wavelength λ of the signal of 5.66 cm (C-band), and a repeat cycle of 35 days, suggested that persistent targets can be identified only for a fraction of the number of buildings or infrastructures larger than a few meters, and the precise position of the SAR scatterers are difficult to discern. With a very-high image resolution (about 3 m) and an acquisition repeatability never available in the past (between 11 and 16 days), the new generation of X-band (λ = 3.11 cm) SAR satellites has improved significantly the possibility of monitoring the movements of single small structures scattered within rural and natural environments. In this work, we present the PSI outcome in the Venice Lagoon using the German TerraSAR-X and the Italian COSMO-SkyMed satellites. The investigated period covers the time interval 2008–2011. The outcomes of the two X-band analyses are compared both at the regional and the local scale in terms of target coverage and distribution and subsidence rates.

Keywords

Land subsidence • Persistent scatterer interferometry • X-band • Venice lagoon

33.1 Introduction

Over the last century, eustacy and land subsidence have produced a relative sea level rise (RSLR) on the Venice coastland ranging from centimetres to meters (Carbognin et al 2010). RSLR seriously increased the vulnerability and the geological hazard (e.g., river flooding, riverbank stability, seawater intrusion, coastline regression) of low-lying environments such as deltas, lagoons, wetlands, and farmlands, a large portion of which lies 3–4 m below the mean sea level. Land subsidence due to natural and anthropogenic activities is the main component of the RSLR in the Venice coastland, and therefore it has been monitored since the end of the 19th century. Over the last decade Persistent Scatterer

P. Teatini (✉)
Dept. of Civil, Environmental an Architectural Engineering, University of Padova, Via Trieste 63, 35121 Padova, Italy
e-mail: pietro.teatini@unipd.it

P. Teatini · L. Tosi
Institute of Marine Sciences, National Research Council, Arsenale - Tesa 104, Castello 2737/F, 30122 Venezia, Italy
e-mail: luigi.tosi@ismar.cnr.it

T. Strozzi
GAMMA Remote Sensing AG, Worbstrasse 225, CH-3073 Gümligen, Switzerland
e-mail: strozzi@gamma-rs.ch

G. Lollino et al. (eds.), *Engineering Geology for Society and Territory – Volume 4*,
DOI: 10.1007/978-3-319-08660-6_33,

Interferometry (PSI) has been effectively used to detect land displacements in the Venice region. In particular ERS1/2 and ENVISAT ASAR provided an excellent subsidence monitoring "network" suitable both for scientific and coastal management purposes over a 20-year acquisition period (Teatini et al. 2012; Tosi et al. 2010).

Recently, the development of new very-high resolution X-band radar satellites, such as TerraSAR-X and COSMO-SkyMed, has increased the applicability of PSI to monitor movements of single structures with millimetric precision and metric spatial resolution (Strozzi et al. 2009). In this work we compare the results of a ground movement analysis at the Venice Lagoon and the Po River Delta obtained by PSI on X-band radar images.

33.2 Materials and Methods

IPTA—Interferometric Point Targer Analysis (Wegmüller et al. 2004), a specific PSI processing chain, was applied on:

- a stack of 31 COSMO-SkyMed stripmap images acquired between September, 2008 and July, 2011 with a revisiting time usually equal to 16 days but sometimes much larger (up to 144 days from February 23, 2011, and 3 July 17, 2011). HH polarization scenes are characterized by a perpendicular baseline ranging from −719 to 695 m with respect to the central reference image of January 19, 2010, a bandwidth equal to 100 MHz and Doppler values from 600 to 800 Hz. The images are characterized by a 35° incidence angle;
- a stack of 30 TerraSAR-X stripmap images acquired between March, 2008 and January, 2009 with a regular 11-day revisiting time. HH polarization images are characterized by perpendicular baseline values between −148 and 190 m with respect to the central reference image of June 15, 2008, and by a 150 MHz bandwidth with good overlap of common reflectivity spectra. The images are characterized by a 28° incidence angle.

IPTA combines data from SAR images acquired from repeat track orbits to exploit the phase differences of the radar signals that are related only to surface displacement occurring between the image acquisitions. The contribution of surface topography to the interferometric phase is systematically removed and the variability in the tropospheric path delay mitigated by utilizing differences in the spatial and temporal characteristics of the phase due to deformation and atmosphere. IPTA-derived deformation measurements are interpreted for a number of point radar-bright and radar-phase-stable targets (PTs) that are coherent over the entire time interval and cover the monitored area as a sparse "natural" benchmark net.

IPTA results have been calibrated using permanent GPS stations to correct the so-called flattening problem (Teatini et al. 2012), i.e. the slight phase tilt resulting by the inaccuracy in estimation of the orbital baseline due to the not perfect knowledge of the satellite position.

33.3 Results

An image of the average land displacement obtained at regional scale by processing the COSMO-SkyMed and the TerraSAR-X scenes is shown in Fig. 33.1.

Note the larger frame of the available COSMO-SkyMed scenes with respect to TerraSAR-X, this letter covering the Venice Lagoon only.

The figure shows displacements ranging from −50 mm/yr of subsidence to a few mm uplift. In general the higher sinking rates are detected in the nearby of the inlets of the Venice Lagoon because of the soil consolidation induced by the MOSE works and in the embankments bounding the inner waterbodies in the Po Delta. Stable areas, i.e. with displacement of a few millimeters per year, are the historical center of Venice, its mainland, and the ancient sandy beach ridges well developed along the inner sector of the Po delta. The regional subsidence generally increases moving southward toward the Po Delta.

Interesting results are obtained at the local scale. As an example, we show the comparison between the displacements obtained using COSMO-SkyMed and TerraSAR-X images on the coastal structures under construction and reinforcement at the Lido inlet (Fig. 33.2). COSMO-SkyMed and TerraSAR-X produce similar outcomes on the littoral strip of Punta Sabbioni and provide a significant pattern of the displacement rates of the coastal structures. The spatial variability of the movements is related to the different consolidation degree of the coastal deposits below the constructions. A lowering on the order of 20–40 mm/yr is detected both by TerraSAR-X and COSMO-SkyMed all along the north structure with the exception of the 10-m deep dock excavated in the central part where TerraSAR-X points out a rising by almost 10 mm/year. This temporary upheaval is caused by the squeezing of the fine-grained subsoil forced by the load of the surrounding structures and the hydraulic overpressure acting on the lock bottom located at about 10 m below the mean sea. The southern jetty shows a substantial stability in the offshore portion and displacement rates down to −70 and −30 mm/yr with TerraSAR-X and COSMO-SkyMed, respectively, toward the root of the structure due to its recent reshape. The smaller displacement rates detected by COSMO-SkyMed is due to the longer period spanned by these images that account for the ongoing consolidation process.

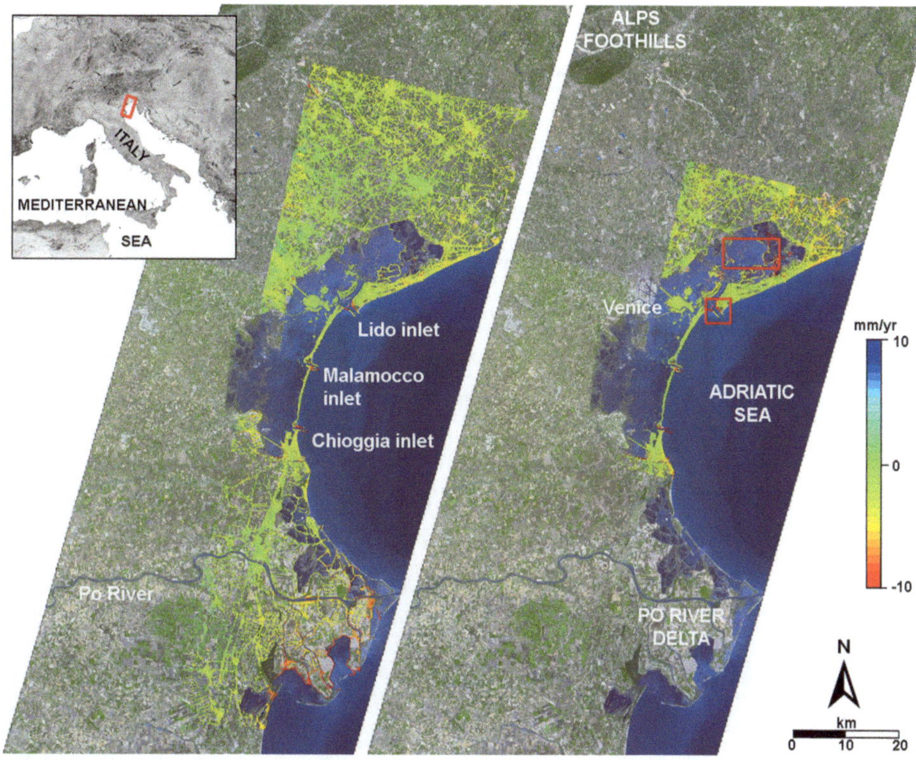

Fig. 33.1 Mean displacement rates of the Venice coastland obtained by PSI on (**a**) COSMO-SkyMed scenes from September 2008 to July 2011 and (**b**) TerraSAR-X scenes from March 2008 to January 2009. The *red boxes* in (**b**) represent the areas shown in Figs. 33.2 and 33.3. Movements are in the satellite line-of-sight direction. Negative values indicate settlement, positive uplift

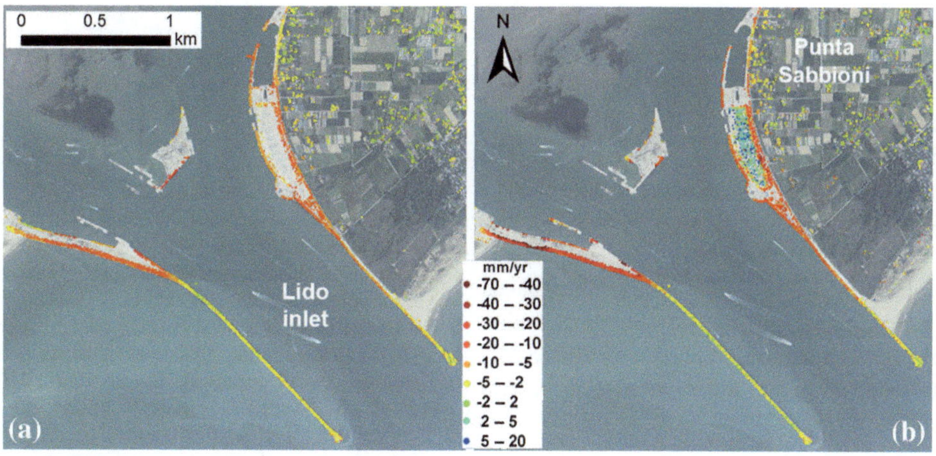

Fig. 33.2 Mean displacement rates at the Lido inlet of the Venice Lagoon obtained by PSI on (**a**) COSMO-SkyMed scenes and (**b**) TerraSAR-X scenes. The location of the area is provided in Fig. 33.1b. Movements are in the satellite line-of-sight direction. Negative values indicate settlement, positive mean uplift

A comparison of the suitability of COSMO-SkyMed and TerraSAR-X to monitor land subsidence in the natural lagoon environment is shown in Fig. 33.3. The area is located on the northern portion of the lagoon. Both the satellites are able to provide information along the embankments bounding Valle Liona and Valle Grassabò on the eastern side. However, TerraSAR-X detects much more measurable PTs at Valle Cavallino, St. Cristina Island and the marshes bounding the Valle di Ca' Zane westward. This higher capability is likely due to the more regular distribution of the scenes and the shorter period spanned by TerraSAR-X.

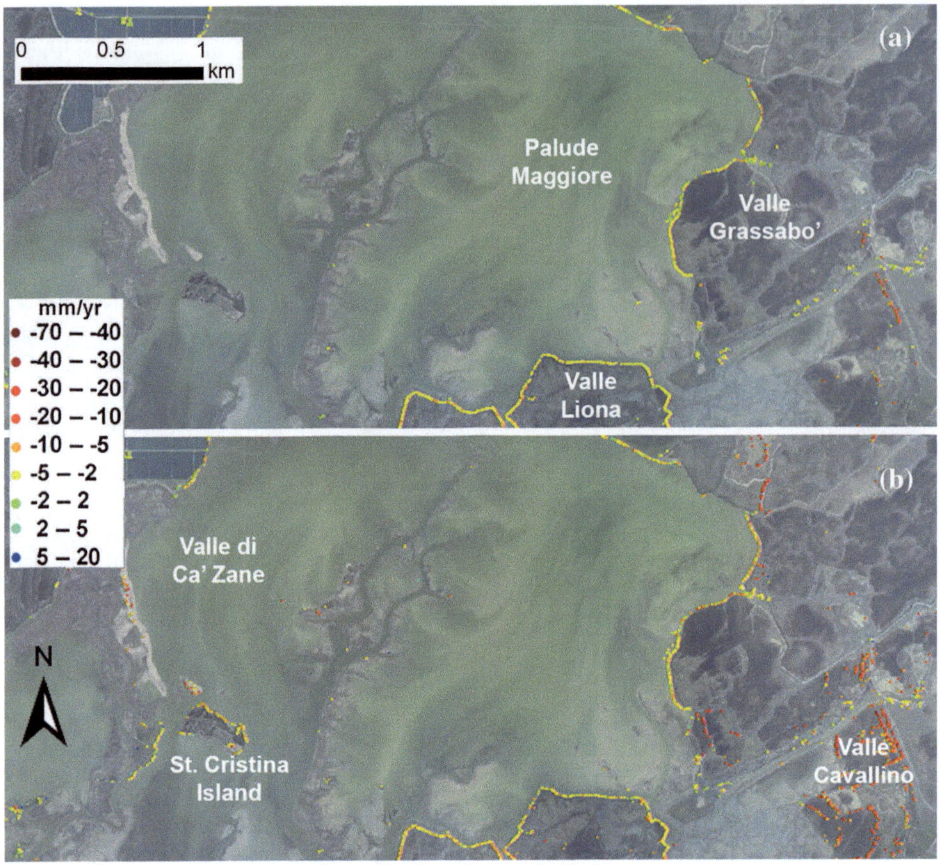

Fig. 33.3 Mean displacement rates on the northern portion of the Venice Lagoon obtained by PSI on (*above*) COSMO-SkyMed scenes and (*below*) TerraSAR-X scenes. The location of the area is provided in Fig. 33.1b. Movements are in the satellite line-of-sight direction. Negative values indicate settlement, positive mean uplift

33.4 Conclusive Remarks

In conclusion the use of X-band radar images to detect ground displacements by SAR interferometry allows the measurement of ground movements with millimetric precision and metric spatial resolution. The comparison between the results obtained by IPTA using COSMO-SkyMed and TerraSAR-X acquisitions shows that both satellite missions provide excellent data for detecting settlements and assessing accommodation of engineering coastal structures at high resolution. However, the shorter revisiting time and more regular acquisition properties of TerraSAR-X allow IPTA to detect coherent reflectors also in natural environments such as the inner Venice Lagoon.

Acknowledgments Data courtesy: (1) COSMO-SkyMed AO, Project ID 2100 © ASI "Land movements in the Venice Lagoon: measuring the natural/regional and anthropogenic/local components by SAR Interferometric Point Target Analysis (IPTA); (2) TerraSAR-X, Project COA0612 © DLR "Assessing vertical movements of natural tidal landforms and anthropogenic structures at the Venice Lagoon inlets".

References

Carbognin L, Teatini P, Tomasin A, Tosi L (2010) Global change and relative sea level rise at Venice: what impact in term of flooding. Clim Dyn 35(6):1055–1063

Strozzi T, Teatini P, Tosi L (2009) TerraSAR-X reveals the impact of the mobile barrier works on the Venice coastal stability. Remote Sens Environ 113:2682–2688

Teatini P, Tosi L, Strozzi T, Carbognin L, Cecconi G, Rosselli R, Libardo S (2012) Resolving land subsidence within the Venice Lagoon by persistent scatterer SAR interferometry. Phys Chem Earth 40–41:72–79

Tosi L, Teatini P, Strozzi T, Carbognin L, Brancolini G, Rizzetto F (2010) Ground surface dynamics in the northern Adriatic coastland over the last two decade. Rendiconti Lincei 21(SUPPL. 1):115–129

Wegmüller U, Werner C, Strozzi T, Wiesmann A (2004) Multitemporal interferometric point target analysis. In: Smits P, Bruzzone L (eds) Analysis of multi-temporal remote sensing images, vol 3. World Sci, Hackensack, pp 136–144

The Impact of Land Subsidence on Preservation of Cultural Heritage Sites: The Case Study of Aquileia (Venetian-Friulian Coastland, North-Eastern Italy)

34

Loredana Alfarè, Sandra Donnici, Mattia Marini, Massimiliano Moscatelli, Luigi Tosi, and Roberto Vallone

Abstract

In the north-western coastland of the Adriatic Sea flooding hazard due to land subsidence and eustasy seriously jeopardizes the many cultural heritage sites present in the area, among which the UNESCO site of Aquileia. Located about 10 km inward from the coastline, Aquileia has recently come to the fore for the severe structural damages caused to its Patriarchal Basilica by differential ground subsidence. This study aims at highlighting the main control factors for subsidence in Aquileia framing it in a regional context. Our investigation is based on the integrated analysis of vertical displacements from Persistent Scatterers Interferometry and levelling surveys from the literature along a N-S transect from Terzo D'Aquileia to Grado. Interpretation of results is based on a geotechnical reference model of the subsoil obtained mainly from borehole data. Beside confirming the regional subsidence scenario affecting the Venetian-Friulian coastland, our study indicates that in Aquileia, differences in subsidence rates between the archeological area and the new city likely relate to diverse consolidation paths induced in the subsoil of the two sites by differential and diachronous man-mad loading over historical time. The smaller scale yet stark variability of sinking rates observed at the Patriarchal Basilica are interpreted to relate to heterogeneities in the shallow subsoil.

Keywords

Land subsidence • Persistent scatter interferometry • Subsoil reference model • Load-induced consolidation • Consolidation history

34.1 Introduction

The north-western coastland of the Adriatic Sea is a subsiding area which during the Holocene high-stand hosted lagoons, deltas and marshes. Many important cultural heritage sites are present along this coastland, some of which are located in lowland areas and seriously jeopardized by flooding hazard due to land subsidence and eustasy (1.2 mm/yr in the northern Adriatic Sea).

Although the impact of such treat has been deeply studied in the past and continuously monitored in major cities such as Venice and Ravenna, there are other sites of great historical and archaeological value that have not been object of specific investigations. Among these, Aquileia is a small city located on the distal reach of the Isonzo fan delta and at the edge of the Grado and Marano lagoons, about 10 km inward from the coastline.

L. Alfarè · S. Donnici · L. Tosi
Istitute of Marine Sciences—National Research Council, Venice, Italy

M. Marini (✉) · M. Moscatelli · R. Vallone
Institute of Environmental Geology and Geoengineering—National Research Council, Rome, Italy
e-mail: mattia.marini@uniroma1.it

G. Lollino et al. (eds.), *Engineering Geology for Society and Territory – Volume 4*,
DOI: 10.1007/978-3-319-08660-6_34, © Springer International Publishing Switzerland 2014

In the antiquity Aquileia was the ninth largest city and one of the most important fluvial harbour of the Roman Empire. Since 1998 It has been inscribed in the UNESCO World Heritage Sites List for its well preserved remnants from the Early Roman Empire and the Patriarchal Basilica Complex from the Middle Age, which still conserves magnificent floor mosaics of the former 4th century basilica. With respect to subsidence-related hazards (i.e. flooding and salt water intrusion), the case of Aquileia has recently drawn the attention of the scientific community for the severe structural damages affecting the Patriarchal Basilica. This threat to the cultural heritage of Aquileia motivated the research action detailed here, which was developed in the frame of the project CHERPLAN (Enhancement of Cultural Heritage through Environmental Planning and Management) under the auspices of the South-East Europe Programme.

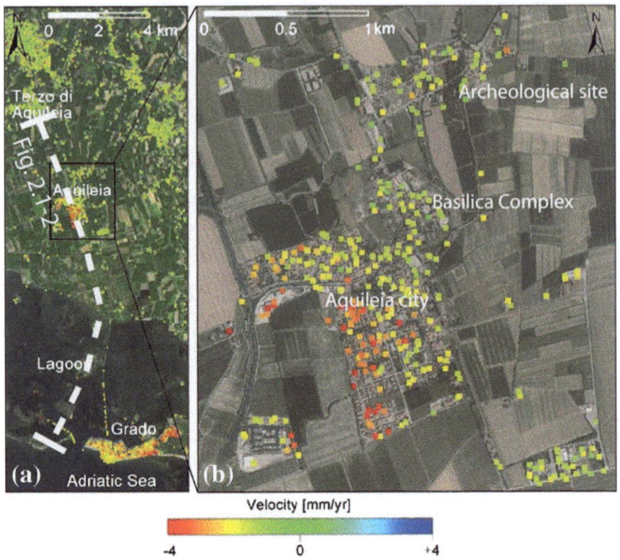

Fig. 34.1 Map of vertical displacement velocity (mm/y) obtained through Persistent Scatter Interferometry on ENVISAT images acquired between 2003 and 2010 for (**a**) the coastland between Terzo di Aquileia and Grado, and (**b**) the surroundings of the city of Aquileia. In (**a**) the dashed line shows the approximate location of Figs. 34.2 and 34.3

34.1.1 Aims, Methods and Materials

This study is intended to achieve a better understanding of subsidence in the Aquileia UNESCO site. Nonetheless, to allow framing our data and interpretations in the regional subsidence scenario, the study area has been extended to cover a N-S transect from Terzo D'Aquileia to Grado.

The approach followed in this research is based on integration of datasets of land vertical displacement from both Persistent Scatterers Interferometry (PSI) and levelling surveys (Marchesini 2006 and references therein), and matching of results with a subsoil reference model mostly based on boreholes.

The analysis of present-day land movements was performed through Persistent Scatterers Interferometry (PSI) on ENVISAT images acquired from 2003 to 2010 and provided by Portale Geografico Nazionale (http://www.pcn.minambiente.it/GN/).

For each persistent scatter (PS), PSI provides the time series of the land displacements along the line-of-sight (LOS) of the satellite, which slopes at 23° from the vertical direction. Since the difference between movements along the LOS and their vertical component is very small, we assume that LOS positive and negative displacement results solely from land subsidence and uplift, respectively (Tosi et al. 2010; Teatini et al. 2012). PS dataset were calibrated using the mean vertical land displacement velocity provided by the permanent GPS stations networks of the Regione Friuli Venezia Giulia and Istituto Nazionale di Oceanografia e Geofisica Sperimentale and levelling carried out by Istituto Geografico Militare Italiano. Finally, the vertical displacement velocities obtained confronting the 2004 and 1997 levelling surveys were utilized as a mean of validation of PSI results (Fig. 34.3).

The subsoil reference model was obtained correlating a number of boreholes from both the literature (Marocco et al. 1984; Arnaud-Fassetta et al. 2003) and public domain repositories. Palaeo-environmental and genetic meaning of the sedimentary units present in the subsoil were reconstructed by framing them in the Pleistocene-Holocene stratigraphic context with the aid of radiocarbon ages reported by Arnaud-Fassetta et al. (2003). Finally, sedimentary units were converted into geotechnical units basing on dominant lithotypes and the shallow subsurface (depth <5 m from ground surface) detailed compiling penetrometer data from technical reports made available by the Consorzio di Bonifica Bassa Friulana and the Aquileia municipality.

34.2 Results

34.2.1 Subsoil Reference Model

In the subsoil of the Terzo D'Aquileia-Grado transect (Fig. 34.2) from top to bottom two main sedimentary units can be recognized consisting of: (i) recent to present-day barrier-island passing landward to peat-bearing clayey-silty lagoon sediments (<10 m), and (ii) a package of alluvial deposits referable to the distal reach of the Isonzo fan delta. In turn, the latter comprises a relatively thin (<10 m) upper unit mostly made of sands and a lower unit of silts bearing thin peat intercalations.

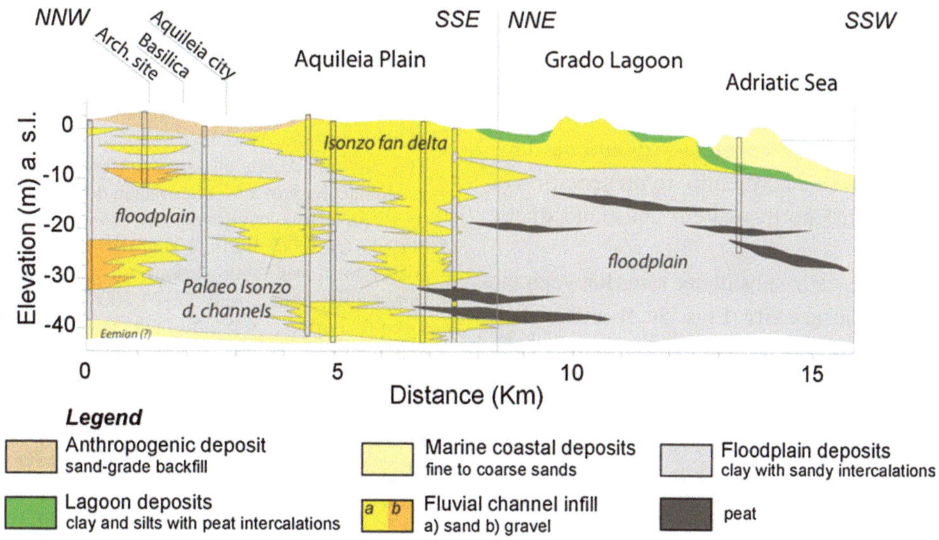

Fig. 34.2 Geological cross section along the Aquileia-Grado transect, modified after Marocco et al. (1984)

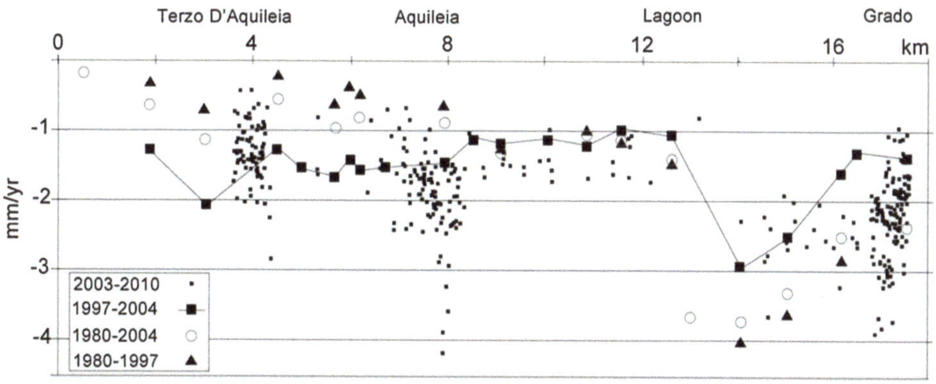

Fig. 34.3 Plot of velocity of vertical displacement (mm/y, negative values indicate subsidence) from levelling surveys (*black triangles, empty circles* and *black squares*) and Persistent Scatter Interferometry (*black dots*) along the Terzo di Aquileia-Grado transect (see Figs. 34.1 and 34.2 for location and geological cross section). Modified, after Marchesini (2006)

In the subsoil of Aquileia (Fig. 34.2), four major sedimentary-stratigraphic units can be recognized based on boreholes and radiocarbon dating from the literature (Arnaud-Fassetta et al. 2003).

From top to bottom these are (i) a relatively thin (<5 m) anthropogenic layer of historical age; (ii) a package of Holocene alluvial sediments of the Isonzo fan delta, comprising gravelly-sandy distributary channel infills (up to 10 m thick in the talvegs and 1 km wide), floodplain clays with rare peaty and sandy intercalations and deltaic sands; (iii) 15–25 m of Last Glacial Maximum (LGM) alluvial deposits of the Palaeo Isonzo river comprised of gravelly-sandy channel infill encased in peat-bearing floodplain clays; (iv) a package of sea shells-bearing sands referable to the Eemian interglacial.

34.2.2 Persistent Scatterers Interferometry

Consistently with levelling data (Fig. 34.3), PSI (Fig. 34.1a) indicate an increase in rates of vertical displacements from inland (1–2 mm/yr) to sea (2–3 mm/yr). Moreover, a broad variability (1–4 mm/yr) is highlighted in the urban areas of Aquileia and Grado, which in the former locality allows recognizing two distinct sectors with sinking rates of 1–2 and 3–4 mm/yr (Fig. 34.1b) corresponding to the archaeological site and the new city, respectively. Also, a smaller scale yet stark variability of vertical displacements (0.5–2.5 mm/yr) is observed at the Patriarchal Basilica, which is located halfway between the archaeological site and the new town.

34.3 Discussion and Conclusions

Our comparison of subsidence rates from PSI and levelling data from the literature (Fig. 34.3) confirms the scenario of regional subsidence affecting the Venetian-Friulian coastland and stresses its significance and criticality to preservation of the cultural heritage of the north-western coastland of the Adriatic Sea.

In Aquileia, the difference in subsidence rates between the archaeological area and the new city (Figs. 34.1b and 34.3) is interpreted to reflect the diverse consolidation paths of the dominantly clayey subsoil (Fig. 34.2), induced by differential and diachronous man-made loading over historical time. Conversely, the smaller scale variability of subsidence rates observed at the Patriarchal Basilica Complex would reflect differential ground surface settlements associated to heterogeneities of the shallow subsoil (Fig. 34.2), such as the presence of sandy-gravelly palaeo-channel infills encased in the high-compressibility peaty-clayey soils of the floodplain.

References

Arnaud-Fassetta G, Brigitte Carré M, Marocco R, Maselli Scotti F, Pugliese N, Zaccaria C, Bandelli A, Bresson V, Manzoni G, Montenegro ME, Morhange C, Pipan M, Prizzon A, Siché I (2003) The site of Aquileia (northeastern Italy): example of fluvial geoarchaeology in a Mediterranean deltaic plain. Géomorphologie: relief, processus, environnement 9(4):227–245

Marchesini C (2006) Vertical land movements in the Grado Lagoon (Italy) measured with various methods. In: 3rd IAG/12th FIG symposium, Baden, May 22–24 2006

Marocco R, Pugliese N, Stolfa D (1984) Some remarks on the origin and evolution of the Grado lagoon (Northern Adriatic Sea). Bollettino di Oceanologia Teorica ed Applicata 2:11–17

Teatini P, Tosi L, Strozzi T, Carbognin L, Cecconi G, Rosselli R, Libardo S (2012) Resolving land subsidence within the Venice Lagoon by persistent scatterer SAR interferometry. Phys Chem Earth Parts A/B/C 40–41:72–79. doi:10.1016/j.pce.2010.01.002

Tosi L, Teatini P, Strozzi T, Carbognin L, Brancolini G, Rizzetto F (2010) Ground surface dynamics in the northern Adriatic coastland over the last two decades. Rend Fis Acc Lincei 21(Suppl 1):S115–S129. doi:10.1007/s12210-010-0084-2

Ground Deformation Monitoring Over Venice Lagoon Using Combined DInSAR/PSI Techniques

35

Penelope Kourkouli, Urs Wegmüller, Pietro Teatini, Luigi Tosi, Tazio Strozzi, Andreas Wiesmann, and Kevin Tansey

Abstract

Synthetic Aperture Radar Interferometric (InSAR) techniques are widely applied to detect ground displacements caused by both natural and anthropogenic effects. Such techniques are a powerful tool for monitoring the ground motion, however, they have limitations such as temporal and geometrical decorrelation. Persistent Scatterer Interferometry (PSI), a pixel based methodology, has been developed to overcome such constraints. Nevertheless, also with PSI poor spatial coverage is observed in many case studies, especially over natural environments. In this contribution, we propose a methodology using elements from InSAR and PSI techniques in order to improve the spatial coverage with valid information over natural landscapes. For testing the applicability of this methodology, we selected as a test area the Venice Lagoon and specifically an area covered by saltwater marshes. The synergistic approach shows a good potential to achieve a good spatial coverage over the study area.

Keywords

SAR interferometry • Venice Lagoon • Marshland • Combined methodology

35.1 Introduction

The environmental and economic impacts of land subsidence can vary from negligible to very severe, depending on the location and features of the affected area, subsidence magnitude, rate and spatial extent (Abidin et al. 2007). Therefore there is a need to detect, measure and monitor land subsidence over a range of spatial and temporal scales. Remote sensing of the land surface from spaceborne satellites offers such a solution.

Since the beginning of 1990s, Differential SAR Interferometry (DInSAR) has been developed providing a new way for low-cost, time-effective, precise ground-motion monitoring, covering both small and extended areas (Goldstein and Zebker 1987; Gabriel et al. 1989). Furthermore, in the following years, Persistent Scatterer Interferometry (PSI), which is a pixel based method, has been developed (Ferretti et al. 2001). DInSAR and PSI both have their advantages and disadvantages. One important limitation, especially for DInSAR, is that it does not perform well over non-urban areas. For this reason, the spatial coverage is a severe limitation in many cases. To overcome or reduce such limitations, we developed a methodology combining components from DInSAR and PSI techniques, aiming at achieving a better spatial coverage and mitigating some limitations which arise when using the two techniques separately (Kourkouli et al. 2013).

P. Kourkouli (✉) · U. Wegmüller · T. Strozzi · A. Wiesmann
Gamma Remote Sensing AG, Worbstrasse 225, 3073 Guemligen, Switzerland
e-mail: kourkouli@gamma-rs.ch

P. Teatini
University of Padova, Department of Civil, Environmental and Architectural Engineering, Padova, Italy

L. Tosi
Institute of Marine Sciences, National Research Council, Venice, Italy

P. Kourkouli · K. Tansey
Department of Geography, University of Leicester, Leicester, UK

G. Lollino et al. (eds.), *Engineering Geology for Society and Territory – Volume 4*,
DOI: 10.1007/978-3-319-08660-6_35, © Springer International Publishing Switzerland 2014

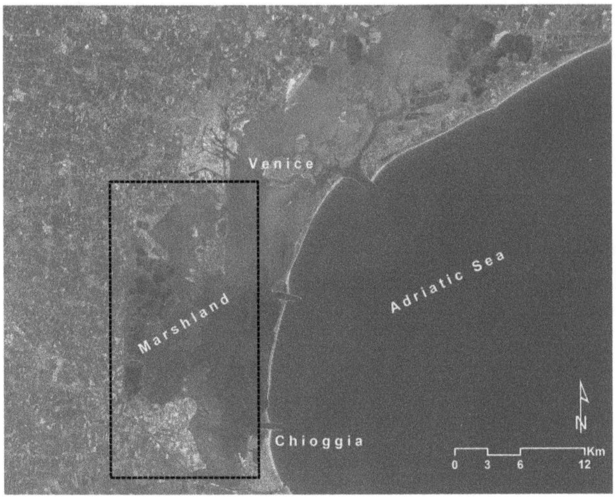

Fig. 35.1 Location map. The *black box* indicates the study area. As background used an ETM-Landsat 7 image

In this work the proposed methodology is further tested over an area in the Venice Lagoon (Italy). Specifically, the aim is to get better information over the natural areas of the lagoon, focusing on the salt marshes or tidal marshes (Fig. 35.1) ("barene" in Italian), which are defined as vegetated lands, a few decimeter above the msl, mostly inundated during high tides. This attempt is challenging as in such land cover types, the coherence is quite low and due to this the interferometric result is quite poor. The objective of this study is to investigate the potential of the proposed methodology and determine how well it performs in natural environments.

35.2 Study Area

Coastal environments are of great interest. Lagoons represent transitional environments between the land and the sea. For the current study, we focus on the Venice Lagoon, which is the largest lagoon in Italy. It is located on the NE part of Italy and covers about 550 km^2. The lagoon is formed by a number of morphologies which include channels, shallows, mud flats and salt marshes. Especially, the salt marshes within the lagoon environments are considered as one of the most valuable ecosystems on Earth. Their existence is vulnerable due to climate change and human inference. Over the last two centuries the area covered by saltwater marshes reduced in the Venice Lagoon from 115 to 33.5 km^2. Some of the reasons which contributed to the retreat are land reclamation, and the change of the lagoon hydrodynamics due to the dredging of navigable channels, ship-induced waves, sea level rise, land subsidence, which induced a severe erosional process to the lagoon morphologies (Amos et al. 2010). Previous studies carried out for monitoring the land deformation of the Venice area using both leveling, and GPS or SAR interferometric techniques (e.g., Teatini et al. 2007; Strozzi et al. 2009; Tosi et al. 2013) have not provide a comprehensive picture of the displacements because no information was retrieved in salt marshes and tidal flats. The deformation rates over the salt marshes (Amos et al. 2010) are still not known, except for a few artificial corner reflectors recently established and monitored by PSI (Strozzi et al. 2013). For this reason, applying the proposed methodology is of particular interest.

35.3 Proposed Methodology and Materials

35.3.1 Data Used

For the present study a dataset of 14 PALSAR scenes acquired from the Advanced Land Observing Satellite (ALOS) was used. Its long wavelength ($\lambda \sim 23$ cm) permits deep penetration over natural environments (e.g. vegetation, sand) resulting in higher temporal coherence over long time intervals. The data acquired in Fine Beam Single (FBS) and Fine Beam Dual (FBD) mode with 10 m spatial resolution, at HH polarization, in ascending orbit (Track: 639, Frame: 900) between February 2007 and July 2010 were used.

35.3.2 Combined DInSAR/PSI Techniques

In order to monitor land displacements over the salt marshes, we propose a synergistic approach (Kourkouli et al. 2013), combining elements from both DInSAR and PSI methodologies. Firstly, a 2-D differential interferometric (DInSAR) technique was applied using a single reference approach. Hence, a total number of 14 differential interferograms with respect to the date 20-July-2008, was calculated. A multi-looking (ML), 4 × 10 looks in range and azimuth respectively, applied on the interferograms to eliminate the speckle noise and keep a number of potential distributed scatterers into the later analysis. The 2-D ML differential phases were extracted into point locations in order to introduce these into the PSI approach.

The PSI approach was applied additionally using the same scene as reference. A candidate point list (PT) was generated regarding two different criteria. The first one is based on the spectral characteristics of the individual point targets and the second one on their temporal variability. Using this point list, the point data interferograms were calculated.

Afterwards, both 2-D ML differential phases and point data interferograms combined into one point data stack, as well as the PT list with the 2-D ML point locations. Hence, we proceeded with the combined point data stack of differential interferograms and the combined point list to the later analysis. A least-squares regression was performed on the point phases to estimate point heights and deformation rate. Filtering was applied on the residual phases, both spatial and temporal, in order to eliminate contributions in atmosphere and non-linear deformation terms. Final unwrapped phases were converted into displacement values. The final result presents relative displacements towards the Line of Sight (LOS).

35.4 Results and Discussion

The combined interferometric methodology permitted to achieve a quite good spatial coverage over the marsh surfaces (Fig. 35.2), which was the main objective. To validate our approach, the combined methodology was tested on the historical city of Venice and we noticed that the displacement rates are similar with those provided by previous studies (Teatini et al. 2007; Strozzi et al. 2009; Tosi et al. 2013). It is worth noting that the final accepted time series were derived after a quality control using a coherence threshold of 0.8 keeping only the most reliable solutions in the result. Several monitoring points were detected within the lagoon. The highest concentration are located on the higher reclaimed areas located south of the industrial port. These artificial islands ("casse di colmata" in Italian) were constructed in the 1960s from material which came from the digging of the "Canale dei Petroli" channel. A significant number are also spread over low lying marshes. The measured displacement rates are largely variable, ranging from a substantial stability to settlements of more than 25 mm/yr. A significant variability of the subsidence rates was also detected by the artificial reflectors (Strozzi et al. 2013), reflecting the heterogeneity of the deposits composing the lagoon subsurface. Specific multi-disciplinary investigations are required to deeply understand these values.

A regional trend was extracted by interpolating the DInSAR/PSI outcome using the Kriging methodology (Fig. 35.3). Overall, the northern part of the study area, as well as the southern part, is quite stable. On the contrary, the central part of the study area, where the natural morphological features are mainly located shows higher subsidence rates up to 20 mm/yr. Additionally, two profiles were plotted on the displacement map (Fig. 35.4). The first one (AA′) is along the coastline (NS direction). In the major part of the profile, the deformation rates vary from −8 to −15 mm/yr. The second profile (BB′) is oriented NW-SE. It shows that the deformation rates near the coastline varies from −2 to −8 mm/yr, while in the central part of the profile is observed the maximum subsidence values (−18 to −20 mm/yr). Moving to SE of the profile, we observed that the deformation rates vary from −9 to −14 mm/yr. High rates are reasonable in this case, due to the fact that the longer L-band wavelength offers a good possibility to achieve fast movements. Furthermore, applying the combined methodology permitted to detect movements of the marsh surfaces. Conversely, the previous PSI measurements have always provided information of anthropogenic "natural" or "artificial" structures that, having a certain foundation depth, neglect the deformation of the shallowest (approximately 1 m thick) deposits.

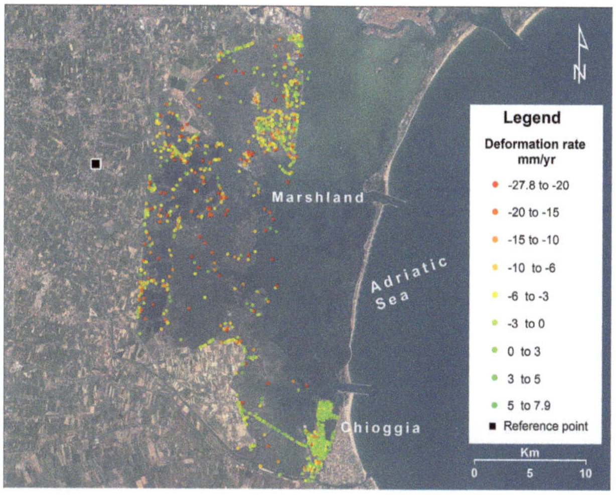

Fig. 35.2 Combined interferometric result showing displacements towards the Line of Sight (LOS). As background used an ETM-Landsat image

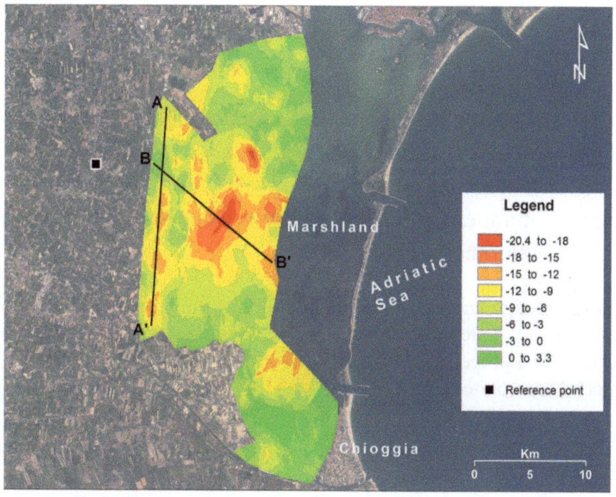

Fig. 35.3 Kriging interpolation of the combined interferometric result. AA′ and BB′ profiles are drawn

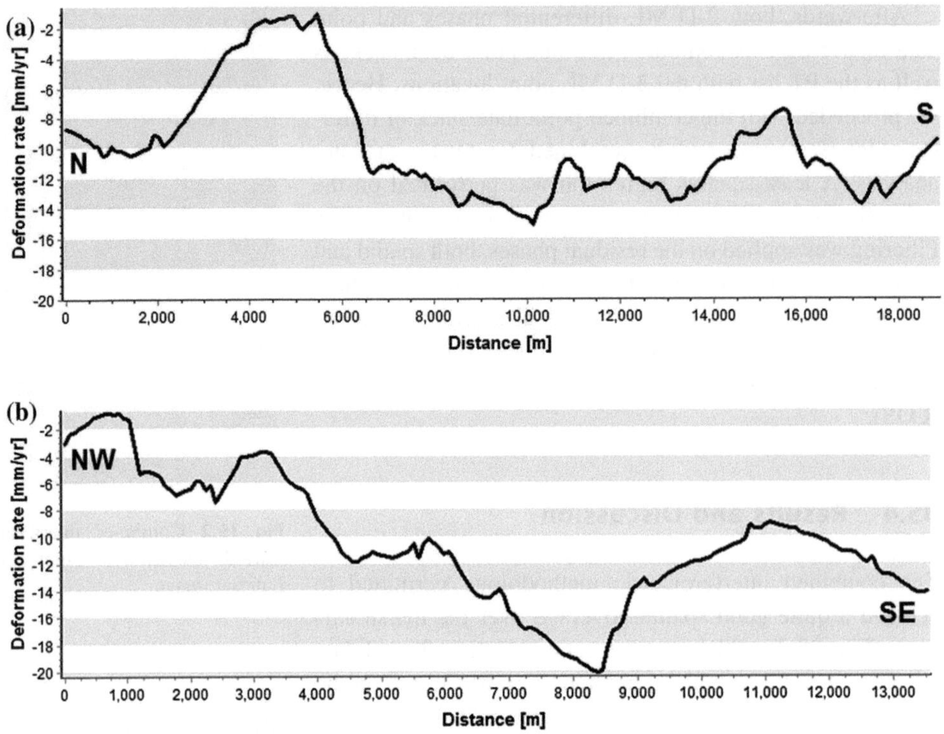

Fig. 35.4 Profiles of the AA′ and BB′ that are drawn in Fig. 35.3

Results show good potential as methodology for this case. The combined method achieved a good spatial coverage over the marshland which permits to achieve information about natural environments whose subsidence cannot detect easily with other techniques.

Acknowledgments Penelope Kourkouli gratefully acknowledges the Marie-Curie fellowship funded by the Marie Curie People Actions under the European Seventh Framework Programme (GIONET project). Data courtesy: PALSAR © Jaxa.

References

Abidin HZ, Andreas H, Djaja R, Darmawan D, Gamal M (2007) Land subsidence characteristics of Jakarta between 1997 and 2005, as estimated using GPS surveys. GPS Solut 12(1):23–32

Amos CL, Umgiesser G, Tosi L, Townend IH (2010) The coastal morphodynamics of Venice lagoon, Italy: an introduction. Cont Shelf Res 30(8):837–846. doi:10.1016/j.csr.2010.01.014

Ferretti A, Prati C, Rocca F (2001) Permanent scatterers in SAR interferometry. IEEE Trans Geosci Remote Sens 39(1):8–20

Gabriel AK, Goldstein RM, Zebker HA (1989) Mapping small elevation changes over large areas: differential radar interferometry. J Geophys Res 94(B7):9183

Goldstein RM, Zebker HA (1987) Interferometric radar measurement of ocean surface currents. Nature 328:707–709

Kourkouli P, Wegmüller U, Wiesmann A, Tansey K (2013) An integration of 2-D multi-looked differential interferograms with persistent scatterer interferometry. Living Planet Symposium, Edinburgh, pp 9–13

Strozzi T, Teatini P, Tosi L (2009) TerraSAR-X reveals the impact of the mobile barrier works on Venice coastland stability. Remote Sens Environ 113(12):2682–2688

Strozzi T, Teatini P, Tosi L, Wegmüller U, Werner C (2013) Land subsidence of natural transitional environments by satellite radar interferometry on artificial reflectors. J Geophys Res Earth Surf 118:1177–1191. doi:10.1002/jgrf.20082

Teatini P, Strozzi T, Tosi L, Wegmüller U, Werner C, Carbognin L (2007) Assessing short- and long-time displacements in the Venice coastland by synthetic aperture radar interferometric point target analysis. J Geophys Res 112(F1):F01012

Tosi L, Teatini P, Strozzi T (2013) Natural versus anthropogenic subsidence of Venice. Sci Reports 3:2710

Coastal Impacts Around Guadiaro River Mouth (Spain)

36

J. Javier Diez, Rodriguez Fernando, and Efren M. Veiga

Abstract

The "Port of Sotogrande" Beach (San Roque, Cadiz, Spain) has suffered significant erosion and changes since the construction of the marina and port of Sotogrande (San Roque, Cadiz, Spain). This paper reviews the dynamical processes on Guadiaro front and establishes relationship between them. It sets from a comparative evolution of the Alboran Sea Coast outlets and bays since the Little Ice Age, which shows that the Guadiaro estuary has remained functional while all other Alboran fluvial estuaries silted to soon become deltas, to analyse the most detailed recent changes on the beach. The study shows the evidences of the coastal impacts around the mouth, even further than mouth littoral barriers, of the port infrastructures; and it provides new elements to understand the dynamical processes on the mouth and surroundings as well. That should be fundamental for shore protection along the whole coastal stretch.

Keywords

Outlets-inlets • Geometric/massif singularities • Refraction-diffraction • Estuary-delta, global changes

36.1 Introduction

The Port of Sotogrande (San Roque, Cadiz, Spain) was designed and built on the middle of old Sotogrande beach that extended southern of Chullera point at both sides of Guadiaro river-mouth (Fig. 36.1). It extends southward since just in front of the left bank of the estuary, where the mouth spit-barrier was originated. The about 10° NNE-SSW coastal alignment southern of Chullera point is edged by sandy beach and finely uniform for several kilometres; its central stretch is shaped by a littoral barrier closing the Guadiaro mouth and the whole beach and shoreline was considered stable till the earlier seventies, as can be noticed from Fig. 36.2.

The Guadiaro estuary opens to a coastal stretch of the Alboran basin, between Iberian and North African coasts, next to the Strait of Gibraltar, western Mediterranean basin, where coastal orientation is NNE-SSW (nearly N-S, in fact). The coastal processes along Iberian Alboran coast as a rule is mainly affected by eastern and western wind and wave actions, but in this west end the action of eastern (Levantes) is totally determinant, as a simple look at the map indicates; The coastal orientation around Guadiaro permits to understand, however, that the obliquity of that action is weak and the transversal wave action dominates. Different studies "with numerical model based on the mild slope equation (waves, currents, and sediment transport) and the coastal evolution (SMC and MIKE 21) (Muñoz et al. 2010) asset a southwest-wards net littoral transport", while the mouth spit remains currently growing northwards even during easterly wind-waves. Those studies have obtained relevant results

J. J. Diez · R. Fernando (✉)
Universidad Politecnica, Madrid, Spain
e-mail: fernando.rodriguezl@upm.es

J. J. Diez
e-mail: josejavier.diez@upm.es

E. M. Veiga
UPM, Madrid, Spain
e-mail: efrenmartin@gmail.com

G. Lollino et al. (eds.), *Engineering Geology for Society and Territory – Volume 4*,
DOI: 10.1007/978-3-319-08660-6_36,

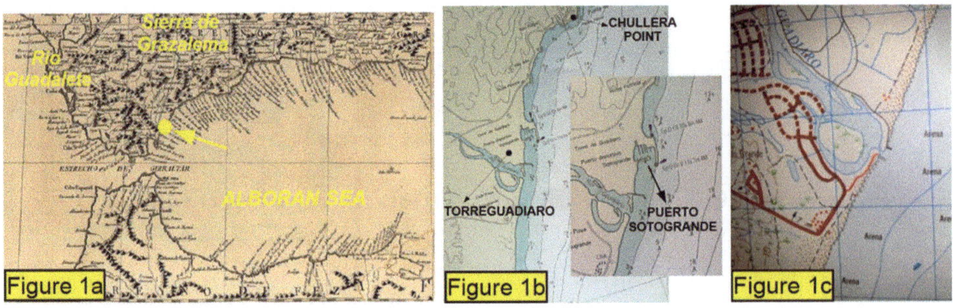

Fig. 36.1 **a** Historical map Guadiaro river mouth situation in Iberian Alboran sea cost. **b** Guadiaro mouth (1 jetty) & Port of Sotogrande. (1989 Spanish Nautical Chart 453). **c** Guadiaro river mouth (2 jetties). (1973 Spanish Topographic map 14-47/1075)

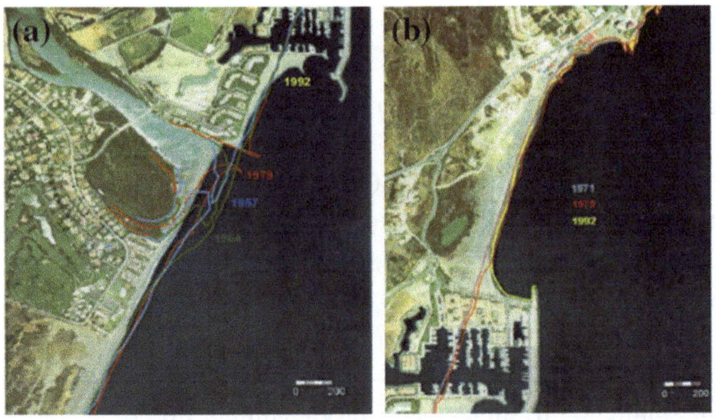

Fig. 36.2 Southern **a** 1957, 1964, 1979, 1992 and Northern, **b** 1971, 1979, 1992 shoreline situation, following University of Cadiz (2009)

on the refraction, though not on the diffraction, analysis. Two successive structural actions around the mouth can have significantly modified coastal processes and their influence was not well analyzed through those numerical models: (1) a couple of jetties on the mouth (1973), one soon removed (1975), and (2) the breakwater of the port of Sotogrande (built in 1986 and extended in 1994) on the left side of the estuary spit barrier.

The Guadiaro river sources of Grazalema Sierra (Cádiz), western Penibetic ridge, are among the highest and most homogeneous rainy in Spain throughout the year. More than regulatory, a recently implemented upstream transfer to the Atlantic through Guadalete's reduced its mouth flow peaks and favoured its low tidal mouth closure, but this transference is posterior to the impacts here discussed.

36.2 Survey Observations

Beside proper works, this survey has taken into account the available researches (Muñoz et al. 2010 and Dirección General de Costas 1996) and the pictorial documentation of the Environmental Departments of both the Spanish National Geographic Information Centre (CNIG) and the Andalucía Cartographic Institute (ICAE).

36.2.1 About Alboran Outlets

Ancient maps of the Iberian Peninsula, since Cantelli's (1676) till last Tomas López' (1785), including many others of just qualitative estimations, mainly the recently discovered geomorphologic Teixeira's of 1634 (King Planet Atlas, Wien Library), indicate that Little Ice Age related mouth transformation, since estuarine bays to sea-advanced deltas could accentuate along the eightieth century. It is congruent with other data (Esteban et al. 2006; Diez 2012).

36.2.2 Geomorphological Research

Coastal evolution was analyzed by comparing aerial orthogonal photos from the years 1946, 1956 ("American flight"), 1964, 1973, 1986, 1989, 1996, 1998, 2000, 2001, 2003, 2004, 2005, 2006, 2007, 2008, 2009, and 2011, and other side ones.

Shoreline changes and evolution could be analyzed since 1957, long before that related developments and port works took place, from just Fig. 36.2, that show both sides of the beach around the mouth. It evidences the alongshore littoral transport barrier role of the port, as to produce heavy river mouth and southern erosions, particularly on the beach stretch immediately southern of the port itself. All the beach forms in it are concordant with a North to South net littoral transport, and show wide "dry beach" (Bores 1978) till port construction in 1986. Figure also shows the supported and the sheltered configurations of the beach beside both jetties (note that the 1975 demolished southern jetty effects disappeared in 1977).

A bathymetric research was carried out from bathymetric documentation of the cited studies and Coastal Administrations (1984, 1989, 1994), up to compare their cross profiles, between the port and the current outlet, with the November 2011 bathymetry facilitated by the Sotogrande Beach Condominium (Fig. 36.3).

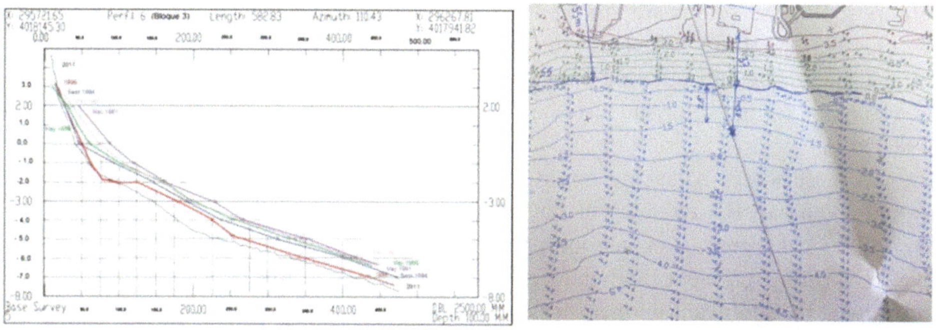

Fig. 36.3 **a** Compared evolution of 6 beach cross profile in the first 300 m from shoreline: 1969 (*red*), 87 (*mauve*), 89 (*green*), 94 (*dark blue*), 96 (*brown*) and 2011 (*black*). It shows the erosive process losing the bar and steepening the slope. **b** November 2011 Bathimetry (surveyed by Port of Sotogrande Beach Condominium)

36.3 Former Results: Guadiaro Estuary Evolution

Nearly all Iberian Alboran bays and river-outlets have been dramatically evolving since/along the Little Ice Age from open mouth geometries, by forming deltas in within the estuaries, to the sea deltaic advance (Diez 1996); likely a consequence of the hydrologic-erosive changes of the Global Change (Diez 2012). Guadiaro's seems to have delayed the processes, however, maybe due to its regular and intense fluvial activity, which avoided its early spit barrier to close the mouth. Teixeira shows a very northern spit on the left bank of the mouth that supported a long beach southern of Chullera point already in 1634. The evolution of the mouth inlet was typical: down-drift migration and narrowing, together with the southward spit barrier progression, till the inner/outer level gradients due to tides, storms and fluvial flow can break a new up-drift inlet branch, more direct and efficient for the water flow. These changes repeat cyclically about. The inlet migration and the spit barrier progression precisely witness the southward sense of the littoral drift around Guadiaro mouth, which is so remaining since already 1634 till the first port works (the couple of mouth jetties in 1972—documented since 1973).

The Guadiaro delta has remained in within the mouth for a long time permitting the slow infilling on both margins and building a wide littoral coastal plain southern of Chullera point. The mouth has so behaved following the Geometric Singularity more than the Massif Singularity (Bores 1978) and the infilling by silting permitted the development of a littoral spit barrier, which, itself favoured the plain increasing. The seaward deltaic deposits are not evidenced till entered twentieth century, mainly as outlet ebb tidal delta, with river and coastal sediments together, and it shown its maximum around 70's more due to both the jetties construction and harbour development than to the reduction of fluvial sediments. One of the mouth jetties, southern, was immediately demolished, probably due to several accidents attributed—not documented yet- to the great and fast movements of delta sands after their construction. As a result, erosion immediately started along the southern coast and is still remaining, after having accentuated since the port breakwater building (1986) and accretion (1995).

36.4 Discussion: Analysis of Works Effects and Reflections about it

Starting development in 1976, first action however was the building in 1972 of two mouth jetties, unexplained and unexplainable (the aiming to better regulate the river does not appear clear, distant nearly 0.4 km is too excessive in a small tide and regular flow mouth, they could be mainly designed in fact as for determining the littoral processes, particularly the alongshore transport regime), whose major aim might have been the observation of the littoral process answer. This answer was immediate, for 1973 (Fig. 36.1b), closing the existing outlet beside the northern jetty and opening a new one south of the southern jetty (one bad quality side pictures, author unknown and dated 1973, shows both jetties and the southern mouth-outlet). The obvious erosion effect south of the southern jetty and the whole mouth sedimentary instability was important and that jetty was demolished before 1977. Jetties construction is also the first moment in which spit barrier progress northwards after having been progressing southwards all time.

The breakwater for the harbor was built opening to South since 1986 on the beach and spit northern of the mouth and splitting it in Torreguadiaro Beach, at the north and Port of Sotogrande Beach, at the south of the port. The significant mouth erosion initiated in 1973 accentuated and extended to the beaches southern of the harbor since 1986. It is curious to notice the 1962 Spanish nautical chart, 453, showing the port in 1989 superimposed on the southward spit of 1962. Topographic maps of 1973 and 1988 (SECEG) (the first showing the jetties and the second still ignoring jetties and port) show the same situation for the spit. And also the 772 British Chart of 1971.

The beach answered these works balancing northward, that is, eroding its half south part (by reducing its bar and steepening its cross profile) and accreting its half north part; balance becomes a port silting problem, and the profile changes are the first signal of significant permanent beach erosion. To correct balancing and avoid silting the breakwater and counter-dyke system were extended since 1995, but no dynamic measure was implemented to correct net erosions, which increased and led to relevant shore damages. Since 1998, several actions for back-passing and nourishments, to retreat and reinforce sea wall, and to rebuild the jetty were taken (2004, 2010).

The ortho-photos before 1973 undoubtedly show that alongshore littoral transport maintains a southward net sense before the beginning of the mentioned works, about 1972 (1964'—its shoreline—shows the maximum advance of the ebb deltaic front). The mouth beach system between points was stable since the spit barrier closed the mouth pushing the fluvial/tidal outlet to the south; the littoral drift became then maximum southward and the sand infilling of the mouth minimum, but cohesive sediments from the river basin deposited mostly in within the mouth fastly extending the estuary delta end reducing the tidal prism, only significant under storm conditions because tide is small here. Therefore the cyclic process of southward outlet migration and opening of new updrift outlet became also faster permitting the formation of a increasingly local and notorious ebb delta, itself increasing the width and stability of the whole beach.

The spit currently grows northward, despite wave direction and alongshore transport occur mainly, and almost permanently, southward, but it remained growing southwards all time (Teixeira 1634) up to the beginning of port of Sotogrande works in 1973. This northern spit growing that closes the current Guadiaro inlet is explained as a north-eastern-wards littoral transport attributed (Muñoz et al. 2010), on the basis of a mild slope equation numerical model, to the (ebb delta refractive) "Finkl effect" though realizing that the "use of wave propagation numerical models by non-experts is prone to error, as small variations in input parameters can induce dramatically different results that may be contradictory in certain bathymetric configurations." However there is total photo evidence of both the wave diffraction consequences, by the jetties first and by the breakwaters later, and its time-space tight relation with the transport and spit reversal, just since 1973. The transport reversal in front of the mouth evidences through the mouth outlet and the spit barrier evolution and changes (e.g., the southern counter-spit is the consequence of the ebb-delta refraction plus northern spit diffraction effects); the change to northward transport around the whole mouth is only dominant, however, since 1973, though the role of the ebb delta refraction is obvious, diffractions on jetty, first, and breakwater, later, are much more determinant, and evident in any space-time analysis as all data support.

36.5 Conclusions

The erosive problems of the Port of Sotogrande Beach are obviously related to the littoral/inlet processes around Guadiaro mouth and mainly affected by eastern wind waves with net southward littoral drift. Wave action has a strong orthogonal effect, however, through cross profile and bar changes. The alongshore transport reversal in front of the mouths only dominates since 1972 and, being explainable by ebb delta refraction, diffractions on jetty and breakwater are more determinant.

The reversal effects of ebb deltas are obvious but very subtle, and for a long time have been confused with inexistent "up-drift migration" of tidal inlets; they need to be rightly managed and the "gross" effects of hard obstacles as jetties and, overall, port breakwater cannot be confused with those subtle adaptations. The determinant role of jetty and port works on the beach erosion, besides the port and also other stretches of the whole beach, cannot be therefore ignored, further that their natural evolution.

References

Bores PS (1978) Shore classification-simple forms with prevailing wind wave action. In: III IAEG International Congress Proceedings, Madrid, pp 150–169

Diez JJ (1996) Guía Física de España: VI Las Costas. Alianza Editorial, Madrid, pp 712

Diez (2012) Hydrological and morphological evidences of flood changes in Iberian Peninsula through XVII and XVIII centuries (COST Deciphering River Floods Symposium), Viena

Dirección General de Costas (1996) Estudio de la dinámica litoral del rio Guadiaro y su entorno. San Roque, Cádiz. INSUB-ERA 3 S.L, p 137

Esteban V, Diez JJ, Fernandez P (2006) Evolution of the Iberian Peninsula coast and recent climatic changes: port facilities and coastal defense in the Muslim domain. J C R. SI 39, Brazil, pp 1839–1842

Muñoz-Perez JJ, Caballero I, Tejedor B, Gomea-Pina G (2010) Reversal in longshore sediment transport without variations in wave power direction. J Coast Res 26(4):780–786. W Palm B (Florida)

Universidad de Cádiz, Gestión Integrada de Áreas Litorales (2009) Gestión integrada de zonas costeras y cuencas hidrográficas: introducción a un caso de estudio. El río Guadiaro. VI report UCA-costas convenio. Universidad de Cádiz Publisher. www.gestioncostera.es

Actual Geological Processes Acting on the Western Coast of Black Sea

37

Mihaela Stănciucu

Abstract

In the last decades the whole border of Black Sea had been tremendously damaged due to a number of causes such as the construction of Iron Gates barrages and other dams on Danube effluents, or global sea level rising. The main objective of the project "Implementation of adequate infrastructure of natural risk prevention in most vulnerable areas. Reduction of coastal erosion" was to propose large scale measures in order to mitigate those effects (erosion and landslides). For this purpose, extensive and complex site investigations have been made by international research teams, among which geological, geotechnical and geophysical, on the whole Black Sea coast. The most interesting results were provided by offshore seismic investigations which depicted the presence of a fault network inside the major structural units, on which vertical tectonic movements can be a major cause of long term costal erosion and large scale landslides. The findings raise the problem of the choice of reference benchmark of all topographical, geological and hydrogeological monitoring systems of the whole South Dobrogea unit, knowing that the very large scale vertical movements are much more difficult to detect with usual survey programs.

Keywords

Paleorelief • Erosion risk

37.1 Introduction

After the damming of the Danube River at Iron Gates I and II (1970) the river sediment discharge diminished by almost 40–45 % and this evolution of solid flow induced a quick erosion of the Black sea border. In the last 30 years the maximum erosion rates of the shoreline ranged between 20 m/year on the northern coast, to 5 m/year on the southern coast. Simultaneously, large landslides developed wherever the shore raised substantially above the sea level (Stănică 2011). Geological, geotechnical and geophysical surveys have been carried out, onshore and offshore, in order to investigate the most critical areas exposed to erosion. Geotechnical boreholes, light penetration tests, longitudinal and transversal refraction seismic profiles were carried out on 25 locations. The subsidence of at least some parts of South Dobrogea unit is a very probable phenomenon (that remain to be proven in further research), due to the following argumentation: (a) this movement is in agreement with the geological trend of this unit, already underlined by stratigraphyc studies (Seghedi 2012). The shape of the tectonic structure defined by near vertical faults predisposes mainly to vertical motions and in addition the proximity of seismogene area of Shabla provides the impulse for these presumptive neotectonic movements. (b) The phenomenon is also caused by groundwater withdrawal produced by intense exploitation of fresh waters for human or industrial use. Consequences of this presumptive process are obviously major for the land management in the costal areas were severe slope instability already occurred. The most

M. Stănciucu (✉)
Faculty of Geology and Geophysics, Department of Engineering Geology, University of Bucharest, Bucharest, Romania
e-mail: stanciucumihaela@yahoo.com

G. Lollino et al. (eds.), *Engineering Geology for Society and Territory – Volume 4*,
DOI: 10.1007/978-3-319-08660-6_37,

instable slopes and beaches erosion rates (that menace even urban communities) superpose over the instable areas revealed by seismic investigations of this study.

37.2 Geomorphologic, Geologic and Tectonic Setting of Black Sea Shore

The morphology of the Black Sea border reflects the deep geological structure which is divided in four sectors with different geological characteristics separated by three crustal faults (Seghedi 2012) (Fig. 37.1). The northern sector Chilia—Sfântul Gheorghe, is a deltaic plain consisting of fossil littoral formations, followed by a deltaic front and maritime deposits (Panin 1996). This area corresponds geologically to the westernmost segment of the Scythian Platform, the Pre-Dobrogean Depression, which consists in thick cover (Cretaceous to Quaternary sediments) disposed over a magmatic basement situated at 1–4 km depth. The basement is highly tectonized and the whole unit was subjected to several oscillatory movements (Seghedi 2012). The area of the Black Sea border between Sfantul Gheorghe and Peceneaga-Camena major faults, define the North-Dobrogea Orogen. Its Hercynian basement and Cimmerian cover (Triassic-Jurassic) deposits are splitted in three units by tectonic lines (Săndulescu 1984) and covered by thin post-tectonic Upper Cretaceous sediments. The last two southern units are separated by Capidava-Ovidiu crustal fault, but both are related to Moesian Platform which has a Precambrian basement (Seghedi 2012). The Central Dobrogea unit is an upland plateau with a maximum elevation of 350 m decreasing in altitude from North to South, to the East towards the Black Sea and to the West towards the Danube. This unit which appears as a horst, compared to the neighboring structural units, represents an uplifted sector of Neoproterozoic Moesian Platform composed by two types of metamorphic rocks. The Paleozoic calcarenites which cover the basement are almost completely removed by erosion caused by the up lifting movements. The southernmost unit extended from Mamaia to Vama Veche, presents two types of littoral morphology features, respectively high cliff areas, separated by limited beaches. This pattern exhibits the geological structure of the South Dobrogea unit, which is considered a subsided Moesian block (over 600 m) along the Capidava-Ovidiu and Intramoesian faults (Seghedi 2012). The basement was intercepted only at the northern limit and the remnant parts of the South Dobrogea Platform contain only the Mezozoic-Cenozoic Moesian cover. Structural and tectonic architecture of southern Dobrogea is extremely complex, being characterized by two major nearly vertical faults systems that affects the cover formations and detach them in seven blocks at the shore line (Zamfirescu et al. 1994).

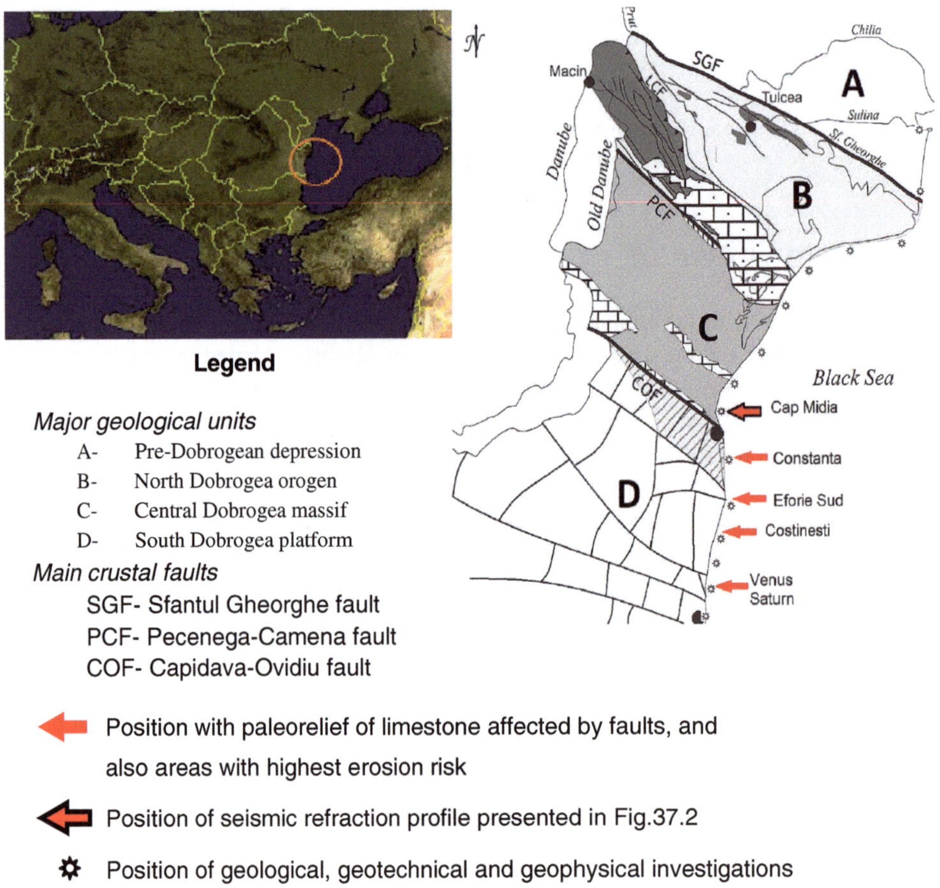

Fig. 37.1 Schematic map of main geological units of the Black Sea Border [compilation after 5, 8]

37.3 Interpretation of Geotechnical Investigations

A synthesis of the geotechnical parameters required for the preliminary design of anti-erosion protection measures has been developed, based on the geotechnical analyses and interpretation of SPT resistance for all lithological complexes found onshore and offshore: loess complex on cliffs, Pleistocene clay complex at the base of the cliffs, Quaternary marine deposits and Sarmatian limestone on beaches. Due to the limited space of this publication, we present only the results obtained for the marine sediments, which are exposed in two major groups: the first one including geological units A, B, C in the northern sector, which is less subjected to erosion processes and the second one in the southern sector composed on unit D, which is, in the author opinion, the area of Romanian seashore with the most severe erosion processes (Fig. 37.1). The analysis of granulometric data allows the following observations: (a) the fine sand is more abundant in the northern sector with about 24 % on shore, and about 32 % offshore; (b) the sediments had much uniform granulometric distribution in the northern sector compared with the southern sector. Based on penetration resistance we use the well accepted empirical relationship developed by Sanglerat (1965), regarding the estimation of bearing capacities of soils as function of resistance to light penetration for marine sediments. For the northern areas, the resistances to penetration vary between 1 KPa and 4.6 MPa, which corresponds to bearing capacities of 100 to 468 KPa. In the southern areas, marine deposits are in a denser state and resistance to light penetration varies between 0.68 KPa and 5.7 MPa, which corresponds to bearing capacities of 45 to 570 KPa. Average values of both characteristics are presented in Table 37.1. The results of SPT resistance obtained on southern locations, have been corrected and normalized according to Eurocode 8. Interpretation of parameters derived from $(N_1)_{60}$ values has been made according to Mayne et al. (2001), Bowles (1996) and Hatanaka and Uchida (1996) relations and synthetically exposed in Table 37.2.

37.4 Interpretation of Geophysical Investigations

For the 25 selected sites along the Romanian Black Sea border, seismic measurements using the refraction method have been made on two profiles along the shore (longitudinal) and offshore (transversal). Based on registered seismic wave velocities (V_p), dynamic stiffness moduli (E_{din} and G_{din}) are calculated assuming proper bulk densities and Poisson coefficients for sands, gravel or bedrock. Geophysical investigations in the Northern area identified one or two seismic reflectors, as derived from primary wave velocities, which defines two or three lithological complexes exposed synthetically in Table 37.3. The transversal profile Cap Midia reveals an intense fractured geological structure characterized by several faults in horts-graben system and a possible contact with green schist as presented in Fig. 37.2. Interpretation of seismic investigations in the

Table 37.1 Light penetration resistance of marine sediments

Sector	Average penetration resistance Rd (MPa)	Average bearing capacity q_{ad}	
		Min (KPa)	Max (KPa)
Northern	2.98	198	298
Southern	2.63	170	343

Table 37.2 Interpretation of SPT resistance of marine sediments

Depth (m)	Dr (-) (Mayne et al. 2001)				E (KPa) (Bowles 1996)				ϕ' (degree) (Hatanaka and Uchida 1996)			
	Onshore		Offshore		Onshore		Offshore		Onshore		Offshore	
	Min	Max	Min	Max	Min	Max	Min	Max	Min	Max	Min	Max
0.5	0.32	0.77	0.38	0.70	3,636	12,357	4,439	10,636	22	43	33	41
1	0.35	0.74	0.33	0.56	3,980	11,669	3,751	7,538	32	42	31	38
2	0.39	0.67	0.37	0.72	4,554	9,947	4,325	11,095	33	41	32	42
3	0.26	0.74	0.5	0.58	3,024	11,592	6,390	7,920	29	42	36	38
4	0.38	0.76	0.83	1	4,389	12,308	14,255	20,022	33	43	45	50
5	0.20	0.6	0.57	0.63	2,554	8,196	7,577	9,021	28	38	38	39
6	0.13	0.58	0.62	0.65	2,123	7,943	8,768	9,333	26	38	39	40

Table 37.3 Specific average values of dynamic parameters in Northern area

Complex	Parameter (MPa)	Onshore			Offshore		
		Min	Av.	Max.	Min	Av.	Max.
A1: upper deposit: unconsolidated marine and deltaic sediments (fine sands in a loose or very loose state)	E_d	–	–	–	258	270	286
	G_d	–	–	–	88	92	97
A2: deeper part of submerse sediments, (fine sands with gravel in medium relative density, normal consolidated)	E_d	524	846	1,512	503	648	707
	G_d	178	288	514	171	220	240
D: bedrock (green schist or compacted limestone)	E_d	1,225	1,567	2,168	–	4,230	–
	G_d	417	533	373	–	1,469	–

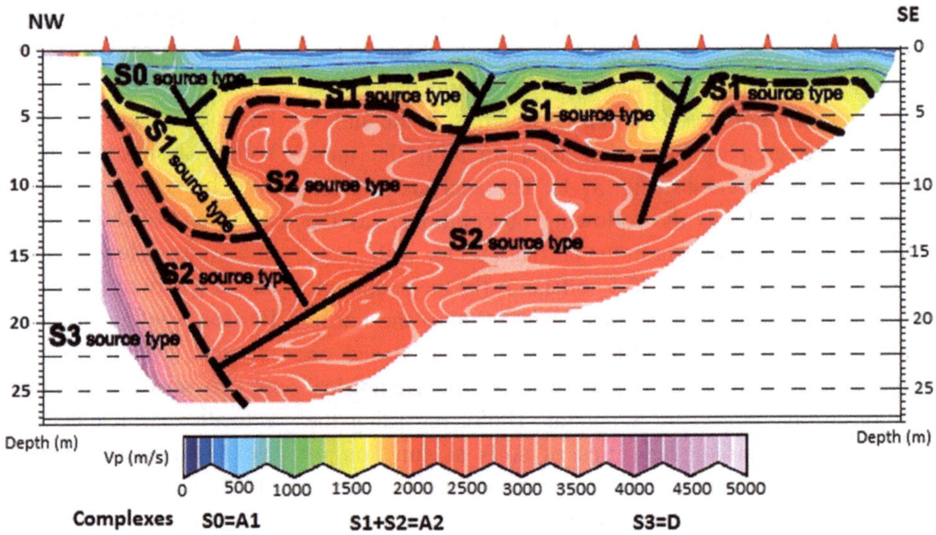

Fig. 37.2 Seismic refraction profile transversal on the Black Sea border at Cap Midia location

Table 37.4 Specific average values of dynamic parameters in Southern area

Complex	Parameter (MPa)	Onshore			Offshore		
		Min	Av.	Max.	Min	Av.	Max.
A1: upper part of the unconsolidated marine sediments (dry or submerse fine sands in loose or very loose state)	E_d	442	–	–	1,209	452	154
	G_d	655	–	–	1,606	773	263
A2: unconsolidated marine deposits (sand with or without gravel, in medium state of relative density)	E_d	1,627	1,519	534	1,596	1,222	423
	G_d	1,679	3,016	1,077	2,063	2,990	1,045
A1+B: undifferentiated marine sediments followed by very soft clay	E_d	504	–	–	–	–	–
	G_d	800	–	–	–	–	–
B: Pleistocene clay with limestone boulders, or cavernous limestone layers interbedded with thin clay	E_d	1,126	2,077	742	1,617	1,828	640
	G_d	1,931	3,990	1,425	1,830	2,361	843
C: un-differentiated limestone strata	E_d	2,535	7,097	2,535	2,479	6,769	2,418
	G_d	3,724	14,839	5,299	3,114	10,406	3,716
C1: cavernous limestone strata	E_d	–	–	–	1,930	3,877	1,380
	G_d	–	–	–	2,698	7,812	2,790
C2: massive compact limestone strata	E_d	–	–	–	2,721	8,175	2,920
	G_d	–	–	–	3,577	13,733	4,905

southern area reveals several lines of contrast in seismic velocities which allow differentiating three main complexes, in different states as presented in Table 37.4. Transversal seismic profile provides some additional information related to the geological structure: (a) limestone has significant paleorelief on profiles Constanta, Tuzla, Olimp, Jupiter, Mangalia, 2Mai and Vama Veche; (b) in certain positions Constanta, Eforie Sud, Costinesti, Venus and Saturn the paleorelief is affected by faults most of them reverse, or thrust reverse.

37.5 Conclusions

Beside the ranges of geotechnical parameters, bearing capacities of soils and the presence of very weak layers in several locations, the most important finding of this study is the confirmation of the intense faulted structure of the South Dobrogea unit, accomplished by interpretation of transversal offshore seismic profiles in five locations. These specific locations correspond with the areas of highest erosion risk recorded in previous marine-costal studies (Stănică 2011). This fact allows to generate another hypothesis (that remain to be proved by specific survey studies), that the latest intense rate of erosion in this areas and on the whole South Dobrogea, may be produced either by vertical subsidence movements of some specific block of the tectonic architecture depicted by Zamfirescu et al. (1994), or of the whole geological structure as underline Seghedi (2012), or both the causes.

References

Bowles JE (1996) Foundation analysis and design. McGraw-Hill, New York

Hatanaka M, Uchida A (1996) Empirical correlation between penetration resistance and effective friction of sandy soil. Soils Found Jpn Geotech Soc 36(4):1–9

Mayne PW et al (2001) Manual of subsurface investigations. National Highway Institute, Washington

Panin N (1996) Impact of global changes on geo-environmental and coastal zone state of the Black Sea, vol 1. Geo-Eco-Marina, Bucharest, pp 7–23

Sanglerat G (1965) The penetrometer and soil exploration: Interpretation of penetration diagrams theory and practice. Elsevier Pub Co, New York

Seghedi A (2012) Palaeozoic formations from Dobrogea and Pre-Dobrogea: an overview. Turk J Earth Sci 21:669–721

Săndulescu M (ed) (1984) Geotectonica Romaniei. Tehnică, Bucureşti

Stănică A (2011) Coastal dynamics and sedimentology studies. Implementation of an appropriate structure for the prevention of natural hazards in the most vulnerable areas. Reduction of coastal erosion. Scientific Report

Zamfirescu Fl et al (1994) Vulnerability to pollution of karstic aquifer system in Southern Dobrogea. In: Proceedings of RAH international symposium impact of industrial activities on groundwater, Constanta, Romania

Part V

Submarine Mass Movements: Hazards and Risk Assessment

This part addresses the critical issues related to estimating submarine mass movements related hazards and risk. Any geohazard study on mass movements must include a demonstated understanding of the slide mechanisms involved, their extent and timing. This session shall therefore provide an opportunity to review the known extent of submarine mass movements, the diversity of the type of failures, their triggering mechanism, and the environmental forcing, if any. From a risk assessment view point, submarine mass movements pose a great challenge since they can take place on very gentle slopes (1–4 °), have volumes of hundreds of cubic kilometers and travel over hundreds of kilometers, while potentialy generating a tsunami. For any final risk analysis, consequences will also have to be evaluated and taken into account, and tools or illustration of such application are also very much welcome.

Submarine Mass-Movements on Volcanic Islands: Examples from the Aeolian Archipelago (Italy)

38

Daniele Casalbore, Alessandro Bosman, Claudia Romagnoli, and Francesco Latino Chiocci

Abstract

High-resolution bathymetry collected around the Aeolian islands allowed the recognition of a large spectrum of geomorphic features referable to mass-movements. Besides large-scale sector collapse scars (recognized only at Stromboli), two main groups of medium-scale submarine slides are distinguished based on their size, geometry and location. The first group is defined by semi-circular slide scars at hundred-meter scale, indenting the outer edges of insular shelves and overlying sedimentary deposits in the first hundred(s) m of water depth (wd). The second group is defined by downslope-elongated scars carving the steep submarine flanks of volcanic edifices and interpreted as retrogressive slope failures. Within this latter group, particular attention has been devoted to the frequent slope failures affecting the submarine portion of the Sciara del Fuoco at Stromboli, and to the nested slide scars affecting the old and extinct Banco del Bagno submarine volcanic center at Lipari. The main predisposing and triggering factors, recurrence times and tsunamigenic potential of the recognized mass-movements are briefly discussed to provide some insights into geohazard assessment of the area.

Keywords

Multibeam • Sector collapses • Retrogressive failures • Triggering mechanisms • Tsunami

38.1 Introduction

Over the last two decades the interest for submarine instability grew up, as well as the need to mitigate hazard related to these phenomena. Volcanic islands pose a major threat in this regard, as demonstrated for instance by the 1998 Sissano events in Papua New Guinea (Tappin et al. 2001). The recent acquisition of high-resolution multibeam bathymetry, side scan sonar, single-channel seismics around the active Aeolian islands enabled us to map and characterize a large spectrum of geomorphic features referable to submarine mass-movements, ranging from small rock-falls up to large-scale sector collapses (e.g. Bosman et al. 2009; Casalbore et al. 2011; Romagnoli et al. 2012, 2013a). Moreover, a submarine landslide affected the NW flank of Stromboli in 2002 (Chiocci et al. 2008), generating tsunami waves, with observed run-up heights up to 10 m (Tinti et al. 2006).

D. Casalbore (✉) · A. Bosman
IGAG-CNR, Istituto di Geologia Ambientale e Geoingegneria, Area della Ricerca di Roma 1, P.le Aldo Moro 5, 00185 Rome, Italy
e-mail: daniele.casalbore@igag.cnr.it

C. Romagnoli
Dipartimento di Scienze Biologiche, Geologiche ed Ambientali, Università di Bologna, P.za Porta S. Donato 1, 40126 Bologna, Italy

F. L. Chiocci
Dipartimento Scienze della Terra, Sapienza Università di Roma, P.le Aldo Moro 5, 00185 Rome, Italy

G. Lollino et al. (eds.), *Engineering Geology for Society and Territory – Volume 4*,
DOI: 10.1007/978-3-319-08660-6_38, © Springer International Publishing Switzerland 2014

The goal of this contribution is to propose a synthetic review of main instability processes affecting the submarine flanks of the Aeolian volcanic edifices, and discuss their extent, recurrence time, tsunamigenic potential, preconditioning and triggering mechanisms according to the available geomorphological and geological constraints. The study can provide useful insights for a first-order hazard assessment of this touristically exploited area as well as for the comparison with similar instability processes in other marine volcanic settings.

38.2 Geological Setting

The study area encompasses the central and eastern sector of the Aeolian Archipelago (Fig. 38.1 inset), representing the younger manifestation of a volcanism related to the NW-directed subduction of the Ionian oceanic lithosphere below the Calabrian Arc (e.g. Argnani and Savelli 1999). The submarine setting of the Aeolian Arc, with volcanic edifices rising from depths comprised between 1,000 and 2,600 m wd, has been described in broad terms by Romagnoli et al. (2013b). The submarine flanks are quite steep, with slope gradients ranging from 20 to 40° in the upper slope to few degree at their base. The morphological continuity between subaerial and submarine flanks is locally interrupted by the presence of insular shelves developed around the older sectors of the volcanic edifices (Romagnoli 2013). The outer edge of insular shelves is mostly located down to 100–130 m wd, i.e. at the lowest level reached by wave erosion during glacial stages and they are commonly overlaid by terrace-shaped depositional wedges (named Submerged Depositional Terraces, SDT hereafter, Chiocci and Romagnoli 2004).

38.3 Instability Features Around the Aeolian Insular Volcanoes

Several instability features were recognized in the central and eastern sector of the Aeolian arc. The most evident features are related to large-scale flank collapses, recognized on the NW and E Stromboli flanks. Main morphologies associated to these events are: (a) large (some km wide and long) and horseshoe-shaped landslide scars, bounded by very steep and few hundred meters-high head-scarps, and (b) related deposits, such as megablocks and fan-shaped bulges with hummocky morphology found at the foot of the submarine collapsed flanks (Fig. 38.1). The estimated volume of the collapses are in the order of 1 km^3 each one (Romagnoli et al. 2009a, b).

The medium- to small-scale instability features can be divided into two main groups according to their size, geometric characteristics, and location:

(1) the more numerous group is represented by few hundred meters wide and long slide scars, indenting the outer edge of the insular shelves and overlying SDTs (depths <200 m wd, Fig. 38.2a). Actually, two types of slide scars can be recognized on swath bathymetry, having low morphological relief or well-marked headwall (scar Ia and Ib in Fig. 38.2a). In both cases, individual deposits related to landslide scars are not recognizable on swath bathymetry or seismics;

(2) the second group is represented by slide scars affecting the submarine volcanic flanks of the edifices below the outer edges of insular shelf. These features are typically longer (several hundreds of meters) than larger (tens or few hundreds of meters). The scars often prolong downslope in channelized features, where related landslide deposits are not recognizable. Within this group, two peculiar cases are represented by slope failures affecting the submarine Sciara del Fuoco slope (SdF hereafter; Fig. 38.1) at Stromboli and the volcanic flanks of the Banco del Bagno volcanic center at Lipari (Fig. 38.2b). In the former case, slope failures are known to occur very frequently within the SdF, i.e. the most volcanically active sector of the Aeolian Islands. The SdF represents, indeed, the remnant of a large collapse scar and nowadays acts as a main channel-way to funnel seaward the huge amount of volcaniclastic material produced by the persistent strombolian activity. In 2002, a submarine tsunamigenic landslide occurred on the upper and steeper part (slope gradients of 30–40°) of the SdF, mobilizing about 9.5×10^6 m^3 (Chiocci et al. 2008).

At Lipari, a nested-scars landslide complex affects the flanks of an eccentric volcanic center, considered to have been active during the early volcanic activity at Lipari (Romagnoli et al. 1989) and whose top (depth <130 m) was completely dismantled by wave action during Late-Quaternary sea-level fluctuations, leading to the formation of a wide insular shelf (IS in Fig. 38.2b). This landslide complex is 2.5–4 km long and 1–2 km wide, extending over areas of 10 km^2 (Fig. 38.2b); the headwall is made up by steep head-scarps with an almost constant relief of 10–20 m. On 1.4 kJ Sparker profile, it is evident that the landslide scars affect the upper part of the sedimentary cover (Fig. 38.2c). In the northern part, the landslide complex is bounded by a wide channel, characterized by coaxial crescent-shaped scarps that can be interpreted as cyclic steps (e.g. Fildani et al. 2006; Paull et al. 2010; Casalbore et al. 2014).

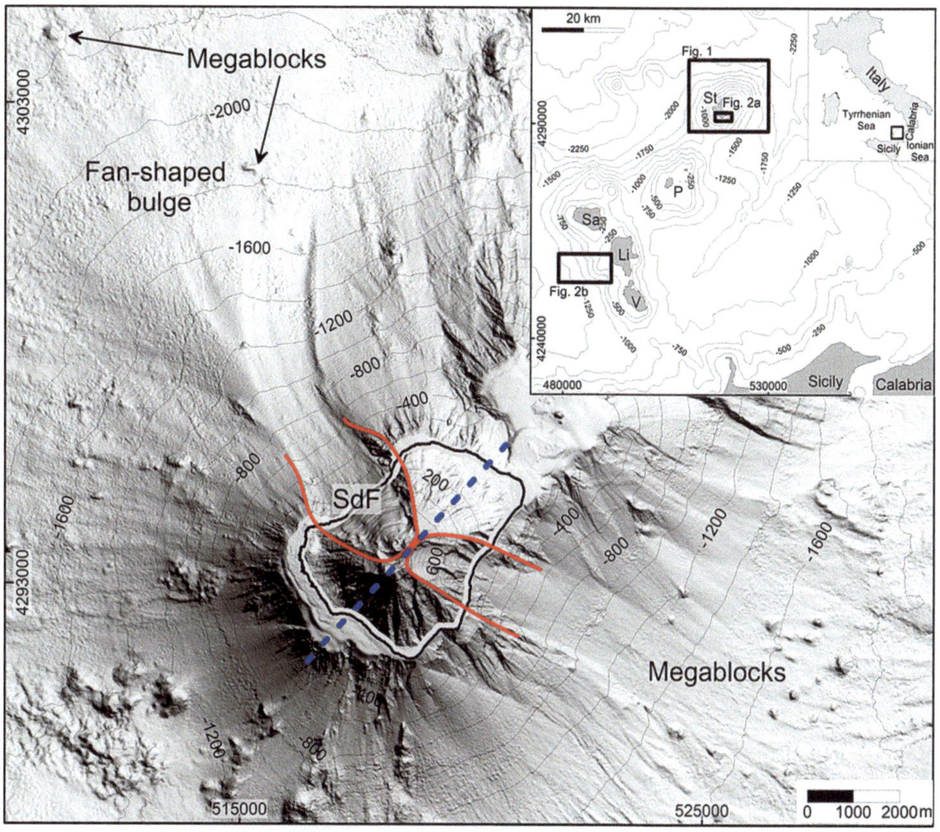

Fig. 38.1 Shaded relief of Stromboli edifice, with the indication of main sector collapse scars (*red lines*) at both sides of the main SW-NE rift axial zone (*dashed blue line*). In the inset, bathymetric map of the study area, with the indication of the following figures; *V* Vulcano, *Li* Lipari, *Sa* Salina, *P* Panarea, *St* Stromboli

38.4 Discussions and Conclusions

The morpho-bathymetric data collected on the submarine extension of the Aeolian Archipelago allowed us to recognize a large spectrum of instability processes, ranging at different spatial scale and not randomly distributed. The most striking evidence is the presence of large-scale flank collapses only on the NW and E Stromboli flanks but not on the other volcanic edifices. Their occurrence is mainly related to the presence of a well- developed axial rift zone, with collapses mainly occurring perpendicularly to it (Fig. 38.1; Tibaldi 2001; Bosman et al. 2009). Given the large volume of material mobilized during these events, their tsunamigenic potential is very high, but their estimated recurrence period is low, i.e. some to tens of thousands years (Casalbore et al. 2011 and reference therein).

Differently, smaller-scale slope failures commonly affect the edge of insular shelves and overlying submerged depositional terraces (SDTs), mobilizing volume of hundreds of thousand cubic meters. Preconditioning factors are represented by steep gradients, leading to over-steepening of sometimes incohesive sediments (in the case of SDTs). Triggering mechanisms can be represented by eruptive and seismic crisis as well as by cyclic load due to storm-waves in shallow water (Locat and Lee 2000; Casalbore et al. 2012). The analysis of the morphological relief of these scars allows to distinguish between two mass-wasting/erosive stages: some scars (see, for instance, Scar Ia in Fig. 38.2a) show a low relief and are partially infilled by the prograding sequences of the SDTs. Contrastingly, other scars (such as Scars Ib in Fig. 38.2a) have well-defined head-scarps and indent both the insular shelf and SDTs, witnessing a younger erosive stage (younger than ~20 ka). Given the relevant number of the latter type of scars (tens recognized in each volcanic edifice), their recurrence time could be statistically estimated in the order of few hundreds of years in the last 20 ka. However, we cannot exclude the possibility that most of landslides occurred at the same time, for instance in relation with sea-level changes. A rough assessment of their related tsunamigenic potential was recently computed at Stromboli, with maximum tsunami wave amplitudes typically higher than 0.5 m (Casalbore et al. 2011).

The other group of medium- to small-scale landslide scars occurs on the steep submarine flanks of the edifices. These features can be interpreted as the result of retrogressive failures (e.g. Prior and Suhayda 1979), i.e. generating on the middle and lower parts of the flanks and migrating upslope, on the basis of the following evidence: (a) their downslope-elongated shape, (b) the presence of channelized features at their foot, and (c) the occurrence of minor scarps coaxial to the main scar (Fig. 38.2a). The lacking of landslide deposits at the foot of these scars

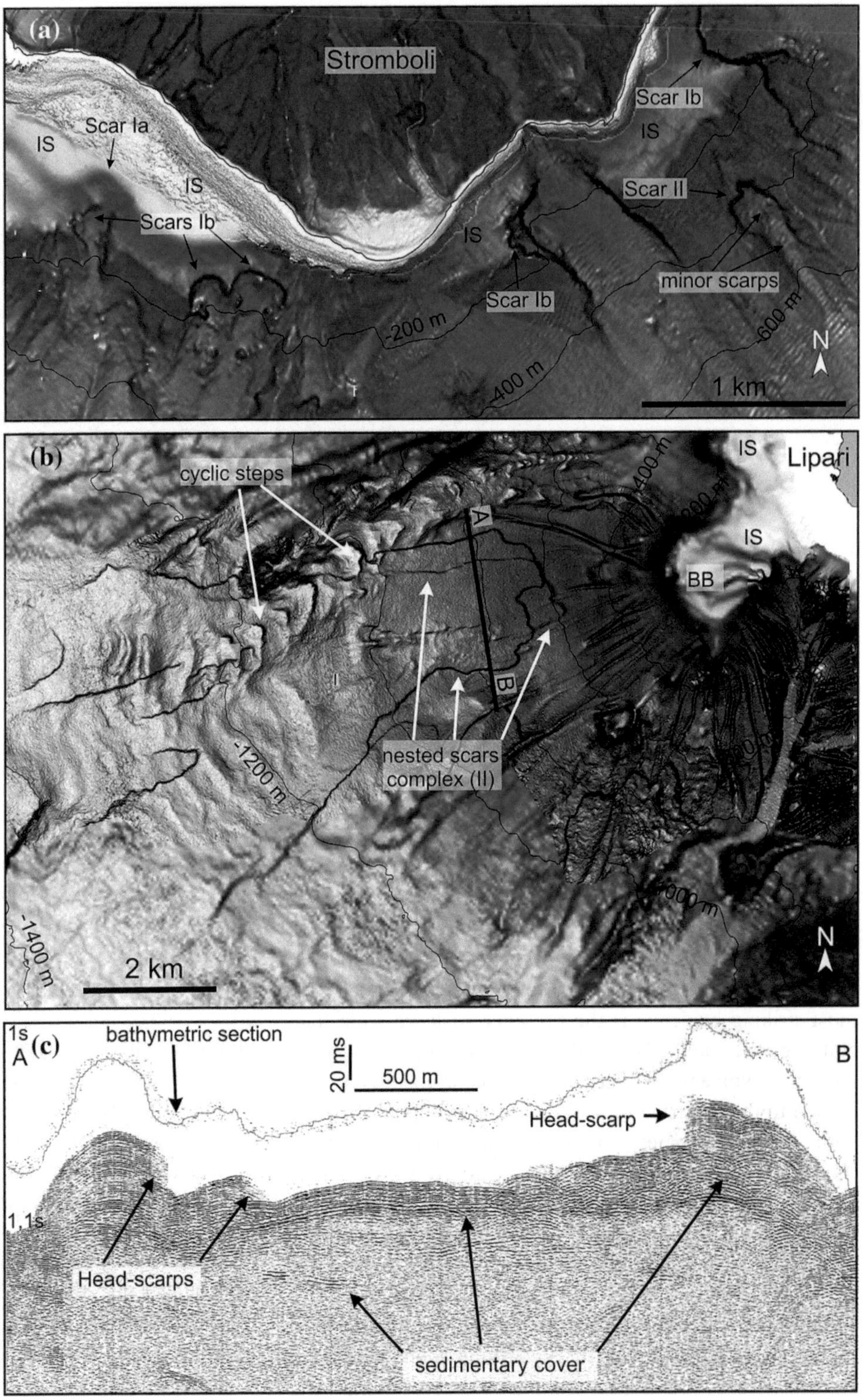

Fig. 38.2 Shaded relief of the southern sector of Stromboli (**a**) and western sector of Lipari (**b**), with the indication of the different types of landslide scars (Ia, Ib and II; see text for details); IS: insular shelf; BB: Banco del Bagno volcanic center. (**c**) 1,4 kJ Sparker profile crossing the slide scar of Banco del Bagno and affecting the sedimentary cover with well-defined head scarps; the *black line* represents the corresponding bathymetric section with a vertical offset of 40 ms

suggests the occurrence of disintegrative failures that may evolve downslope into debris flows and/or turbidity currents able to carve the seafloor. The geo-hazard related to these features is low due both to their low recurrence period (i.e. thousands of years in relation to the low number of recognized features) and to their low tsunamigenic potential, in relation to the higher depths where they formed. A striking exception is represented by the very frequent submarine slope failures affecting the upper part of the SdF slope, such as the 2002 event (Romagnoli et al. 2009b). This setting is related to the concomitant presence of several predisposing and triggering mechanisms, such as high sedimentation rates, steep slope gradients and internal heterogeneity (Chiocci et al. 2008). Given the relatively high frequency and tsunamigenic potential of the slope failures affecting the SdF slope, the related geohazard is the highest among all the recognized mass-movements in the Aeolian islands.

A final consideration deals with the nested-slides landslide complex recognized on the flanks of the Banco del Bagno volcanic center in the western Lipari. These slope failures affects the sedimentary cover draping the steep flanks of this volcanic center, as witnessed by seismic profiles (Fig. 38.2c) and hemipelagic mud recovered from cores (unpublished data). The scar morphology is very different as well, being more similar to translational slides affecting the submarine flanks of old and extinct volcanoes, such as Ventotene in the Central Tyrrhenian Sea (e.g. Casalbore et al. 2014) or open-slope of continental margins. In this case, the main preconditioning mechanisms for the slope failures are likely related to the bedding attitude of the sedimentary infilling that mimic the steep slope of the underlying volcanic flanks as well as to erosion at the foot of the slope by turbidity currents, as suggested by the presence of a wide channel with cyclic steps immediately northward of the landslide complex (Fig. 38.2b).

Acknowledgments This research was funded by the National MaGIC (Marine Geohazards along the Italian Coasts) Project and carried out in the Ritmare flagship project (Ricerca ITaliana per il MARE). We would also acknowledge crews of R/V Urania, Universitatis and Thetis along with those people taking part in the surveys as well as an anonymous reviewer for useful comments.

References

Argnani A, Savelli C (1999) Cenozoic volcanism and tectonics in the southern Tyrrhenian sea: spacetime distribution and geodynamic signifiance. J Geodyn 27:409–432

Bosman A, Chiocci FL, Romagnoli C (2009) Morpho-structural setting of Stromboli volcano, revealed by high resolution bathymetry and backscatter data of its submarine portions. Bull Volcanol 71:1007–1019

Casalbore D, Romagnoli C, Bosman A, Chiocci FL (2011) Potential tsunamigenic landslides at Stromboli Volcano (Italy): insights from marine DEM analysis. Geomorphology 126:42–50

Casalbore D, Bosman A, Chiocci FL (2012) Study of recent small-scale landslides in geologically active marine areas through repeated multibeam surveys: examples from the Southern Italy. In: Submarine mass movement and their consequences, "Advances in natural and technological hazards research", vol 31 Springer International Publishing, pp. 573–582

Casalbore D, Bosman A, Martorelli E, Sposato A, Chiocci FL (2014) Mass wasting features on the submarine flanks of Ventotene volcanic edifice (Tyrrhenian Sea, Italy). In: Submarine mass movements and their consequences, vol 37. Springer International Publishing, pp. 285–293

Casalbore D, Romagnoli C, Bosman A, Chiocci FL (2014). Large-scale seafloor waveforms on the flanks of insular volcanoes (Aeolian Archipelago, Italy), with inferences about their origin.Marine Geology. http://dx.doi.org/10.1016/j.margeo.2014.06.007

Chiocci FL, Romagnoli C (2004) Terrazzi deposizionali sommersi nelle Isole Eolie. Memorie Descrittive della Carta Geologica d' Italia 58:81–114

Chiocci FL, Romagnoli C, Tommasi P, Bosman A (2008) Stromboli 2002 tsunamigenic submarine slide: characteristics and possible failure mechanisms. J Geophys Res 113:B10102. doi:10.1029/2007JB005172

Fildani A, Normark WR, Kostic S, Parker G (2006) Channel formation by flow stripping: large-scale scour features along the Monterey East Channel and their relation to sediment waves. Sedimentology 53:1265–1287

Locat J, Lee H (2000) Submarine landslides: advances and challenges. Can Geotech J 39:193–212

Paull CK, Ussler W, Caress DW, Lundsten E, Covault JA, Maier KL, Xu J, Augenstein S (2010) Origins of large crescent-shaped bedforms within the axial channel of Monterey canyon, offshore California. Geosphere 6:1–20

Prior DB, Suhayda JN (1979) Submarine mudslide morphology and development mechanism, Mississippi Delta: 11th offshore technology conference, vol 2. Houston, Texas, pp 1055–1061

Romagnoli C, Calanchi N, Gabbianelli G, Lanzafame G, Rossi PL (1989) Contributi delle ricerche di geologia marina alla caratterizzazione morfostrutturale ed evolutiva dei complessi Vulcanici di Salina, Lipari e Vulcano (Isole Eolie). Boll. GNV-CNR 2:971–978

Romagnoli C, Casalbore D, Chiocci FL, Bosman A (2009a) Offshore evidence of large-scale lateral collapse on the eastern flank of Stromboli, Italy, due to structurally controlled, bi-lateral flank instability. Mar Geol 262:1–13

Romagnoli C, Kokelaar P, Casalbore D, Chiocci FL (2009b) Lateral collapses and active sedimentary processes on the northwestern flank of Stromboli volcano. Mar Geol 265:101–119

Romagnoli C, Casalbore D, Chiocci FL (2012) La Fossa Caldera breaching and submarine erosion (Vulcano Island, Italy). Mar Geol 303–306:87–98

Romagnoli C, Casalbore D, Bosman A et al (2013a) Submarine structure of Vulcano volcano (Aeolian islands) revealed by high-resolution bathymetry and seismo-acoustic data. Mar Geol 338:30–45

Romagnoli C, Casalbore D, Bortoluzzi G, Bosman A, Chiocci FL, D'Oriano F, Gamberi F, Ligi M, Marani M (2013b) Bathy-morphological setting of the Aeolian Islands. In: Lucchi F, Peccerillo A, Keller J, Tranne CA, Rossi PL (eds) The Aeolian Islands volcanoes, vol 37. Geological Society, London, Memoirs, pp 27–36

Romagnoli C (2013) Characteristics and morphological evolution of the Aeolian volcanoes from the study of submarine portions. In: Lucchi F, Peccerillo A, Keller J, Tranne CA, Rossi PL (eds) The Aeolian Islands volcanoes, vol 37. Geological Society, London, Memoirs, pp 13–26

Tappin DR, Watts P, McMurty GM, Lafoy Y, Matsumoto T (2001) The Sissano, Papua New Guinea tsunami of July, 1998-offshore evidence on the source mechanism. Mar Geol 175:1–24

Tibaldi A (2001) Multiple sector collapses at Stromboli volcano, Italy: how they work. Bull Vulcanol 63:112–125

Tinti S, Maramai A, Armigliato A, Graziani L, Manucci A, Pagnoni G, Zaniboni F (2006) Observations of physical effects from tsunamis of 30 December 2002 at Stromboli volcano, Southern Italy. Bull Volcanol 68:450–461

39 Evolution of a Submarine Mass-Transport Complex in Space and Time

William C. Haneberg and Kerry J. Campbell

Abstract

Submarine mass-transport deposits (MTDs) can be petroleum reservoirs, drilling hazards, or indicators of seafloor instability that need to be considered in deep-water engineering projects. In 3-D seismic reflection data, MTDs are typically characterized by laterally extensive but relatively thin layers of debris with weak and incoherent reflectors and, in many cases, little or no seafloor geomorphic expression. Because of their great lateral extent, it is unusual to see proximal and distal portions of large MTDs in a single data set. We infer that spatial variation in the geometry of a large and exceptionally well-imaged MTD can serve as a proxy for the temporal evolution of the mass transport event at a point. Near its source, the MTD is characterized by disaggregation of originally intact strata. As distance from the source increases, the complex becomes increasingly chaotic the manifestation of the original structure weakens. Each facies also has a characteristic seafloor geomorphologic expression. The spatio-temporal evolution is particularly evident when the fabric is visualized using 3-D seismic volume amplitude attributes that display the lateral continuity of reflectors.

Keywords

Mass transport • Slope stability • Submarine Geohazards • 3D seismic

39.1 Introduction

Submarine mass transport deposits can—depending on their depth, stratigraphic location, and lithology—be petroleum reservoirs, drilling hazards, and manifestations of past seafloor slope instability that must be considered when planning deep-water seafloor infrastructure such as pipelines and anchors for production platforms. Often referred to as MTDs, mass transport deposits are the sedimentary records of debris flow- or debris avalanche-like mass transport events typically characterized by great lateral extent along with the ability to flow across gentle slopes and to transport intact blocks with characteristic dimensions of tens or even hundreds of meters (e.g., Shipp et al. 2011).

Sawyer et al. (2009) described the seismic, geophysical log, and geotechnical property signatures of MTDs in the Ursa Basin, Gulf of Mexico, and postulated a mechanism of formation involving retrogressive failure of submarine slopes and disaggregation of initially intact strata as the failure progressed. They defined two seismic facies typical of the MTDs they studied, which were in general characterized by low amplitude reflections above a high amplitude base. The two facies were *Chaotic* and *Discontinuous Stratified*. Sawyer et al. (2012) produced similar results in scale model experiments for some ratios of sediment shear stress to yield strength. Other authors have described, and we have observed in many proprietary 3-D seismic data sets

W. C. Haneberg (✉) · K. J. Campbell
Fugro GeoConsulting, Inc, 6100 Hillcroft, Houston, TX 77081, USA
e-mail: whaneberg@fugro.com

K. J. Campbell
e-mail: kcampbell@fugro.com

G. Lollino et al. (eds.), *Engineering Geology for Society and Territory – Volume 4*,
DOI: 10.1007/978-3-319-08660-6_39,

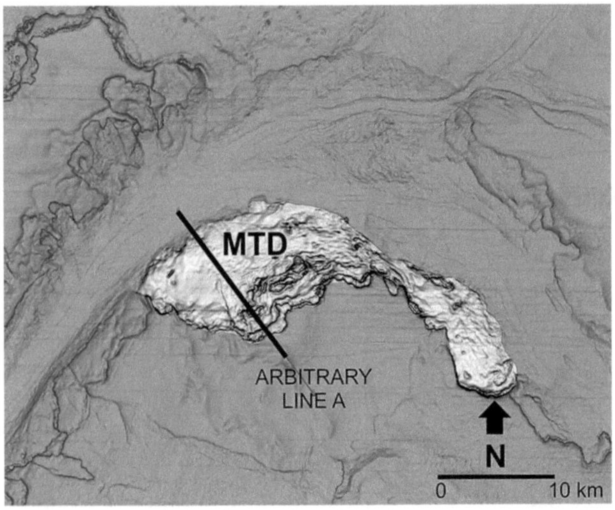

Fig. 39.1 Seafloor rendering showing the arbitrary seismic line A and surficial expression of the MTD described in this paper highlighted. Image is a slope-shade map produced from the regional 3-D seismic reflection seafloor pick and converted to depth. Dark shades correspond to steep slopes whereas light shades correspond to gentle slopes. Many large submarine slope failures are apparent along the edges of the channel complex passing through the middle part of the area

encountered in our consulting practice, many of the same characteristics.

Using an example from the Carnarvon Basin, offshore Australia, we show how information from a 3-D exploration seismic volume can be used to illustrate the evolution of a deep-water MTD in an area where both the proximal and distal portions of the deposit are discernable and the failures have well developed seafloor expressions (Fig. 39.1). This example offers an integrated surface (geomorphological) and subsurface (sedimentological) archetype of a MTD ranging from retrogressive failure of intact strata to initial disaggregation to form a discontinuous stratified facies to, finally, complete disruption of the original sedimentary fabric to create a chaotic facies. Spatial changes in the fabric, structure, and stratigraphy of the MTD can be interpreted as a representation of the temporal evolution of the mass transport event at the observer's location.

39.2 Geologic Setting

The example described in this paper comes from the Northern Carnarvon Basin off the northwestern coast of Australia. Space does not permit a lengthy summary of the basin geology as described in detail by authors such as Hocking (1988). However, we do note two salient points: First, the axes of prominent regional folds (e.g., the Kangaroo syncline and the Resolution arch) and major normal faults, all or most of which are likely related to a long history of continental rifting and syndepositional sedimentation associated with the breakup of Gondwana, strike approximately NE-SW throughout the region. Second, although we have no age control, comparison of the stratigraphy shown in our project area with published cross sections suggests the sediments that evolved into the MTD that we describe are likely of Miocene age.

39.3 3-D Seismic Data

Figure 39.1 is an index map showing location of 3-D seismic line A depicted in subsequent figures. The image is a slope-shade map produced from the regional 3-D seismic reflection seafloor pick and converted to depth. Dark shades correspond to steep slopes whereas light shades correspond to gentle slopes. Many large submarine slope failures are apparent along the edges of the large channel complex passing through the middle part of the area. Water depths in the area range from about 1000 to 1500 m.

Inline (E–W) and crossline (N–S) spacing for the post-stack time migrated 3-D seismic volume were 25 and 12.5 m respectively, and the sampling rate was 0.002 s. The dominant frequency of the upper half-second of the seismic data is approximately 45 Hz. For typical near-seafloor sediment velocities, a frequency of 45 Hz yields a vertical limit of separability, which is the thinnest bed for which the top and bottom can be separately imaged, of 8 to 10 m. It is a first estimate of the vertical resolution of the data. The data were available to the authors in 8 bit format.

As illustrated in Fig. 39.2, the MTD occupies the uppermost portion of the section directly below the seafloor, and can be divided into three components: (1) Largely intact and gently dipping strata, (2) Partially disaggregated blocks created during the initial stages of slope failure (discontinuous stratified facies of Sawyer et al. 2009), and (3) Completely disaggregated strata characteristic of MTDs commonly interpreted in 3-D seismic volumes (chaotic facies of Sawyer et al. 2009). There are indications of extension of the largely intact strata near the headscarp and buckling or shortening or the disaggregated strata near the toe of the MTD.

The distribution of materials within the MTD is further illustrated in Fig. 39.3, which shows the seafloor geomorphic expression of the MTD along with three instantaneous lateral continuity (ILC) time slices. ILC is one of several measures of reflector continuity or similarity, which typically have different commercial names in 3-D seismic analysis programs; such measures help to quantify and enhance edges such as faults and the edges of MTD blocks.

White portions of the ILC slices in Fig. 39.3 indicate zones where the lateral continuity or similarity of seismic amplitudes is high (e.g., intact blocks) whereas black portions indicate little or no continuity (e.g., highly brecciated

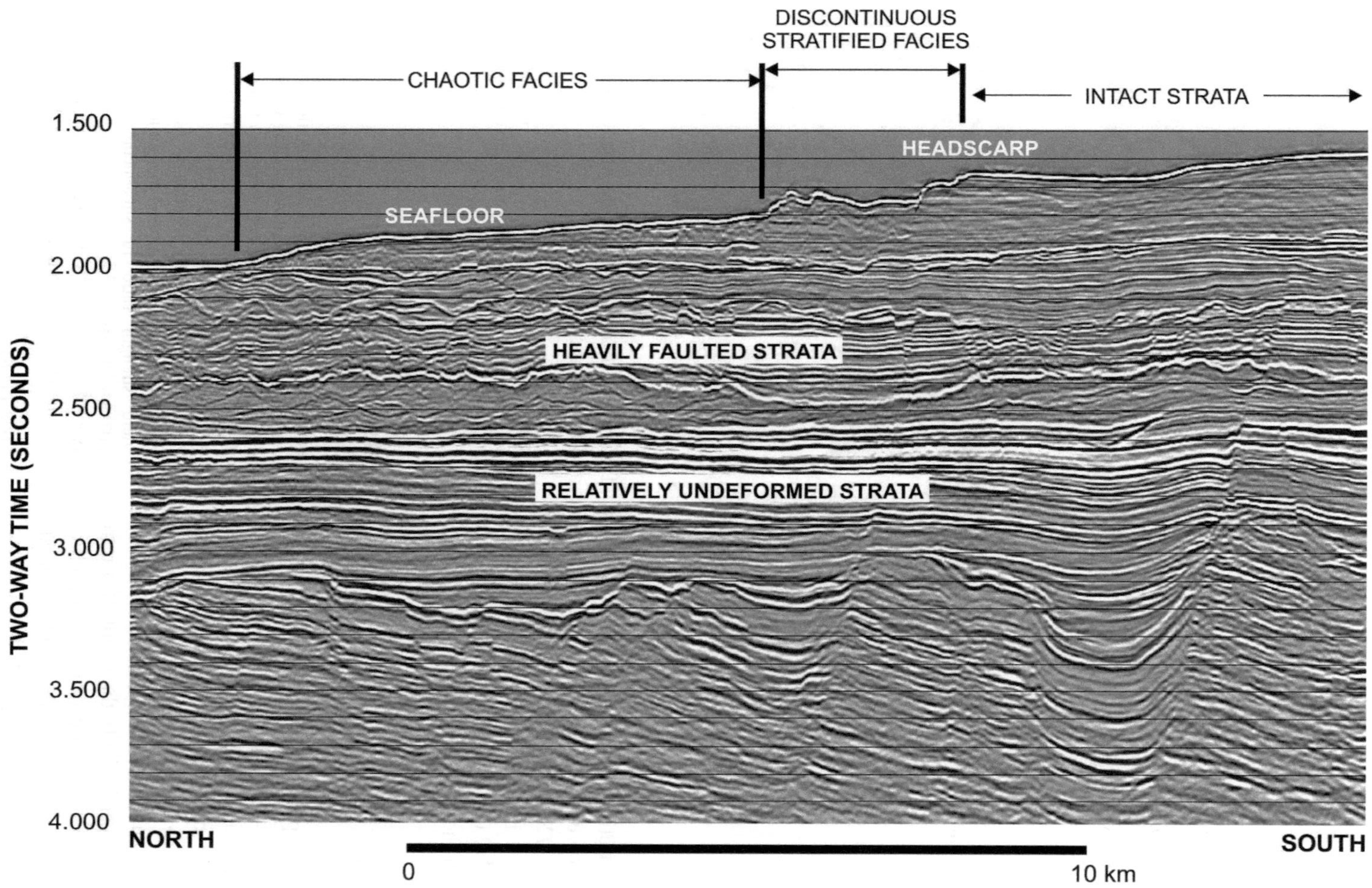

Fig. 39.2 Close-up of a portion of arbitrary seismic line A showing interpreted faults and base of the MTD. Vertical axis shows two-way travel time (TWT) in seconds and positive impedance contrasts are shown in white. Shallow sediments such as those shown here typically have seismic velocities of 1,550 to 1,600 m/s, which can be used to very approximately convert two-way time to depth

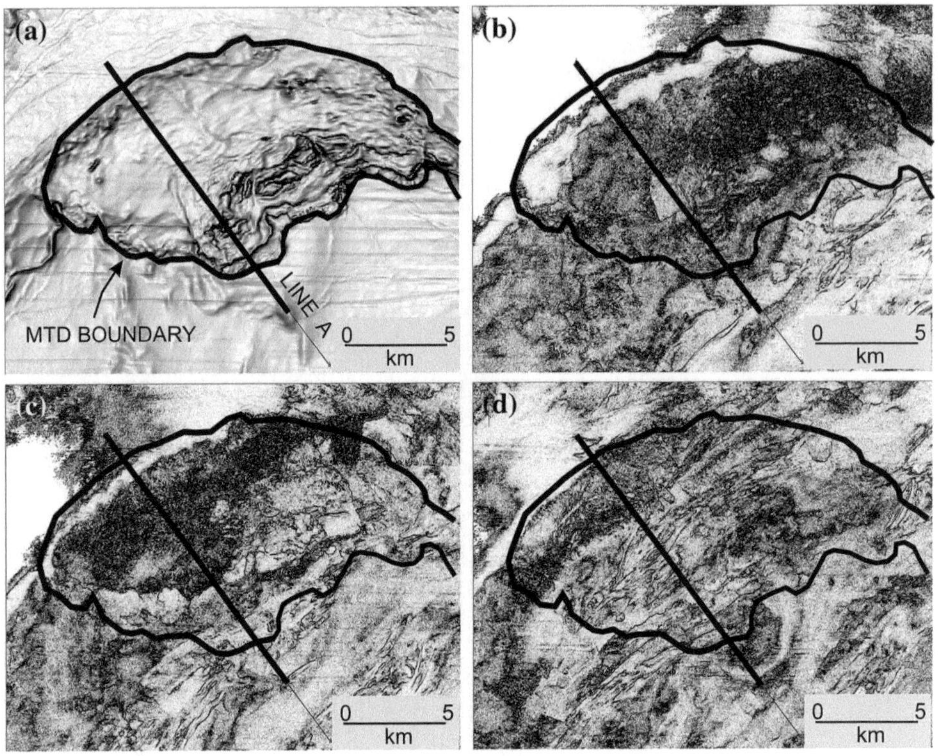

Fig. 39.3 Close-up images showing seafloor geomorphic expression of the MTD and instantaneous lateral continuity (ILC) time slices at depth. **a** Seafloor slopeshade image. **b** 1.920s TWT ILC slice showing intact blocks within disrupted or chaotic matrix within the MTD. **c** 1.960s TWT slice showing intact blocks within disrupted or chaotic matrix within the MTD. **d** 2.000 s TWT ILC slice showing regional structural trends beneath the MTD. The portion of Line A shown in Fig. 39.2 is indicated by the *thick black line*

or broken zones), illustrating the disaggregation of intact strata as the MTD evolved in space and time. The two uppermost ILC slices (1.920 and 1.960 s) cleary show large intact blocks surrounded by a disaggregated matrix. The lowermost ILC slice (2.000 s) illustrates the regional structural grain beneath the MTD.

39.4 Conclusions

The short length of this paper limits the amount of data that can be presented and discussed. Based upon careful interpretation of the 3-D seismic volume, knowledge of similar features in many other data sets, and experience both mapping and quantitatively analyzing smaller scale onshore analogs (e.g., Haneberg and Tripp 1991), we infer that the MTD described in this paper is likely the result of a retrogressive failure. The spatial variation in facies, which are similar to those defined by Sawyer et al. (2009), illustrates the temporal evolution of the mass as it slid, disaggregated, began to flow, and finally came to rest. Thus, this MTD offers an unusually complete example of a large mass failure now frozen in time and offers insights into its spatial and temporal evolution.

Acknowledgments Fugro N.V., owner of the seismic data upon which this paper is based, generously gave permission to publish the seafloor image and data examples.

References

Haneberg WC, Tripp G (1991) An irrigation-induced debris flow in northern New Mexico. Bull Assoc Eng Geol 28:359–374

Hocking RM (1988) Regional geology of the northern Carnarvon Basin. In: The North West Shelf, Australia. Proceedings of PESA symposium, Perth, WA, pp 21–43

Sawyer DE, Flemings PB, Buttles J, Mohrig D (2012) Mudflow transport behavior and deposit morphology: role of shear stress to yield strength ratio in subaqueous experiments. Mar Geol 307–310:28–39

Sawyer DE, Flemings PB, Dugan B, Germaine JT (2009) Retrogressive failures recorded in mass transport deposits in the Ursa Basin, Northern Gulf of Mexico. J Geophys Res 114(B10). doi: 10.1029/2008JB006159

Shipp RC, Weimer P, Posamentier HW (2011) Mass-transport deposits in deepwater settings, vol 96. SEPM Special Publication, Tulsa

High-Resolution Studies of Mass Transport Deposits: Outcrop Perspective for Understanding Modern Submarine Slope Failure and Associated Natural Hazards

40

K. Ogata, G. A. Pini, A. Festa, Ž. Pogačnik, G. Tunis, J. J. Mountjoy, K. Senger, and M. Strasser

Abstract

Micro- to meso-scale outcrop studies on selected field analogues allow direct calibration and testing of geophysical interpretations performed on mass transport deposits in modern continental margins, in terms of genetic processes and sliding dynamics. This comparative approach provides important information for forecasting and mitigating submarine landslide-related geohazards. The comparison of fieldwork studies (i.e., siliciclastic, carbonate and mixed, seismic-scale Eocene-Oligocene submarine mass transport deposits of the Northern Apennines in Italy, central Pyrenees in Spain and the north-western Dinarides in Italy/Slovenia) with multibeam bathymetric, seismic and drill core data from some modern analogues (i.e., offshore of New Zealand, Japan and Svalbard in Arctic Norway) is proposed in order to upscale the outcrop observations and downscale the geophysical features. Our results show that slide mass mobility is a function of the degree of internal liquefaction/fluidization, mainly achieved at the basal sliding interval and within the slide body. This is due to undrained shearing/loading-unloading of poorly-lithified sediments, and consequent development of fluid overpressure that is able to accommodate deformations at high strain rates. Structures related to these processes are observable at all scales, and represent diagnostic criteria to recognize potentially catastrophic mass transport events.

Keywords

Exhumed mass transport deposits • Geophysical analogues • Shear-induced liquefaction

K. Ogata (✉)
Dipartimento di Fisica e Scienze della Terra, Università degli Studi di Parma, 43124 Parma, Italy
e-mail: kei.ogata@gmail.com; kei.ogata@unipr.it

G. A. Pini · G. Tunis
Dipartimento di Matematica e Geoscienze, Università di Trieste, 34010 Trieste, Italy
e-mail: gpini@units.it

G. Tunis
e-mail: tunis@units.it

A. Festa
Dipartimento di Scienze della Terra, Università di Torino, 10125 Turin, Italy
e-mail: andrea.festa@unito.it

Ž. Pogačnik
Salonit Anhovo, Building Materials, Joint-Stock Co, 5210 Deskle, Slovenia
e-mail: zeljko.pogacnik@salonit.si

J. J. Mountjoy
National Institute of Water and Atmospheric Research, 14901 Wellington, New Zealand
e-mail: joshu.mountjoy@niwa.co.nz

K. Senger
EMGS Electromagnetic Geoservices ASA, 0250 Oslo, Norway
e-mail: senger.kim@gmail.com

M. Strasser
Geological Institute, ETH Zurich, 8092 Zurich, Switzerland
e-mail: michael.strasser@erdw.ethz.ch

G. Lollino et al. (eds.), *Engineering Geology for Society and Territory – Volume 4*,
DOI: 10.1007/978-3-319-08660-6_40, © Springer International Publishing Switzerland 2014

40.1 Introduction

Our inability to directly observe and monitor submarine landslide mass emplacement is a major impediment to understanding landslide mechanics. Deformation mechanisms acting within a submarine landslide during its downslope evolution can be better understood by combining submarine geophysical observations and detailed outcrop studies on modern and exhumed continental margins, respectively. Modern seafloor and subsurface imaging techniques provide the gross morphology, areal extent and internal character of single mass transport deposits (MTDs) and composite mass transport complex (MTCs). However, even where single-point ground-truthing of geophysical data by coring exists, marine geological data cannot resolve internal details and lateral complexity to any fine scale. Conversely, outcrop studies allow analyses from the microscopic scale up to the cartographic/map scale (comparable with the geophysical scale), but rarely preserve complete, source-to-deposit examples of large MTCs. Nonetheless the different boundary conditions and scales, also on-land studies on subaerial landslides would benefit of the knowledge acquired on submarine failures through this integrated approach.

In this study we integrate geophysical, drill core and outcrop data in order to characterize the deformation processes critical to the emplacement of submarine landslides through a continuity of observation across scales. Two key datasets are used to bridge the outcrop detail to the seismic scale: (1) observations and analytical data from large-scale (10 s-100 s of m-thick and >100 m^2-extensive) MTD/MTCs exhumed in the Early Oligocene Epiligurian succession (Northern Apennines, Italy; Ogata et al. 2012a), the Eocene Jaca Basin (central Pyrenees, Spain; Labaume et al. 1987), and the Paleogene Friuli basin of the northwestern Dinarides (Italy and Slovenia; Tunis and Venturini 1992), and (2) published and unpublished multichannel seismic reflection and drill core data from MTD/MTCs on the Hikurangi margin of New Zealand (Mountjoy and Micallef 2012), the Nankai wedge in Japan (Strasser et al. 2011) and the inner Isfjorden, Svalbard (Arctic Norway) (Fig. 40.1). We synthesize observations from these data sets to develop a scale-independent model representing the mechanical process, which control the mobility of large-scale submarine mass movements.

40.2 Synthesis of Key Observations and Results

The analysed units are composite sedimentary bodies made up by different mass-transport units (i.e. structural-sedimentary facies associations) which vary, laterally and vertically, between three main end-members identified as: (1) slump-type, (2) blocky flow-type and (3) debris flow-type (Fig. 40.2). These end members can coexist within the same unit as a result of: (1) downslope evolution of the slide mass, (2) internal pulses and reactivations within the same event, (3) amalgamation/stacking of different events, and/or (4) vertical preservation of the original (i.e. pre-failure) sedimentary column (Ogata et al. 2012b).

A key observation from outcrop and drill core data is that highly fluidized sediments are widely identified where shear localization occurs. The typical product encompassing such processes is a sedimentary matrix represented by an unsorted, hyper-concentrated mixture of loose and poorly consolidated fine-grained sediments, which sustains multi-sized, rigid and plastic elements (i.e. from mm-sized particles to hundreds of m-sized slide blocks/olistoliths). Analysis on drill cores (i.e. samples and X-ray Computed Tomography scan data) from a satellite basin atop the Nankai wedge demonstrates that MTD intervals are plastically deformed by meter to tens of meter-scale folds, and offset by centimetre to meter-thick mesoscopic ductile shear zones. The shear zones are loci of concentrated deformation by grain (i.e. clast) flow and clay mineral iso-orientation, also responsible for systematic changes in the sediment compaction/porosity (Pini et al. 2013).

At the entire deposit scale, concentrated pore fluid zones can be imaged in 3D seismic reflection data as velocity and amplitude anomalies (Plaza-Faverola et al. 2010). Seismic reflection data from the Hikurangi Margin (Mountjoy and Micallef 2012) indicate high amplitude reflections within regions of MTD over-riding, indicative of shear-zone fluid pressures. Thrust faulting is also characterized by relatively high amplitude reflectivity where shearing is expected to be localized. Multi-scale observations indicate that the internal arrangement of the final mass transport deposit is characterized by low- to high-angle thrusts, isolated and/or rooted in the basal shear plane, sometimes with sea-floor expressions (i.e. pressure ridges).

40.3 Discussion and Conclusions

Liquefaction, fluidization and hydro-plastic deformation are common processes in the internal structural evolution of MTCs and are interpreted as a critical process controlling the mobility of seafloor mass transport. Meso- and micro-structural and sedimentologic analyses suggest a deformation mechanism controlled by un-drained simple shearing of water-saturated, poorly-consolidated to loose sediments due to the dragging forces acting at high strain rates along boundaries of internal elements (i.e. internal differential movements) and at the basal slide interval (i.e. general

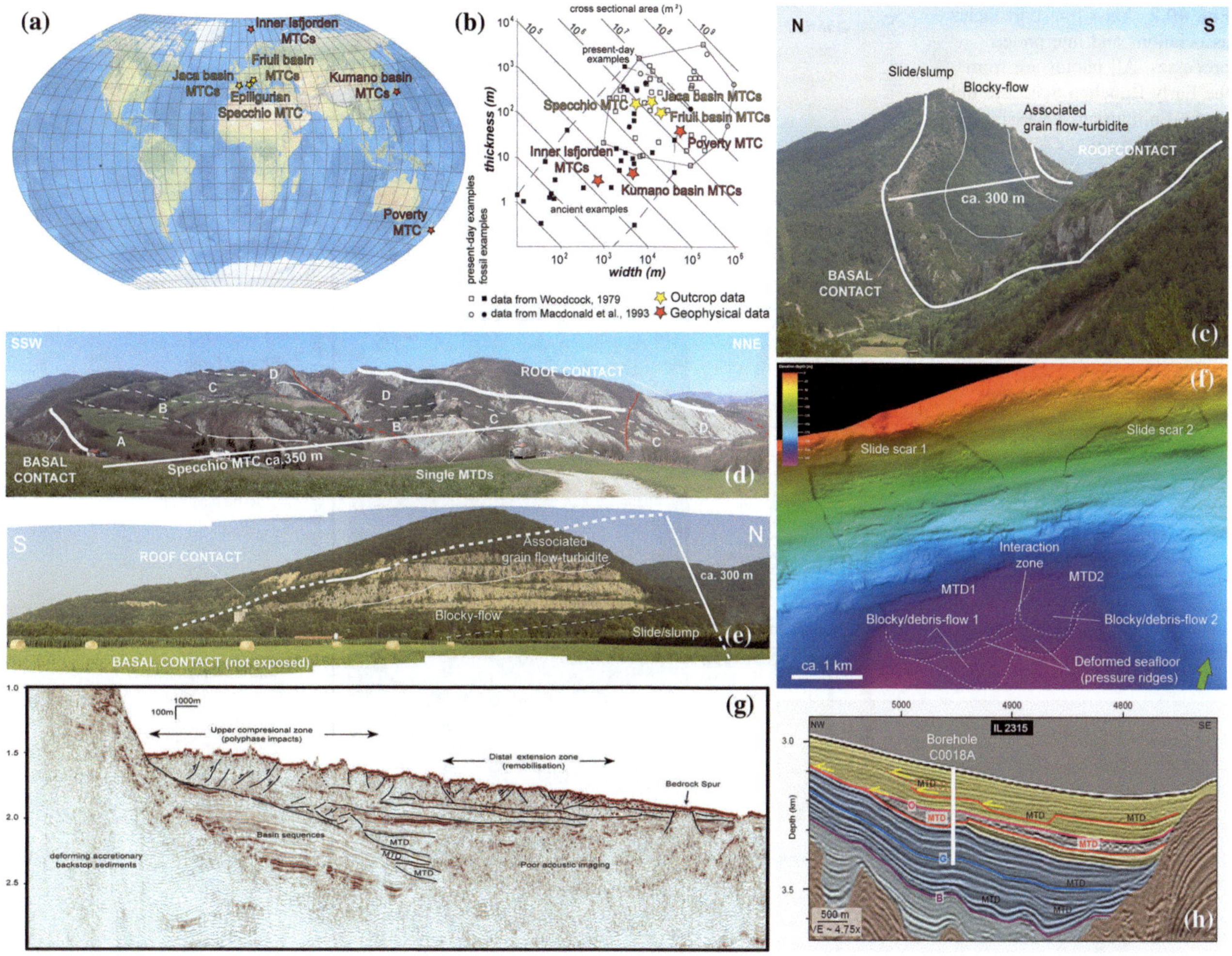

Fig. 40.1 **a** Locations of the studied examples. **b** Thickness versus width diagram showing relative sizes of ancient (*dashed line* envelope) and present-day (*solid line* envelope) MTDs and MTCs. The relative dimensions of the studied examples are plotted. Modified from Lucente and Pini (2003). **c** Panoramic overview of a carbonate MTD of the Eocene Jaca basin (from Ogata et al. 2012b). **d** Panoramic overview of the Specchio MTC of the Early Oligocene Epiligurian succession. **e** Panoramic overview of a carbonate MTD of the Paleogene Friuli basin. **f** Multibeam image of the inner Isfjorden MTDs. **g** Seismic profile of the Poverty MTC of the Hikurangi Margin. **h** Seismic profile of the MTDs of the Kumano basin (from Strasser et al. 2011)

down-slope movement). Most of the shearing achieved during the slide movement is likely accommodated within the basal interval, resulting in a sort of "overpressured carpet" that mechanically separates the slide mass from the substrate due to hampered hydraulic diffusivity (Ogata et al. 2012b). Frictional heating is another possible mechanism able to increase pore pressure up to liquefaction (Goren and Aharonov 2007).

When the mechanical coupling becomes strong enough (e.g. during slide mass deceleration and freezing, and/or impacts against topographic highs), the momentum can then be partly transferred downward into the substratum, with the consequent involvement of the underlying sediments. This deformation is likely caused by dynamic/static overloading and rear push of the sliding mass, causing substrate deformation and incorporation (i.e. erosion in sedimentological meaning). Compressional stress is mainly located at the front of the slide mass and then, transferred to the surrounding sediments, causing folding and thrusting at various scales. During the early post-depositional compaction, the internal fluid overpressure may be dissipated through developing of fluid-escape structures, sometimes reaching the slide surface as mud/sand volcanoes (Strachan 2008; Moernaut et al. 2009), or can be retained for a relatively long time interval, favouring slow differential movements of the entire mass (Major 2000).

The diagnostic product of mass transport-related liquefaction, fluidization and soft-sediment deformation processes can be found in outcrop (or drill cores) as a sedimentary matrix represented by an unsorted, hyper-

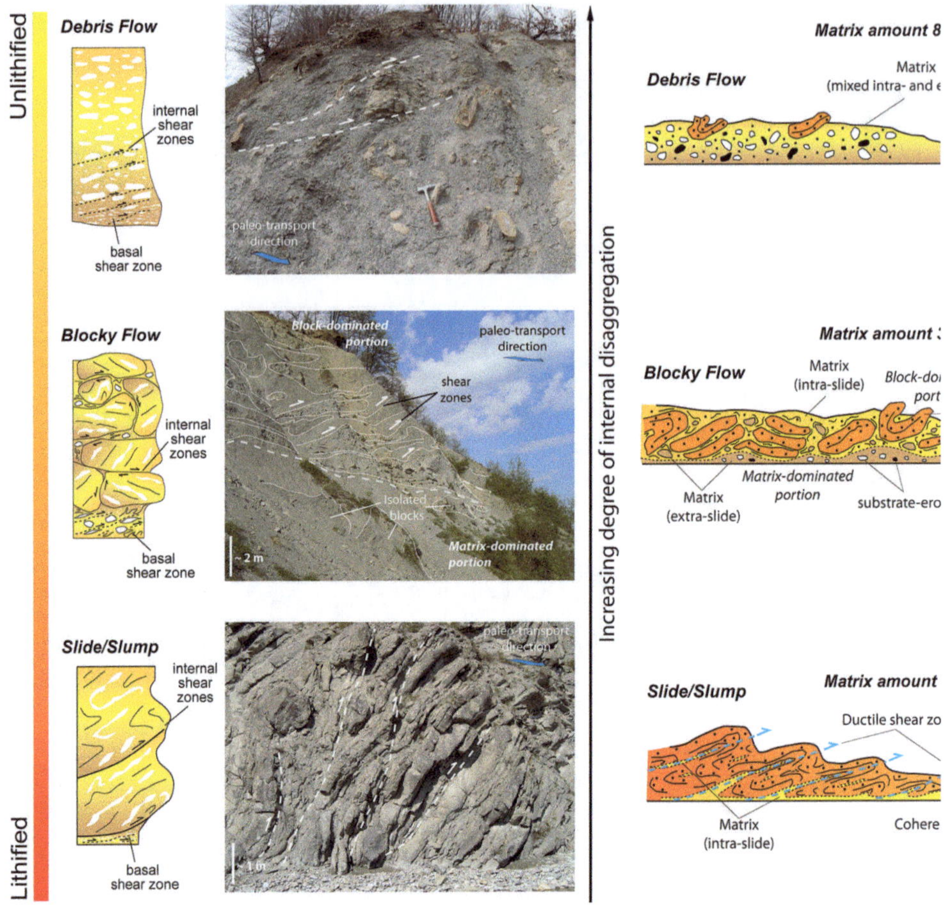

Fig. 40.2 Mass transport facies associations and interpreted processes. All photos are from the Early Rupelian Specchio unit of the Epiligurian succession, Northern Apennines (Ogata et al. 2012a, b)

concentrated mixture of loose and poorly consolidated fine-grained sediments. Shear zones with concentrated fluid excess pore pressure can be visualized in seismic reflection profiles and drill cores, especially at the very base of the unit where horizons of trapped high fluid pressure may induce high amplitude reflectivity and even negative polarity. Such reflectors can be observed also within the slide mass, and depending on their lateral continuity, may indicate the amalgamation surface/interval between two subsequent bodies, therefore representing a powerful tool to distinguish single depositional units (MTDs) from composite, multiple accumulation complexes (MTCs).

Fluidized sediment is a fundamental component of a slide body along with other discrete parts that behave coherently (e.g. slide blocks, un-dissociated masses, etc.), and it typically occurs with the internal and basal shear zones as cm- to m-sized elongated lenses and bands. In this framework the relative amount of composing sedimentary matrix may indicate the catastrophic or episodic/periodic character of the slide event (Fig. 40.2).

The systematic integration of the large amount of available geophysical data with detailed, high-resolution outcrop studies represents the most reliable method for the study of submarine landslides. This synergic approach permits observations covering all the scales, overcoming the intrinsic resolution limits of the single methods. This is particularly so for fine scale, shear-zone mechanical processes that exert a major influence on the mobility of extremely large continental margin mass transport processes, representing the key concepts for forecasting and mitigation of submarine landslide-related geohazards.

Acknowledgments Norwegian Hydrographic Survey provided multibeam data from Svalbard. Joshu Mountjoy supported by NIWA under Coasts and Oceans Research Programme 1 (2013/14 SCI). This work is part of the MIUR-PRIN n. 2010AZR98L_002 program.

References

Goren L, Aharonov E (2007) Long run-out landslides: the role of frictional heating and hydraulic diffusivity. Geophys Res Lett 34:1–7

Labaume P, Mutti E, Seguret M (1987) Megaturbidites: a depositional model from the Eocene of the SW-Pyrenean Foreland Basin, Spain. Geo-Mar Lett 7:91–101

Lucente CC, Pini GA (2003) Anatomy and emplacement mechanism of a large submarine slide within a Miocene foredeep in the Northern Appennines, Italy: a field perspective. Am J Sci 303:565–602

Major JJ (2000) Gravity-driven consolidation of granular slurries: implications for debris-flow deposition and deposit characteristics. J Sediment Res 70(1):64–83

Moernaut J, De Batist M, Heirman K, Van Daele M, Pino M, Brümmer R, Urrutia R (2009) Fluidization of buried mass-wasting deposits in lake sediments and its relevance for paleoseismology: results from a reflection seismic study of lakes Villarrica and Calafquén (South-Central Chile). Sed Geol 213:121–135

Mountjoy JJ, Micallef A (2012) Polyphase emplacement of a 30 km3 blocky debris avalanche and its role in slope-gully development. In: Yamada Y, Kawamura K, Ikehara K, Ogawa Y, Urgeles R, Mosher D, Chaytor J, Strasser M (eds) Submarine mass movements and their consequences. Advances in natural and technological hazards research. Springer, Netherlands, pp 213–222

Ogata K, Tinterri R, Pini GA, Mutti E (2012a) The Specchio Unit (Northern Apennines, Italy): an ancient mass transport complex originated from near coastal areas in an intra-slope setting. In: Yamada Y, Kawamura K, Ikehara K, Ogawa Y, Urgeles R, Mosher D, Chaytor J, Strasser M (eds) Submarine mass movements and their consequences. advances in natural and technological hazards research. Springer, Netherlands, pp 595–605

Ogata K, Mutti E, Pini GA, Tinterri R (2012b) Mass transport-related stratal disruption within sedimentary mélanges. Tectonophysics 568–569:185–199

Pini GA, Strasser M, Boschetti A, di Blasi F, Ogata K, Panieri G, Festa A (2013) Structural characterization of a large-scale mass transport deposit: origin of the internal features and possible depositional mechanism. IODP Exp. 333 second post-cruise meeting

Plaza-Faverola A, Bünz S, Mienert J (2010) Fluid distributions inferred from P-wave velocity and reflection seismic amplitude anomalies beneath the Nyegga pockmark field of the mid-Norwegian margin. Mar Pet Geol 27(1):46–60

Strachan LJ (2008) Flow transformations in slumps: a case study from the Waitemata Basin, New Zealand. Sedimentology 55:1311–1332

Strasser M, Moore GF, Kimura G, Kopf AJ, Underwood MB, Guo J, Screaton EJ (2011) Slumping and mass transport deposition in the Nankai fore arc: evidence from IODP drilling and 3-D reflection seismic data. Geochem Geophys Geosysst (G3) 9:Q0AD13

Tunis G, Venturini S (1992) Evolution of the southern margin of the Julian Basin with emphasis on the megabeds and turbidites sequence of the Southern Julian Prealps (NE Italy). Geologia Croatica 45:127–150

Landslide Hazard Assessment of the Black Sea Shelf Adjoined to the Western Caucasus

41

Vsevolod Yu. Ionov, Ernest V. Kalinin, Igor K. Fomenko, and Sergey G. Mironyuk

Abstract

The presented manuscript aims to reveal the peculiarities of submarine landslides formation within the Caucasian shelf of the Black Sea and to perform landslide hazard assessment by means of slope stability calculations. For the proposed scope there were studied three adjoining areas in the north-eastern sector of the basin between Cape Mali Utrish and Hosta Bay. Analysis of the new geotechnical and geophysical survey data clearly indicates that submarine landslides occur on the shelf of the Black Sea adjoined to the Western Caucasus and may be a potential hazard for submarine installations which cross the shelf break and continental slope area of this region. 1D and 2D slope stability analyses revealed that gravity force is not sufficient for landslide formation in the shelf area and should be supplemented by additional triggers such as storm-wave loading and seismicity.

Keywords

Submarine landslides • The Black Sea • Slope stability • Shelf • Geohazards

41.1 Introduction

One of the most important objectives of the survey works along submarine pipelines and other oil and gas development installations is to perform the geohazard assessment. Geohazards may result in damage of the installations and as opposed to terrestrial conditions it is almost impossible to develop mitigation measures from their action for submarine environments. For the shelf and continental slope areas the most typical geohazards are related to submarine mass movement processes (Ionov 2012).

Submarine slopes of the Black Sea are not an exclusion—mass transport deposits were discovered here in the middle of the XX century (Arhangelsky 1930). According to further studies (Safyanov et al. 2001; Sorokin and Luksha 2002) it may be asserted that submarine landslides are widespread within the Caucasian continental slope of the Black Sea. On the other hand some studies show (Aibulatov 2002) that submarine landslides almost do not occur in the shelf area adjoined to the Western Caucasus and are not hazardous for submarine installations. Analysis of new geotechnical and geophysical survey data clearly indicates that submarine landslides are widespread not only on the continental slope but also within the shelf area of the Black Sea.

The presented manuscript aims to reveal the peculiarities of submarine landslides in the Caucasian Black Sea shelf sediments and to perform the landslide hazard assessment of the study area by means of slope stability calculations.

For the proposed scope there were studied three adjoining areas (Fig. 41.1) in the north-eastern sector of the basin.

V. Yu. Ionov (✉) · E. V. Kalinin
Lomonosov Moscow State University, Moscow, Russia
e-mail: seva.ionov@yandex.ru

I. K. Fomenko
Gulf Interstate Engineering Company, Moscow, Russia

S. G. Mironyuk
Peter Gaz LLC, Moscow, Russia

G. Lollino et al. (eds.), *Engineering Geology for Society and Territory – Volume 4*,
DOI: 10.1007/978-3-319-08660-6_41, © Springer International Publishing Switzerland 2014

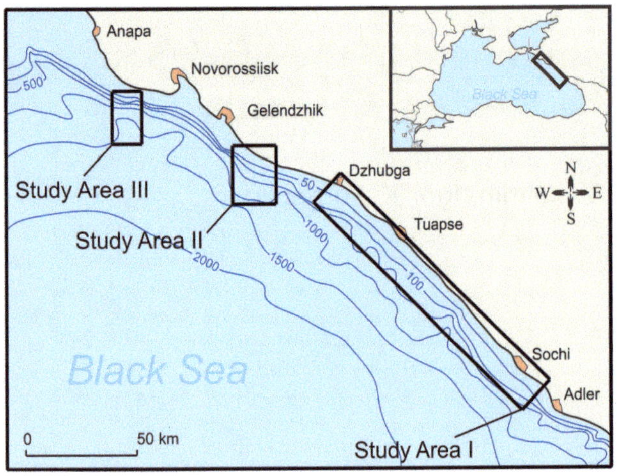

Fig. 41.1 Schematic map showing locations of the study areas (*black boxes*)

41.2 Geological Setting of the Study Areas

Geological setting of the study area was analyzed on the basis of Peter Gaz LLC repository survey data (cruises took place in 2008–2011) and on the basis of repository survey data which was performed for the purposes of governmental geological maps preparation (cruises took place in 1983–2008).

Shelf and continental slope may be marked out within the relief of the study areas. Length of the shelf varies from 1.5 to 14 km. In the shelf area the following elements may be isolated: wave-cut bench zone, inner and outer shelf.

The wave-cut bench forms a zone of abrasion type relief, which stretches as a narrow belt along the shoreline. The inner and outer shelf forms a zone of relief with accumulative type. Inner shelf is under the frequent influence of wave loading and is bounded by the 10 and 30 m depth contours. The Outer shelf is situated beyond the area of frequent wave loading influence and it is bounded by the shelf break. The shelf area is rather flat with slope inclinations of less than 1° towards the shelf break. Beyond the shelf break a continental slope is situated which may be characterized by slope angle 10–60°.

All geomorphological elements are characterized by different geological conditions. It should be mentioned that the profile was studied up to 20 mbsf. According to the drilling data within the wave-cut bench zone terrigenous and terrigenous-carbonate flysch deposits may be observed. Within the inner shelf predominantly different types of sand, gravels, shells and shell debris are deposited. And in the outer shelf mostly clayey sediments are presented. Quaternary sediments are deposited subhorizontally with slight inclination towards the shelf break with an average thickness of 20 mbsf. Within the upper part of the continental slope (which is the steepest) flysch deposits are outcropped. In general the most widespread sediments in the study area are very soft organic clay and very soft organic silty clay. Soil properties of samples which were taken during the surveys were determined in the offshore and onshore laboratories separately for every study area. Subsequently all the laboratory data was statistically reprocessed by means of standard approach (GOST20522-96 1996). As an example physical and mechanical properties of soils and rocks which were discovered within the study area II are presented in the Table 41.1.

Within the *study area I* the authors discovered four submarine landslides in the inner shelf by analyzing the digital terrain model (DTM). These landslides are about 1 m in thickness, up to 1,000 m long and 750 m wide. Inclination of the surface within the areas affected by landslides is in the range between 0.4–0.8°. A detailed description and location of these landslides may be seen in the previous publications (Ionov et al. 2014). According to the sediments sampled closest to the landslide locations, we assume that landslide deposits are predominantly presented by sandy soils.

Apart from the inner shelf of the Black Sea landslides were also discovered in the outer shelf sediments. Close to the shelf break within the *study area III* landslides may be clearly identified on the seabed (Fig. 41.3). Their width is up to 1,000 m, length up to 500 m and thickness up to 6 m. All the discovered landslides are located on the surface with inclinations from 1° to 10°. Slope angle in the headwall area of these landslides reaches 12°. Landslides are composed from over consolidated clayey sediments. Slip surfaces of the discovered landslides coincide with the beds of the quaternary sediments which may be concluded from the geophysical data interpretation (Fig. 41.2).

All the observed landslides may be classified as translational according to classification presented by Locat and Lee (2002).

41.3 Methods

Presence of landslides within the study areas draws a necessity of slope stability assessment of the Caucasian shelf of the Black Sea. As long as geotechnical data was available only for the study areas I and II slope stability assessment was performed just for these two areas.

A standard 1D and 2D approaches based on limit equilibrium methods were utilized for the purposes of slope stability analyses. 1D modelling was implemented for different inclinations of the seabed surface of the modeled profiles (1–40°), thickness of sliding mass (1–20 mbsf), strength parameters (in total, effective stresses and with c_u) and for additional loading or triggering mechanisms.

Table 41.1 Physical and mechanical properties of quaternary sediments on the Caucasian shelf of The Black Sea (study area II on Fig. 41.1)

Soil type	Water content (%)	Density, (g/cm^3)	Density of solid particles, (g/cm^3)	Plastic limit (%)	Liquid limit (%)	Undrained shear strength (lab vane) (KPa)	Undrained shear strength (triaxial) (KPa)
Very soft organic clay (gas saturated)	85.0–143.0 108.2(0.15)	1.28–1.49 1.40(0.04)	2.68	35.69	89.13	1.0–9.0 4.6(0.67)	–
Very soft organic clay	41.0–85.0 56.9(0.15)	1.50–1.71 1.63(0.04)	2.69	26.46	58.46	1.0–11.0 6.1(0.54)	–
Soft to firm clay	24.0-61.0 44.2(0.22)	1.60–2.06 1.77(0.07)	2.67	25.83	59.83	14.0–39.0 21.5(0.39)	23.0
Silty clay to silt	31.0–39.0 34.5(0.11)	1.76	2.72	22.00	39.00	7.0	–
Fine sand	27.0	1.94	2.73	–	–	–	–
Stiff clay	18.0–38.0 27.0(0.25)	1.80–2.13 2.00(0.09)	2.68	25.67	50.00	–	198.0
Shells and shell debris	115.00	1.97	2.79	–	–	–	–

The numerator gives minimum and maximum values for individual soil types; the denominator shows mean values and the coefficient of variation (in brackets)

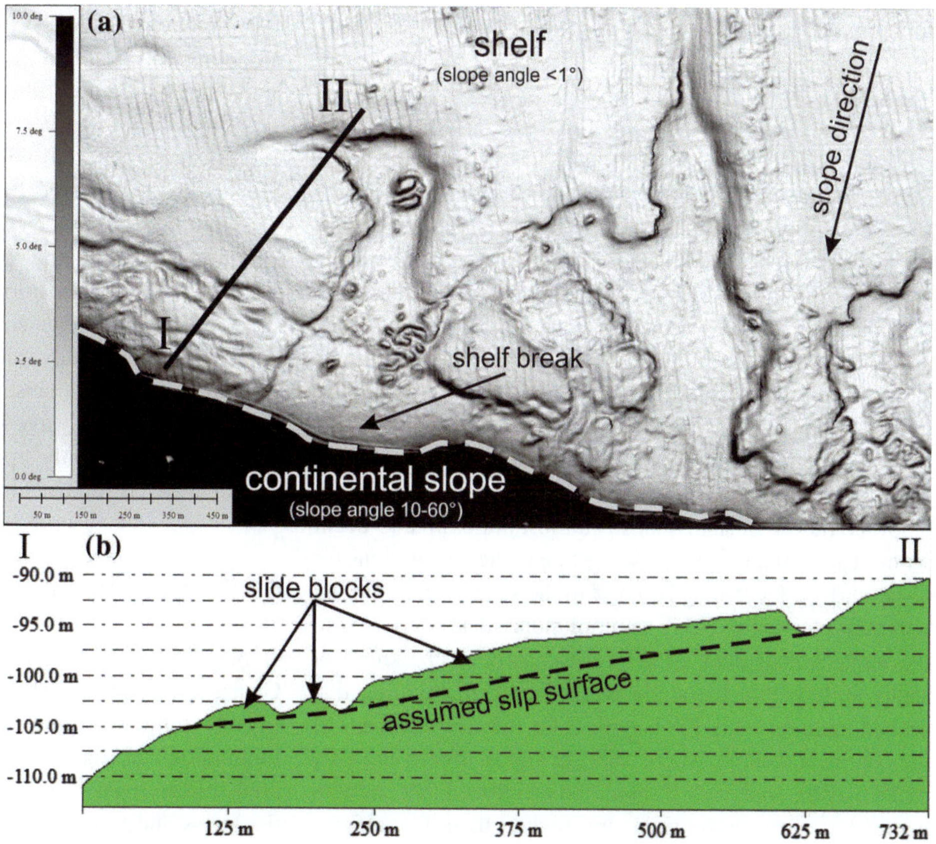

Fig. 41.2 **a** Fragment of the DTM (Global Mapper slope shader) showing the landslides within the study area III. **b** Profile across the landslide area (see **a** for profile location) (Ionov 2013)

Seismicity and storm wave loading were implemented in the 1D model. 2D modeling was performed for static conditions and with seismic loading. Seismic forces were implemented in the analyses by means of pseudostatic approach with the help of seismic coefficients. Slope stability assessment was realized for seismic loading with intensity VII to X (degrees according to MSK-64 scale). Storm wave loading was applied in the modeling process using the equations presented by Seed and Rahman (1978), Lee and Edwards (1986). Only extreme wave parameters

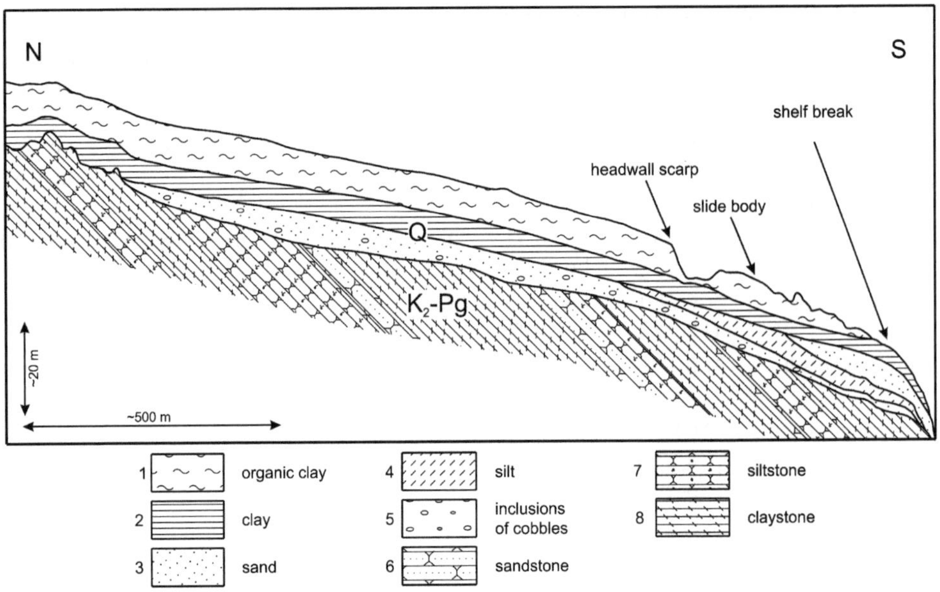

Fig. 41.3 Schematic interpretation of the geophysical profile across the Caucasian shelf of the Black Sea (near Cape Mali Utrish)

for 10 and 100 years periods were used during the calculations. A more detailed description of the methods may be found in the previous publications (Ionov et al. 2012, 2014).

41.4 Results

Slope stability assessment results indicate that soils in the shelf area stay stable for static conditions (without additional triggers). Closer to the shelf break with increasing surface inclination, landslide formation, however, becomes possible even for static conditions. The most unstable soils are very soft and soft clayey sediments. In static conditions slope failure in very soft or soft clayey soils with their thickness about 20 m will occur with slope angle 2–4°. Such inclination of the seabed is typical for the outer shelf close to the shelf break.

Slope failure in sandy soils is almost impossible to occur without additional triggers. According to the calculations for a slip surface located at 10 mbsf in the sandy soils a failure will take place with a 35° angle of slope. And according to the geotechnical data 10 m is the maximum thickness of sandy soils within the study areas. Worth mentioning that seabed inclination of 35° is typical for the steepest areas of the continental slope, where soils are almost absent and only rocks may be observed. It should be mentioned that strength parameters for sandy soils are higher than for clayey. Probably, this difference is caused by the fact that sandy soils occur predominantly in the inner shelf zone and these soils are subjected to frequent storm wave loading.

Slope stability assessment with applied seismic loading shows that landslides are likely to occur in very soft and soft clayey soils both on the shelf and on the continental slope with seismic intensities more than VII (Fig. 41.4). In sandy soils within the shelf area, landslides may appear with seismic intensities XI–X, and on the continental slope with seismic intensities VII–VIII and more. According to the 2D modeling with VII ball seismic intensity the most extensive landslides would be ~280 m long with slip surfaces located at 20 mbsf. With higher seismic intensities the landslides' length increases and with 10 ball seismic intensity it exceeds the geotechnical models lengths.

According to the modeling with applied storm wave loading, landslides may predominantly occur in very soft and soft clayey soils. Formation of landslides triggered by storm waves in sandy soils is only possible in the shelf break area, especially close to the heads of submarine canyons which are incised in the shelf. The calculations revealed that storm wave loading influences the stability of submarine slopes only up to 60–80 m water depth and in deeper water is almost insufficient.

41.5 Conclusion

Geological setting of the study areas shows that: within the Caucasian Black Sea shelf submarine landslides take place in soils in the outer and inner shelf areas on the slopes with inclination starting from 0.4°. Landslides in the outer shelf area differs from the landslides in the inner shelf area in thickness and type of the landslides' sediments (clayey vs. sandy correspondingly) but their mechanism is similar. These landslides may be classified as translational slides.

For the purposes of submarine slope stability assessment in the study area a 1D model may be utilized as long as a 2D. 1D model allows to reveal the critical relations between slopes inclinations, landslides' sediments thicknesses and different soil types which set conditions for landslide

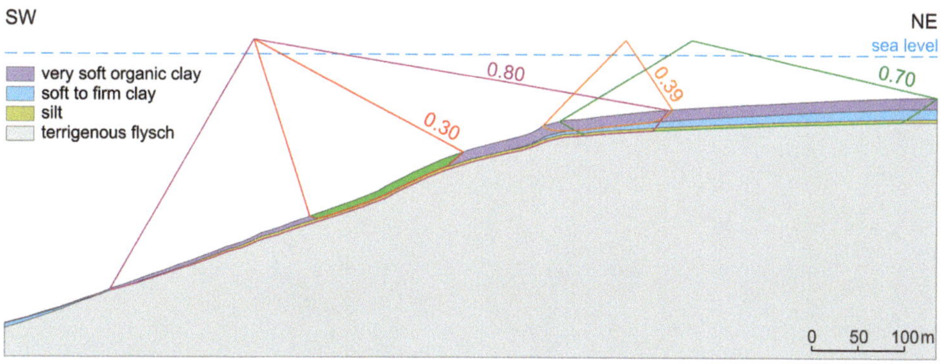

Fig. 41.4 2D Slope stability assessment results within the study area II with seismic loading considered (MSK-64—VII intensity). The numbers mean corresponding factors of safety and *green* area indicates a critical slip surface

formation within the Caucasian Black sea shelf area. 2D model helps to ascertain the most probable locations of the unstable areas of the submarine slopes and to identify their morphometric features.

Calculation results in static conditions shows that in the outer shelf area close to the shelf break on the sections of clayey sediments occurrence and where the soil thickness reaches 20 mbsf or more, according to the modeling landslides may be formed just under gravity force loading. It should be mentioned that the whole thickness of soil mass (20 mbsf and more) may be involved in the landslide process.

Slope stability assessment with additional loading considered shows that the main conditions and triggering mechanisms for the landslide formation on the continental margin of the Black Sea adjoined to the Western Caucasus are relief (slope inclination), soil properties, and different additional loadings (natural and human related). According to the modeling for the formation of landslides in the shelf area the gravity force should be supplemented by storm-wave loading and seismicity.

References

Aibulatov NA (2002) Gravitacionnyi perenos osadochnogo materiala na kontinental'nom sklone i bezopasnost' stroitel'stva i ekspluatacii gazoprovoda Rossiya—Turciya. IRC Gazprom, Moscow

Arhangelsky A (1930) Opolzanie osadkov na dne Chernogo morya i geologicheskoe znachenie etogo yavleniya. Byull Mosk obsch ispyt pri (otd geolog) (1–2):32–80

Ionov VY (2013) Injenerno-geologicheskie usloviya formirovaniya podvodnih opolzney v ollozheniyah shelfa Chernogo moray (mejdu misom Mal.Utrish I buhtoi Hosta): Dis…. kand Geol-min nauk:25.00.08—Zashishena 20.12.2013.—M., 2013.—192 s.—Bibliogr.: s. 156–170

Ionov V, Kalinin E, Fomenko I, Mironyuk S (2012) Submarine landslides formation in the shelfbreak sediments of the Black Sea near Archipo-Osipovka. Injenernaya geologiya 5:36–46

Ionov V (2012) Review of submarine landslides studies as applied to construction of marine oil and gas installations. Injenernye izyskaniya 6:52–63

Ionov VY, Kalinin EV, Fomenko IK, Mironyuk SG (2014) Regional slope stability assessment along the Caucasian shelf of the Black Sea. In: Krastel S et.al (eds) Submarine mass movements and their consequences. In: Advances in natural and technological hazards research/6th international symposium, vol 37. Springer International Publishing, Switzerland, pp 201–212

Lee HJ, Edwards BD (1986) Regional method to asses offshore slope stability. J Geotech Eng 112:489–509

Locat J, Lee H (2002) Submarine landslides: advances and challenges. Can Geotech J 39:193–212

Safyanov GA, Menshikov VL, Peshkov VM (2001) Podvodnye kan'ony—ih dinamika i vzaimodeistvie s beregovoi zonoi okeana. Izd-vo VNIRO, Moscow

Seed HB, Rahman MS (1978) Wave-induced pore pressure in relation to ocean floor stability of cohesionless soils. Mar Geotechnol 3:123–150

Sorokin VM, Luksha VL (2002) Osadki i osadochnye processy v podvodnoi doline Paleokubani v Chernom more. Vestnik Mosk. un-ta. Seriya 4. Geologiya 2:13–24

Recent Morphological Changes of the Nice Continental Slope

42

Maëlle Kelner, Sébastien Migeon, Emmanuel Tric, Françoise Couboulex, Alexandre Dano, Thomas Lebourg, and Alfredo Taboada

Abstract

Time-series bathymetric data acquired between 1991 and 2011 have been used to evaluate the recent morphological evolution of the Nice upper continental slope (SE France, Ligurian Sea). Small-scale landslides lead to a retrogressive evolution of the continental shelf/upper slope transition but their frequency, size and impact are not well known. Mapping was undertaken to identify the morphology of landslide scarps and the location of the shelfbreak. Map comparisons were performed using ArcGIS "raster calculator". Sediment remobilization on the upper slope (up to depths of 200 m) is fast and significant; landslide scars with volumes greater than 25,000 m^3 can appear with a frequency less than 8 years. Shelfbreak migration toward the coastline can reach rates of 60 m over 7–8 years where the continental shelf is over 200 m wide. Furthermore, this quantitative analysis highlights alternations between periods of strongly erosive events and sedimentation periods. On the upper slope, eroded volumes can be multiplied by 10 during periods of enhanced landslide patterns (1999–2006). Such cycle-like landslide activity raises the issue of the triggering processes. On the Nice continental slope thick poorly consolidated beds rapidly deposited on a steep slope, earthquakes and rainfall leading to fresh water circulation below the shelf were identified as potential triggers. Our 4D bathymetric study suggests that over the last 20 years the greatest impacting factor may be freshwater outflows.

Keywords

Ligurian sea • Geohazards • Bathymetry • Continental slope • Geomorphology

42.1 Introduction

Morphological changes, such as erosion of submarine continental slopes, through time are linked to sedimentary processes such as submarine landslides. At a human timescale, low volume, high frequency landslides are largely predominant. They can be responsible for numerous hazards especially when triggered at shallow water depths. Due to the large number of submarine landslides observed on continental slope bordering the Nice coast (SE France, Ligurian Sea; Migeon et al. 2012), we decided to study the present-day mass-wasting processes remoulding the seafloor. Multibeam sonar technology provides an opportunity to

M. Kelner (✉) · S. Migeon · E. Tric · F. Couboulex · A. Dano · T. Lebourg
UMR Géoazur, Université de Nice Sophia-Antipolis, UPMC-CNRS-IRD, Observatoire de La Côte D'Azur, 250 Rue Albert Einstein, Sophia Antipolis, 06560 Valbonne, France
e-mail: kelner@geoazur.unice.fr

A. Taboada
UMR Geosciences Montpellier, Place Eugene Bataillon, 34000 Montpellier, France

G. Lollino et al. (eds.), *Engineering Geology for Society and Territory – Volume 4*,
DOI: 10.1007/978-3-319-08660-6_42, © Springer International Publishing Switzerland 2014

assess morphological change on short time scale (5–8 years) and contribute to the understanding of shelf erosion throughout time. This paper presents the result of the repeated bathymetric survey of the Nice shelf/slope limit between 1991 and 2011. We propose an interpretation of the significant morphological changes observed over 20 years to provide insights on possible triggering mechanisms.

42.2 Geological Settings

Our study focused on the narrow continental shelf ($\sim$100 m from the coastline) and the upper slope, offshore the city of Nice. These settings are affected by a large number of submarine landslides (Migeon et al. 2012). The most recent, mobilizing approximately 8×10^6 m^3, initiated during October 1979. It is responsible for the destruction of the new (in 1979) harbour-airport complex of Nice and the triggering of a 2–3 m high local tsunami (Genesseaux et al. 1980; Ioualalen et al. 2010). The Var river, which reaches the coastline just west of the airport, is impacted by a few but violent Mediterranean rainfalls. These precipitations generate floods and groundwater outflows circulating below the airport, emerging on the continental shelf (Stegmann et al. 2011). Moreover, poorly consolidated sediments deposited on steep slopes are affected by moderate seismicity (Courboulex et al. 2007) which may impact their stability.

42.3 Bathymetric Analysis

To better constrain the frequency and volumes of recent slides including their impact in terms of morphological changes, we compared four high-resolution bathymetric maps compiled from datasets acquired in 1991, 1999, 2006 and 2001 with horizontal resolutions of 10, 2, 2 and 1 m, respectively. The surveys were undergone using multibeam technology. The survey dating from 1991 used a Syledis 5 m resolution terrestrial radio positioning system. The following three used a GPS positioning system. The covered zone varies form one survey to another but portions of the dataset are common in an area extending laterally from the Var river estuary to the Paillon Canyon. The map interpretations and comparisons were performed using GIS tools. Individual scarps and shelfbreak location were identified on each bathymetric map. Then, raster comparisons were performed using the subtraction tool of ArcGIS "raster calculator" in order to calculate volume of missing deposits from slope variations through time. The precision of the calculated volumes is relative to the poorest data resolution of the two periods compared as well as the precision of xyz location. Using the coastline, optical map inspection and excluding depth difference values rang from −2 m to +2 m, we assessed and greatly reduced the positioning errors.

42.4 Morphological Evolution of the Shelfbreak Over the Last 20 Years

Comparison of time-series bathymetric maps provided the specific background to locate and to quantify morphological changes over the last 20 years at shallow water depths (0–200 m). Between 1991 and 2011, an alternation between periods of low morphological changes and periods of active landsliding events or deposition/infillings involving significant volumes can be clearly discriminated.

From 1991 to 1999, due to the poor overlap between surveys, only a clear accretion of about $+6.14 \times 10^6$ m^3 was detected on an area of 541×10^3 m^2 mainly corresponding to the 1979 scar (Fig. 42.1a). Particles were deposited on 61.2 % of the area. Erosion by small-scale landslides affected 4 % of the area, mainly on the western side of the 1979 scar.

During the 1999–2006 period, sediment remobilization on the upper slope was significant; 71 landslides with a mean volume of 1.3×10^5 m^3 were initiated (Fig. 42.1b), representing a total missing volume of 9.21×10^6 m^3 over a 8.3×10^5 m^2 area (27 % of the total area). Similar small landslides with a return frequency of 1 year or less have been described in other environments like fjords, canyon flanks or upper slope (Hill 2012; Hughes-Clarke et al. 2012; Smith et al. 2007). Our observations show that the open Nice continental slope can also be affected by fast erosional processes. During this period, parts of the shelfbreak migrated toward the coastline following the head of retrogressive scars. The shelf was reduced by 20 m along the 130-m-wide western edge of the airport (Fig. 42.1b). Map comparison can also be used to identify areas where sediment deposition prevailed. Sediment infill smoothed older scar and chute morphologies. Landslide scars representing 15×10^4 m^3 of missing deposits were filled-up and disappeared from the sea floor in less than 7 years (zoom Fig. 42.1b). Biscara et al. (2012) made similar observations on the upper Gabon slope and estimated that 15–20 years were necessary to bury scars (17×10^4 m^3).

During the 2006–2011 periods, only subtle morphological changes were observed (Fig. 42.1c) indicating that landslide activity significantly decreased (Fig. 42.1c). Compared to the previous period (1999–2006), the remobilized volumes are 10 times smaller. Sediment remobilization mainly corresponds to retrogressive erosion within pre-existing scars. Only one significant new landslide (about 25×103 m^3) was triggered at 14-m water depth, leading to an instantaneous local landward migration of the

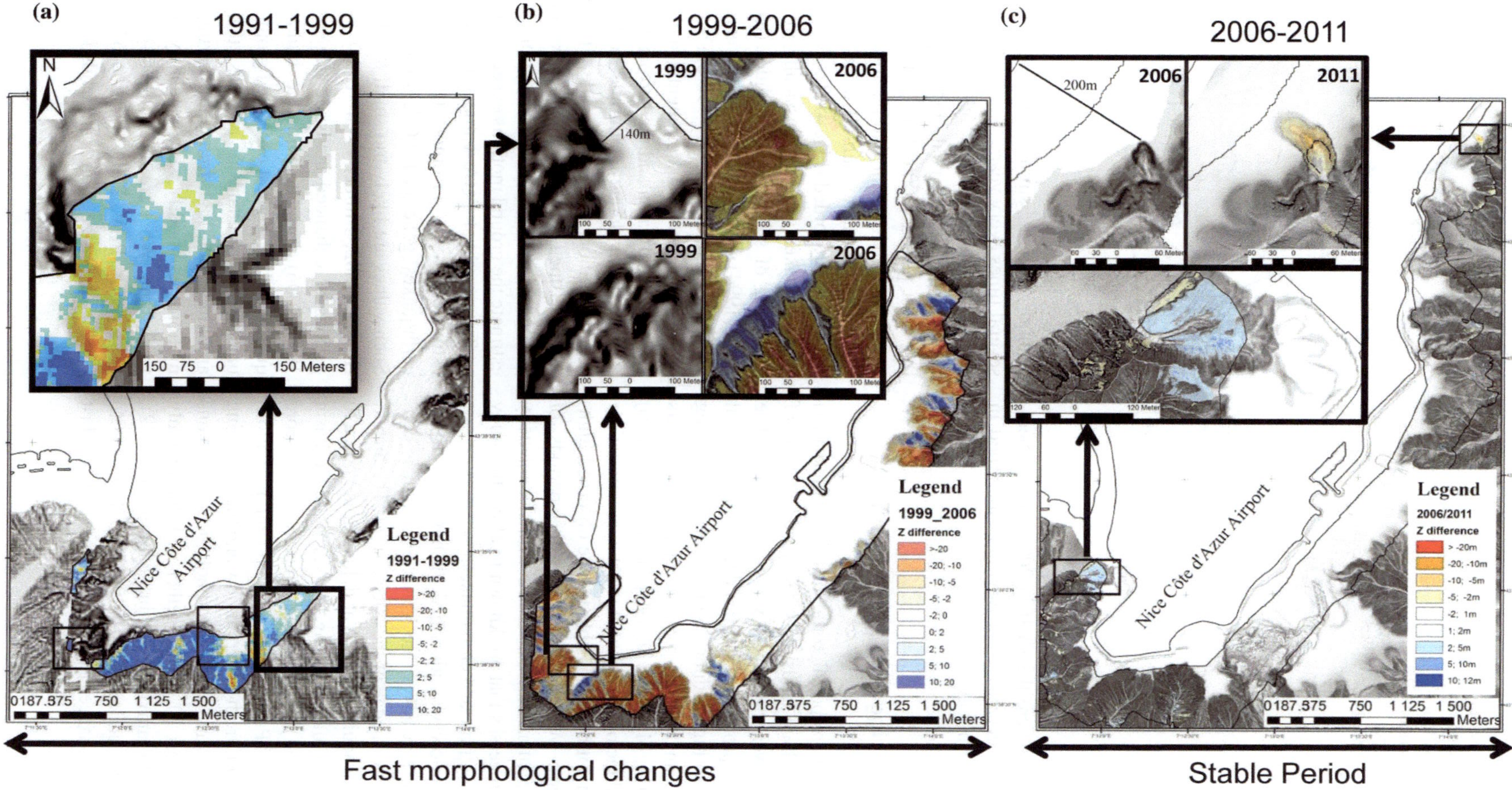

Fig. 42.1 Bathymetric comparisons of the uppermost continental slope: 1991–1999, 1999–2006, 2006–2011. Zoom on erosion/deposition areas: in B&W theoldest map and in colour the erosive area from *yellow* to *red* and the deposition area from clear to *dark blue*

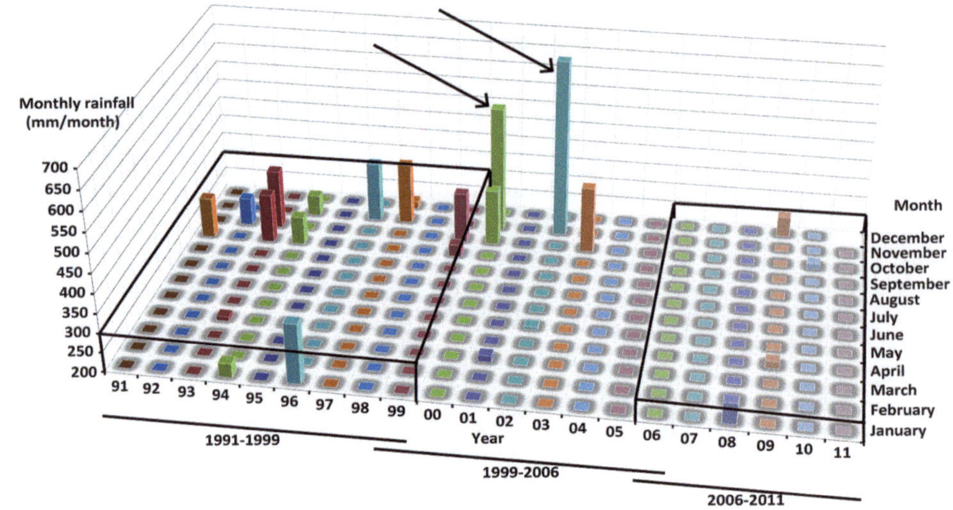

Fig. 42.2 Monthly rainfall events in mm/month, between 1991 and 2011. *Arrows* indicate the 2000 and 2002 strongest rainfall events (500 and 650 mm) of the 1999–2006 time period. Between 2006 and 2011, no significant rainfall events were recorded

shelfbreak of 60 m, where the shelf is 200-m wide. Sediment deposition also decreased and was only identified in front of the Var-river mouth.

42.5 Possible Triggering Mechanisms

The triggering factors responsible for such rapid alternation between periods of high landslide activity and periods of relative quietness over the last 20 years are unclear. The Nice upper slope is prone to landslides due to specific local conditions such as high-sedimentation rates, overloading and critical slope angles, and presence of gas-rich layers (Migeon et al. 2012; Mulder et al. 1994; Sultan et al. 2010) but these potential factors are thought to act over longer time scales and should not be responsible for the observed high-frequency changes. We thus compared the periods of occurrence of the submarine landslides with the intensity and of external processes that are known to affect the study area: fresh-water outflow and earthquakes.

Fresh-water outflow was identified offshore the airport (Dan et al. 2007). Nowadays, the Var-river mouth is constrained by the western edge of the Nice airport. The airport was built above the aquifer of the Var delta that is known to react rapidly with rainfall affecting the Var drainage basin (Stegmann et al. 2011): outflow can generate excess pore pressure of about 10 kPa. To determine a potential relationship between rainfall, fresh-water outflow and the triggering of landslides, we examine the morphological variations identified over the last 20 years together with the monthly rainfall record (>200 mm) on the drainage basin (Fig. 42.2). During the first two periods (1991–1999 and 1999–2006) characterized by an enhanced triggering of landslides and fast seafloor changes, strong rainfall events occurred once or several times every years. The 2000 and 2002 extreme rainfall episodes of respectively 500 and 650 mm unambiguously induced pore pressure peaks in the aquifer. In contrast, during the latest period (2006–2011), characterized by a reduced landslide activity, little morphological variations and sediment deposition, no rainfall events greater than 250 mm/month were recorded (Fig. 42.2).

Earthquakes are also classically invoked to explain the triggering of submarine landslides (Keefer 2002). Mulder et al. (1994) showed that centennial earthquakes with a Peak Ground Acceleration PGA = 98 cm/s^2 can trigger submarine landslides on the Nice continental shelf. Between 1991 and 2011, the study area and the surrounding 100 km were only affected by moderate-magnitude earthquakes (Mw $\leq$ 3–5). Their location and magnitude (BCSF database www.franceseisme.fr) were used in the Akkar and Bommer (2010) Ground Motion Prediction Equations models (GMPEs) to estimate the ground motions expected in the studied area.

The maximal PGA value offshore the Nice airport is estimated to be 17 cm/s^2 (Fig. 42.3) and most of the earthquakes generated PGA values lower than 5 cm/s^2. Such values are likely too small to affect the slope stability (Keefer 2002; Mulder et al. 1994). Nevertheless the two periods of enhanced landslide triggering (1991–1999 and 1999–2006) correlate with the highest number of earthquakes and the highest estimated ground motion values. The underlying mechanism remains unclear, perhaps low-amplitude but high-frequency earthquake-related ground motion can modify slope stability and induce sediment weakening. What are the minimal ground motions parameters (PGA, PGV, PGD, duration of shaking) necessary to trigger small-size submarine landslide?

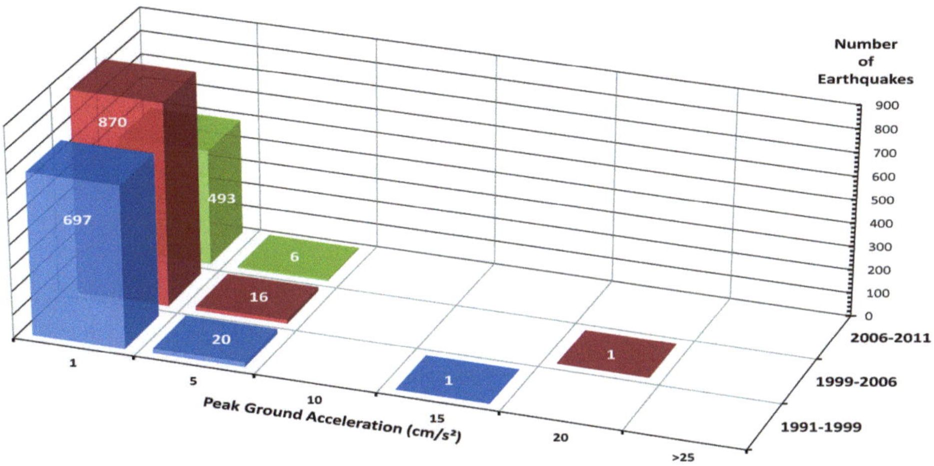

Fig. 42.3 Number of earthquakes ordered by their PGA ranges (cm/s^2) that occurred during the last 20 years. The 1991–1999 and 1999–2006 intervals underwent higher numbers of earthquakes with higher peak-ground accelerations than the 2006–2011 periods

42.6 Conclusion

Analysis of 4D bathymetric data from the Nice continental shelf reveals large variations in landslide frequency over short time periods (less than 7–8 years). Periods of enhanced landsliding can increase erosion rates by a factor of 10. Only a few previous studies (Biscara et al. 2012; Hill 2012; Hughes-Clarke et al. 2012; Smith et al. 2007) discussed the high-frequency occurrence (peri-annual) of small landslides with a $V > 100$ m^3. The present study is a next step in the understanding submarine-slope stability and related hazard, which requires the precise identification of landslides volumes, frequency and triggering factors. Over the last 20 years, the fresh-water outflow circulating in the Var river delta deposits is likely to be the dominant factor in triggering small-size landslides. Despite previous work (Keefer 2002) indicating that earthquakes with magnitude lower than $Mw = 5$ may not trigger submarine landslides, we show that there is a correlation between number of earthquakes and frequency of mass wasting events on the continental shelf. Landslide triggering is also affected by other factors acting over longer timescales such as high-sedimentation rates and the presence of gas-rich layers.

Acknowledgments Maelle Kelner is a Ph.D. student funded by the French Ministry of Research. This work was partly funded by the French Action Marges program. The authors acknowledge the IAEG organizers for encouraging this review. J. S. L'Heureux and J. Locat are thanked for constructive and valuable comments to this paper. Finally, the authors appreciated the help of M. Hassig and M. Paulatto in improving the manuscript's English.

References

Akkar S, Bommer JJ (2010) Seismol Res Lett 81:195
Biscara L, Hanquiez V, Leynaud D, Marieu V, Mulder T, Gallissaires JM, Crespin JP, Braccini E, Garlan T (2012) Mar Geol 299–302:43
Courboulex F, Larroque C, Deschamps A, Kohrs-Sansorny C, Gélis C, Got JL, Charreau J, Stéphan JF, Bethoux N, Virieux J, Brunel D, Maron C, Duval AM, Perez J-L, Mondielli P (2007) Geophys J Int 170:387
Dan G, Sultan N, Savoye B (2007) Mar Geol 245:40
Genesseaux M, Mauffret A, Pautot G (1980) Comptes Rendus L'académie Des Sci. Paris 290:959
Hill PR (2012) Int. Assoc. Sedimentol. 44:47
Hughes-Clarke JE, Brucker S, Muggah J, Hamilton T, Cartwright D, Church I, Kuus P (2012) In: Eberhardt E et al (eds.) Landslides and engineered slopes: protecting society through. improved understanding. Taylor & Francis Group, London, pp 1091–1096
Ioualalen M, Migeon S, Sardoux O, (2010) Landslide tsunami vulnerability in the Ligurian sea: case study of the 1979 October 16 Nice internationalairport submarine landslide and of identified geological mass failures. Geophys J Int. doi:10.1111/j.1365-246X.2010.04572.x
Keefer DK (2002) Investigating landslides caused by earthquakes—a historical review. Surv Geophys 23:473–510
Migeon S, Cattaneo A, Hassoun V, Dano A, Casedevant A, Ruellan E (2012) Submar. In: Yamada Y et al (eds.) Mass movements their consequences. Springer, Amsterdam, pp 451–461
Mulder T, Tisot J-P, Cochonat P, Bourillet J-F (1994) Mar Geol 122:29
Smith DP, Kvitek R, Iampietro PJ, Wong K (2007) Mar Geol 236:79
Stegmann S, Sultan N, Kopf A, Apprioual R, Pelleau P (2011) Mar Geol 280:168
Sultan N, Savoye B, Jouet G, Leynaud D, Cochonat P, Henry P, Stegmann S, Kopf A (2010) Can Geotech J 47:486

The Case of Ischia Underwater Debris Avalanche (Italy, Tyrrhenian Sea) and Its High Mobility

43

G. de Alteriis, Anna Scotto di Santolo, F. L. Chiocci, M. Ramondini, and C. Violante

Abstract

Ischia volcanic island is one of the few cases in the Mediterranean where a clear relationship between constructional volcanism, volcano-tectonic uplift and cyclic multi-scale gravity failures is evident. One of these collapses involved the island's southern flank and led to the emplacement of a submarine debris avalanche spreading over an area of 150 km^2 with an estimated volume of 1.5 km^3. C^{14} stratigraphy suggests that it may have occurred during pre-historical times or even during the Greek colonisation of the island (around 7th century BC). This very large mass movement i.e. the "Ischia debris avalanche" is still apparent on the seafloor with a mega-blocks field detectable until 40 km from the island along the lower continental slope. Its high mobility is proved by the very low H/L ratio when compared to other submarine debris avalanches. Here we report the preliminary results of a back-analysis in which we modeled a land-sea collapse based on some simplified assumptions. The implemented code used in this study assumes a fluid-solid mixture with given rheological behaviors and was already used for simulate the mobility of some terrestrial debris flows in volcanic areas.

Keywords

Ischia volcano • Debris avalanche • Back-analysis

43.1 Introduction

Ischia island is located to the west of the Bay of Naples (Fig. 43.1) and is considered an active volcano based on the wide range of magma-related phenomena like shallow seismicity, ground uplift/subsidence and diffuse hydrothermal vents all suggesting the occurrence of a magmatic reservoir at relatively shallow depths.

Volcanic activity dates back to 150 ky and mostly consisted of medium-scale explosive eruptions with emplacement of pyroclastic deposits and lava domes (Gillot et al. 1982). A major ignimbritic eruption occurred 55 ka with the emplacement of the Mount Epomeo Green Tuff deposit and possibly caldera formation. The latest volcanic event occurred in historical times (AD 1302) as a small lava flow emplaced in the island's eastern sector. A peculiarity of this volcano is the great extent of volcano-tectonic and co-seismic uplift of Mt. Epomeo, the central relief of the island, presently

The original version of this chapter was revised: Incorrect chapter author name has been corrected. The erratum to this chapter is available at https://doi.org/10.1007/978-3-319-08660-6_44

G. de Alteriis (✉) · C. Violante
Istituto per l'Ambiente Marino Costiero (IAMC), CNR, 80133 Naples, Italy
e-mail: giovanni.dealteriis@iamc.cnr.it

A. Scotto di Santolo · M. Ramondini
Dipartimento di Ingegneria Civile, Edile e Ambientale (DICEA), Università Federico II, 80125 Naples, Italy

F. L. Chiocci
Dipartimento di Scienze della Terra, Università La Sapienza, 00185 Roma, Italy

G. Lollino et al. (eds.), *Engineering Geology for Society and Territory – Volume 4*,
DOI: 10.1007/978-3-319-08660-6_43, © Springer International Publishing Switzerland 2014

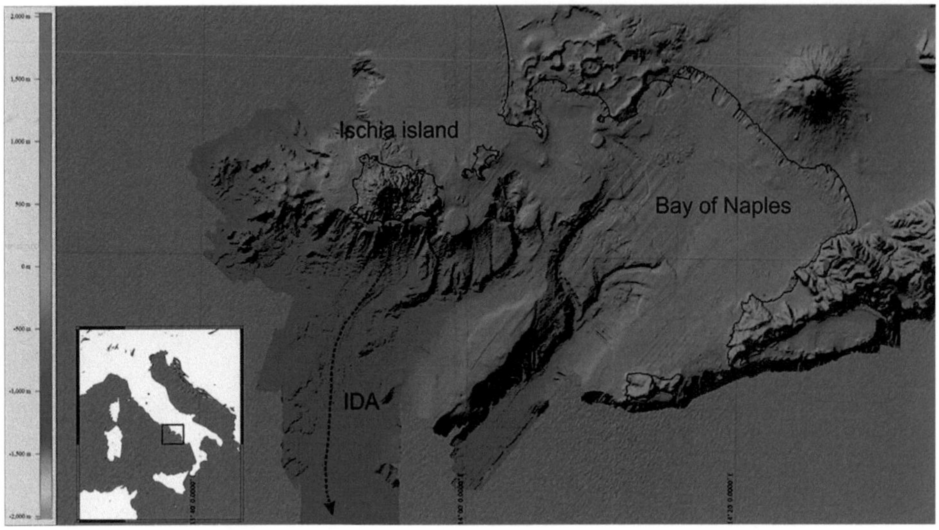

Fig. 43.1 Digital Terrain model of the Naples bay and dispersal axis of Ischia Debris Avalanche (*dashed line*)

attaining 785 m asl. Geological data indicate that this block has risen at least 800 m in the past 30 ky at an average rate of 20 mmy^{-1} but likely higher over shorter time spans. A clear correlation exists between uplift and gravity failures as confirmed by the abundance of subaerial landslides over Mt. Epomeo slopes also favoured by the strong hydrothermal weathering. These mass movements include debris slides, debris flows, mud flows and rock falls with volumes in the order of 10^3 m^3.

Marine geophysical surveys all over the island's offshore recently revealed larger mass movements with volume in the order of 10^9 m^3 radiating out from Mt. Epomeo and entering the sea. Over the western and northern continental shelves two debris field with an estimated volume of 0.2–0.5 km^3 show typical features of debris avalanches although their dating and emplacement mechanism are still matter of study (de Alteriis and Violante 2009).

To date the Ischia debris avalanche (Chiocci and de Alteriis 2006), IDA hereafter, occurring along the island's southern flank, is the best known despite its greater depth also due to extensive coring and radiometric datings (de Alteriis et al. 2010). Consequently the litho-stratigraphy of the deposit and its timing are quite well constrained while little is known about its dynamics as for the almost totality of submarine mass movements. A first attempt of modeling the IDA event was made by Tinti et al. (2011) with the main purpose of simulate the propagation of tsunami waves in the Bay of Naples.

In this paper, taking in consideration the IDA runout, we attempt a numerical simulation based on shallow flow equation in analogy with terrestrial high mobility debris flows (Scotto di Santolo and Evangelista 2009) and neglecting, in first approximation, the role of sea water column.

43.2 IDA Morphological Analysis

The entire IDA area can be divided into: (A) a source or failure upper section; (B) a depositional and transit section with relatively low aspect ratio and peculiar hummocky topography; (C) a deeper depositional section with relatively flat topography and less pronounced hummocks. These features are similar although not identical to many other submarine analogues (see Locat et al. 2004).

Direct sampling along the main dispersal axis has shown the occurrence of a very etherogeneous and ethero-metric assemblage of debris, ranging from gravel to boulder embedded in a relatively stiff, clayey-sandy matrix between the toe of slope at 9 km and about 23 km from source. This prevailing matrix area raises 10–15 m above the surrounding seafloor as a low aspect ratio elongated ridge and corresponds to the (B) depositional section. The latter evolves distally and laterally towards (C): a few metres thick, mud supported debris flow including mostly finer, gravel-size debris.

Blocks occur along (B) and (C) sections and have size from metric to tens of metres with several ones attaining a size of 150 m and raising 35 m above seafloor. They are extremely abundant in the (B) matrix section while are generally not embedded into the (C) debris flow.

Presently the sub-aerial slope is 20–12° steep; this slope can be considered only partly representative of the original failure plane due to intense post-collapse erosion. The underwater slope varies from 30–10° downslope according a typical concave profile. These values can be considered as quite representative of the original failure profile apart from modification due to a post-avalanche depositional marine terrace fed by intense erosion along Mt. Epomeo (Fig. 43.3).

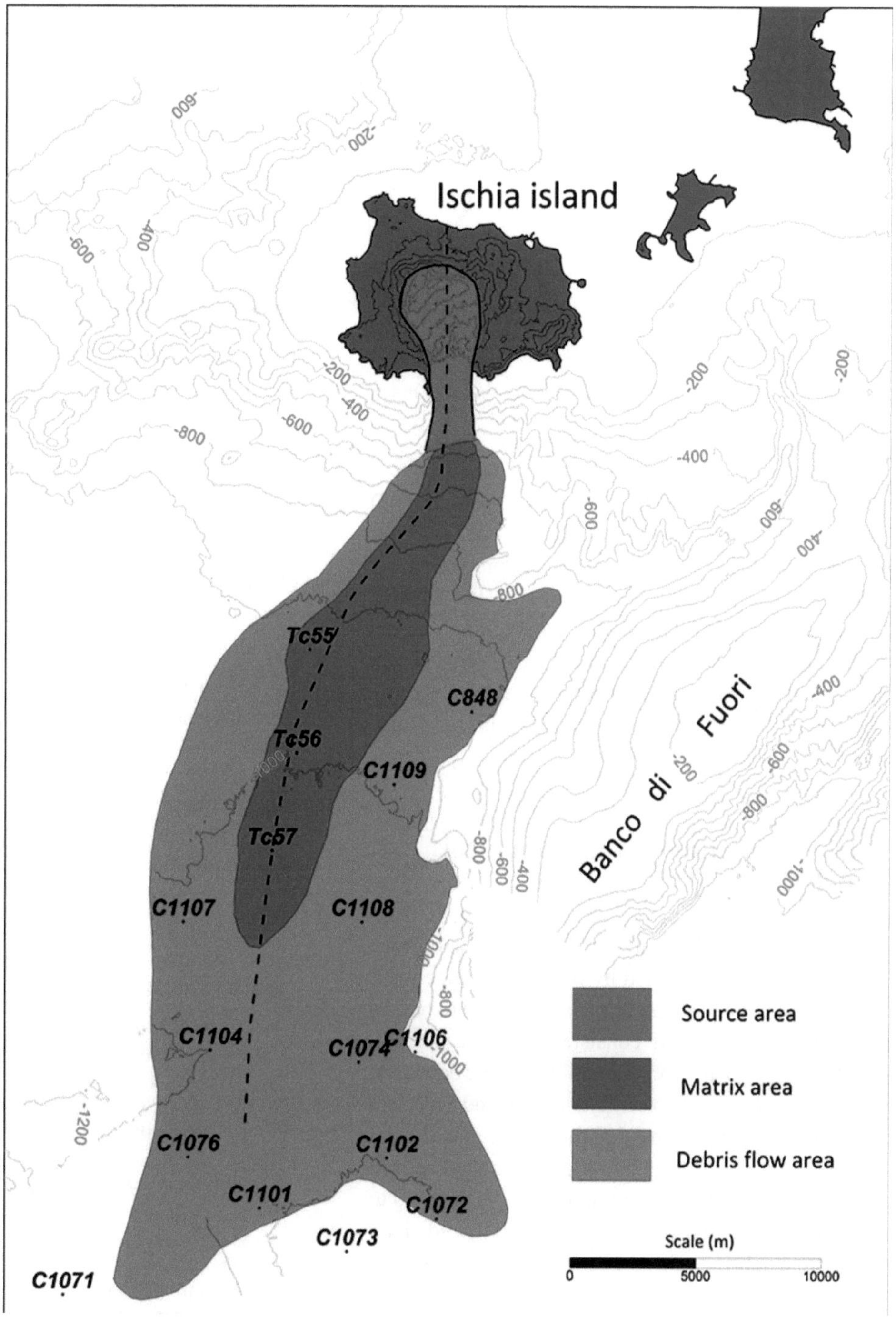

Fig. 43.2 Plan view of IDA dispersal area. The main source and depositional sections are outlined. Labels indicate coring sites. *Dashed line* is the section along the dispersal axis in Fig. 43.3

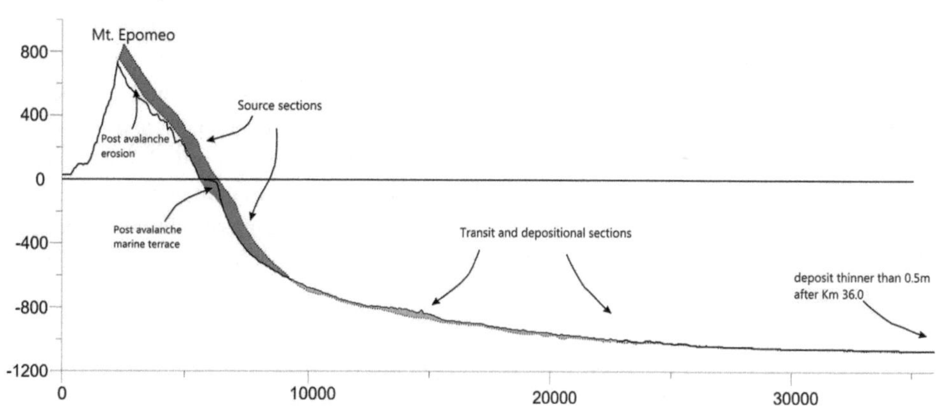

Fig. 43.3 Pre collapse profile reconstruction along IDA main dispersal axis. The *orange area* is the missing volume while *blue area* is the whole deposit including matrix and debris flow

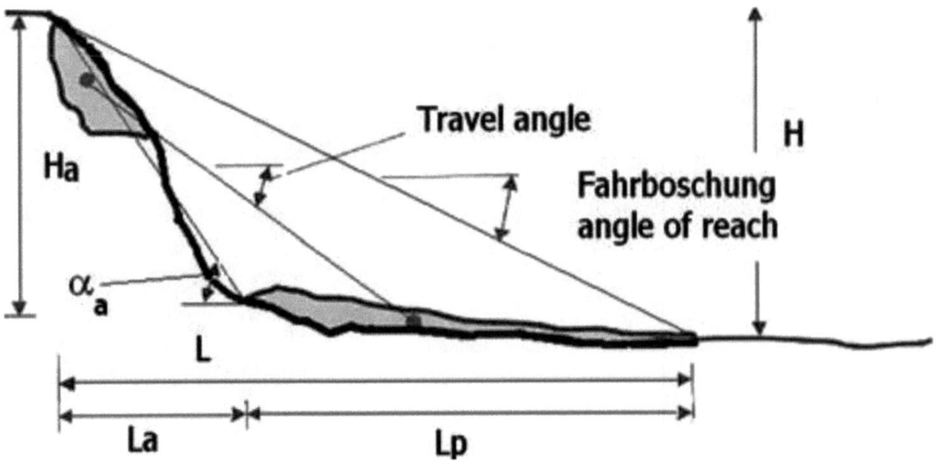

Fig. 43.4 Analysis of the motion: empirical approach (from Scotto di Santolo and Evangelista 2009)

The start of the depositional (and transit) (B) and (C) sections can be set between the toe of upper slope at 600 m down to 1,100 m as far as 40 km from Mt. Epomeo summit here defined as the zero reference point. The IDA deposit spreads over an area of 150 km^2 and is proved until at least 36 km based on the available core stratigraphy (Fig. 43.2). This runout distance increases of 5–6 km when considering the farthest, loose blocks (outliers) detected through multibeam bathymetry or side-scan-sonar records.

Table 43.1 Main geometrical parameters used for numerical simulation

Slope angle of failure, αa	12°
Travel angle between source and deposit centres of mass	4°
Maximum runout, L	33700 m
Extent of the deposit section Lp	26700 m
Mean and maximum deposit thickness (locally constrained by core stratigraphy or inferred by seismic data)	5 and 30 m
Difference between the highest source section and the toe of landslide, H	1900 m
Inferred volume of the failed mass (or trigger volume)	1.5–3.0 Km^3

43.3 Main Assumptions and Numerical Analysis

In our collapse scenario the summit of Mt. Epomeo, once higher than present, was truncated by the failure plane itself that involved also the underwater upper slope down to 600 m depth. The collapsed material consisted primarily of Mt. Epomeo Green Tuff i.e. a trachytic ignimbrite as the dominant lithotype and secondarily of softer pyroclastic and vulcanoclastic deposits. We firstly assume that this relatively coherent rock assemblage rapidly disintegrated and was channelised towards the south whilst a minor amount might be directed to north. After interaction with sea-water the collapsed mass transformed into a debris avalanche and distally into a debris flow i.e. a flowing solid-liquid mixture with different degrees of interactions between the solid and the liquid fraction. A second necessary, although limiting, assumption is that its behavior can be treated in analogy to terrestrial debris flows of known rheology (Scotto di Santolo et al. 2011). Although core sampling results indicate that the deposit is significantly mixed with hemipelagic sediments suggesting that seabed was likely involved during IDA emplacement we have not considered a changing mass in the model.

IDA landslide was therefore analysed using the DAN-W program i.e. Dynamic Analysis of landslides (Hungr 1995) which can be applied to a wide range of rheological behaviors. Among these the most suitable for solid-liquid mixtures i.e. debris flow-debris avalanche types are the frictional, Voellmy and Bingham described by Eqs. (43.1), (43.2) and (43.3) respectively.

$$\tau = \gamma H_i\left(\cos\alpha_i + \frac{a_c}{g}\right)\cdot(1 - r_u)\tan\varphi \tag{43.1}$$

$$\tau = \gamma H_i\left(\cos\alpha_i + \frac{a_c}{g}\right)\cdot\mu + \gamma\frac{v^2}{\xi} \tag{43.2}$$

$$\tau = \tau c + \mu\cdot v \tag{43.3}$$

In (43.1) and (43.2) γ is the unit weight of the flowing material; $a_c = v_i^2/R$ is the centrifugal acceleration resulting from the vertical curvature of the flow path R; r_u is the pore-pressure coefficient (ratio of pore pressure to total normal stress at base of boundary block); φ is the friction angle; μ is a friction coefficient and ξ is a turbulence coefficient with dimensions of [m/s^2]. In (43.3) τ_c is the constant yield strength and μ is the Bingham viscosity. The sliding mass is divided into a number of discrete blocks that interact each other while sliding according to different relationship (Hungr 1995).

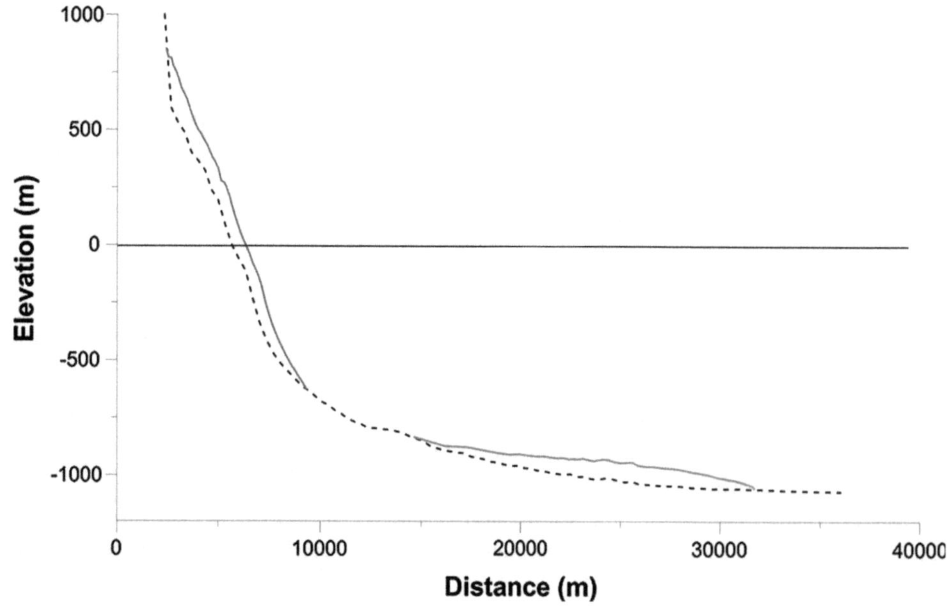

Fig. 43.5 Initial and final profile; Voellmy model ($\mu = 0.03$ and $\xi = 2000$ m/s^2). *Dashed line* is the present day profile. *Red line* is the modeled pre-collapse section, *blue line* the post-collapse depositional section

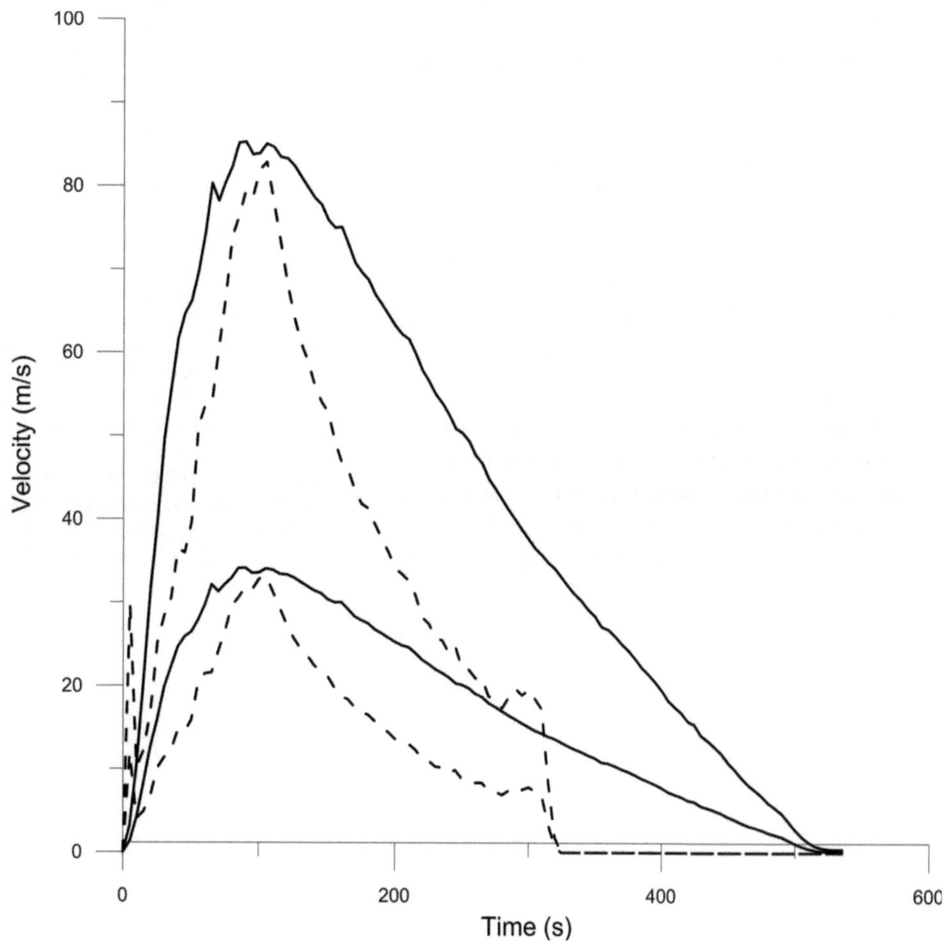

Fig. 43.6 Velocity plot of the mass movement front and rear (continuous and dashed lines respectively) Voellmy model. Velocities were reduced from ~85 m/s to ~35 m/s by lowering the turbulence parameters of one order of magnitude (ξ from 2000 to 100 m/s^2)

The main morphological parameters that were input to modeling shown in Fig. 43.4 are reported in Table 43.1. Most of these parameters are relatively well constrained by multibeam bathymetry and marine geophysics although still many uncertainties remain especially as regards the thickness of the deposits which is hardly detectable through geophysical data.

It is widely accepted that a convenient measure of the mobility of subaerial or submarine marine mass movements is the H to L ratio known also as apparent mobility. The results of the 2D numerical simulation were evaluated by matching the calculated vs. observed maximum runout and thickness of the deposit along IDA main dispersal axis. The best fits were obtained with those models based on a Voellmy rheological behaviour with $\mu = 0.03$ and $\xi = 2000$ m/s^2 (Fig. 43.5).

The velocity estimate which is one of the outputs of modeling such type of mass movements is still a very conjectural issue especially for submarine cases. It worth to recall that, to date, the few documented debris avalanches are terrestrial; for instance the peak values in the order of 50–70 m/sec observed at Mt. St Helens 1980 (Voight et al. 1983) occurred in concurrence with a magma-related lateral blast. Such values should be considered unrealistic in the case of purely gravity driven collapse.

In our case peak velocities in excess of 80 m/s were obtained at around 100 s after initiation of failure which is also in good agreement with similar values obtained by Tinti et al. (2011) with a different approach. Such peak velocities seem overestimated however. A more realistic value of 35 m/sec was obtained reducing the turbulence ($\mu = 0.03$ and $\xi = 100$ m/s^2) but at the expenses of lower mobility and thicker deposit (Fig. 43.6).

It also appears that if the parameters for an entirely sub-aerial movement were utilized this results in a lesser than measured run-out. The presence of higher—than—predicted pore water pressures into the sliding mass is therefore necessary but possibly still not sufficient to explain the very high mobility of IDA.

While high mobility for sub-aerial landslides can be explained either with the increase of the pore water pressure or with a solid to liquid transition, in the case of submarine movements the presence of sea-water i.e. of a drag force should decrease mobility but enhance the hydroplanning (Hurlimann et al. 2000). Our preliminary results indicate the necessity of adopting relatively high turbulence parameters to account for the extreme mobility of such mass movements since yet a conclusive understanding of the phenomenon does not exist despite several working hypotheses.

References

Chiocci FL, de Alteriis G (2006) The Ischia debris avalanche: first clear submarine evidence in the Mediterranean of a volcanic island pre-historic collapse. Terra Nova 18:202–209

de Alteriis G, Violante C (2009) Catastrophic landslides off Ischia volcanic island (Italy) during pre-history. In: Violante C (ed) Geohazard in Rocky Coastal Areas special issue of Journal of the *Geological Society of London.*Spec Pubbl 322:73–104.

de Alteriis G, Insinga D, Morabito S, Morra V, Chiocci FL, Terrasi F, Lubritto C, Di Benedetto C, Pazzanese M (2010) Age of submarine debris avalanches and tephro-stratigraphy offshore Ischia Island, Tyrrhenian Sea, Italy. Marine Geology 278:1–18

Gillot P-Y, Chiesa S, Pasquarè G, Vezzoli L (1982) 33.000 yr K/Ar dating of the volcano-tectonic horst of the isle of Ischia, gulf of Naples. Nature 229:242–245

Hungr O (1995) A model for the runout analysis of rapid flow slides, debris flows, and avalanches. Can Geotech J 32:610–623

Hurlimann M, Garcia-Piera JO, Ledesma A (2000) Causes and mobility of large volcanic landslides: application to Tenerife, Canary Islands. J Volcanol Geoth Res 103:121–134

Locat J, Lee HJ, Locat P, Imran J (2004) Numerical analysis of the mobility of the Palos Verdes debris avalanche, California, and its implication for the generation of tsunamis. Mar Geol 203:269–280

Scotto di Santolo A, Evangelista A (2009) Some observations on the prediction of the dynamic parameters of debris flows in pyroclastic deposits in the Campania region of Italy. Nat Hazards 50:605–622

Scotto di Santolo A, Pellegrino AM, Evangelista A, Coussot P (2011) Rheological behaviour of reconstituted pyroclastic debris flow. Géotechnique 62(1):19–27

Tinti S, Chiocci FL, Zaniboni F, Pagnoni G, de Alteriis G (2011) Numerical simulation of the tsunami generated by a past catastrophic landslide on the volcanic island of Ischia, Italy. Mar Geophys Res 32:287–297

Voight B, Janda RJ, Glicken H, Douglass PM (1983) Nature and mechanics of the Mount St Helens rockslide-avalanche of 18 May 1980. Géotechnique 33-3:243–273

Erratum to: The Case of Ischia Underwater Debris Avalanche (Italy, Tyrrhenian Sea) and Its High Mobility

44

G. de Alteriis, Anna Scotto di Santolo, F. L. Chiocci, M. Ramondini, and C. Violante

Erratum to:
Chapter 43 in: G. Lollino et al. (eds.), *Engineering Geology for Society and Territory – Volume 4*, https://doi.org/10.1007/978-3-319-08660-6_43

The original version of the book was inadvertently published with incorrect chapter author name "Anna Santolo di Scotto" which has to be changed as "Anna Scotto di Santolo" in Chapter 43 and Author Index. The erratum book has been updated with the change.

The updated original online version for this chapter can be found at https://doi.org/10.1007/978-3-319-08660-6_43

G. Lollino et al. (eds.), *Engineering Geology for Society and Territory – Volume 4*, https://doi.org/10.1007/978-3-319-08660-6_44, © Springer International Publishing Switzerland 2018

Author Index

G. Lollino et al. (eds.), *Engineering Geology for Society and Territory-Volume 4*,
DOI: 10.1007/978-3-319-08660-6, © Springer International Publishing Switzerland 2014

MIX
Papier aus verantwortungsvollen Quellen
Paper from responsible sources
FSC® C105338

If you have any concerns about our products,
you can contact us on
ProductSafety@springernature.com

In case Publisher is established outside the EU,
the EU authorized representative is:
Springer Nature Customer Service Center GmbH
Europaplatz 3, 69115 Heidelberg, Germany

Printed by Libri Plureos GmbH
in Hamburg, Germany